INTERNATIONAL ARCHITECTURE OF COMPETITIONS 2

国际建筑竞赛 2

（德）本杰明·胡思巴赫（德）克里斯蒂安·雷姆豪斯 编

王晨晖 译

辽宁科学技术出版社

沈阳

图书在版编目 (CIP) 数据

国际建筑竞赛 2 / (德) 本杰明·胡思巴赫, (德)
克里斯蒂安·雷姆豪斯编; 王晨晖译. — 沈阳:辽宁科
学技术出版社, 2017.6
ISBN 978-7-5591-0149-5

Ⅰ. ①国… Ⅱ. ①本… ②克… ③王… Ⅲ. ①建筑设
计 – 作品集 – 世界 – 现代 Ⅳ. ① TU206

中国版本图书馆 CIP 数据核字 (2017) 第 072859 号

出版发行:辽宁科学技术出版社
　　　　　(地址:沈阳市和平区十一纬路 25 号 邮编:110003)
印 刷 者:辽宁新华印务有限公司
经 销 者:各地新华书店
幅面尺寸:225mm×285mm
印　　张:17
插　　页:4
字　　数:100 千字
出版时间:2017 年 6 月第 1 版
印刷时间:2017 年 6 月第 1 次印刷
责任编辑:宋丹丹
封面设计:李　莹
版式设计:李　莹
责任校对:周　文

书　　号:ISBN 978-7-5591-0149-5
定　　价:298.00 元

编辑电话:024-23280367
邮购热线:024-23284502
E-mail: 1207014086@qq.com
http://www.lnkj.com.cn

Lehrecke Architekten

Glass Dairy in Münchehofe

玻璃奶品厂

st Architekten

en

Glass Dairy Münchehofe
玻璃奶品厂 Munchehofe

Invited competition preceded by an application procedure
邀请赛，开始前有申请程序。

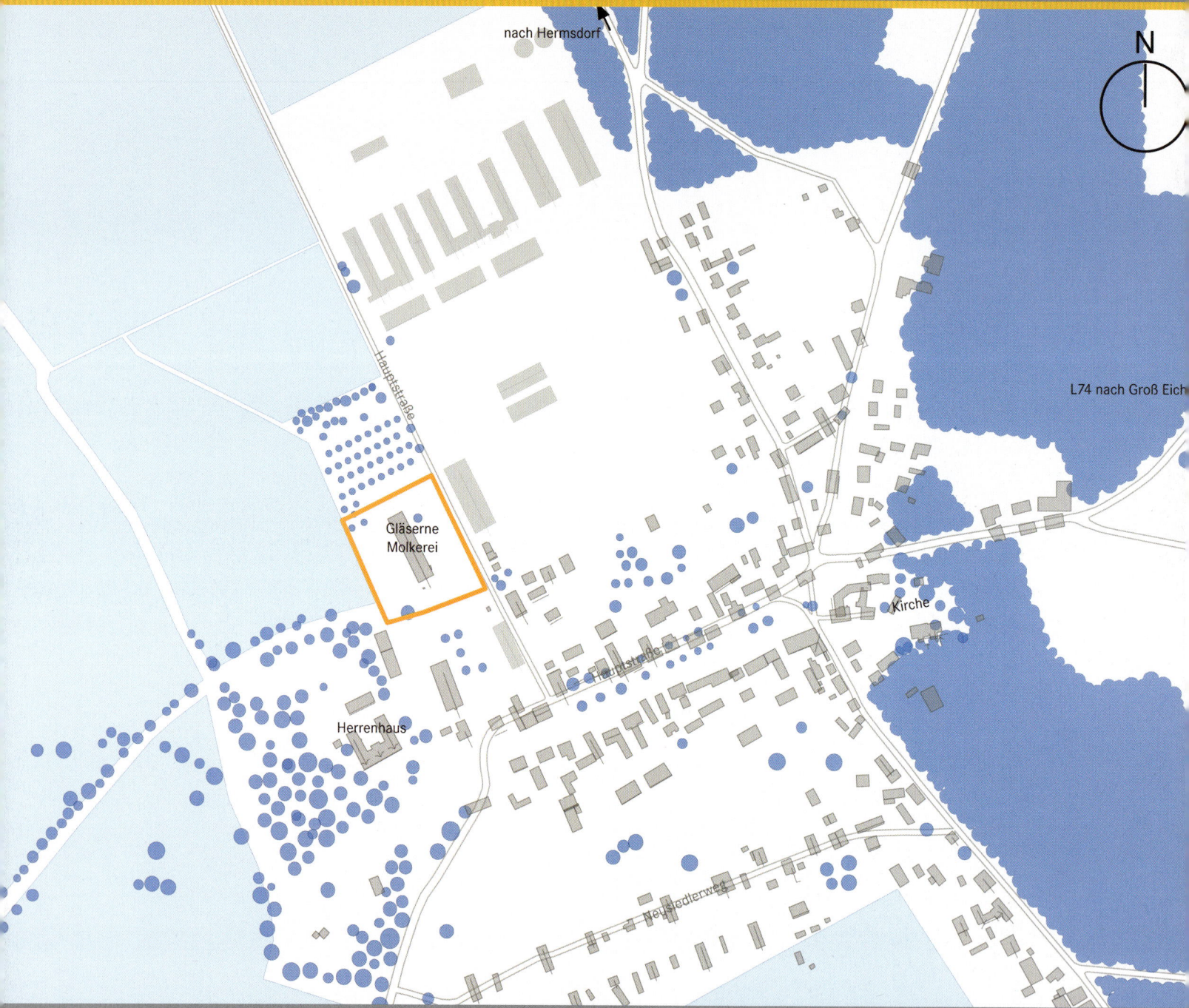

nach Hermsdorf

L74 nach Groß Eich

Hauptstraße

Gläserne Molkerei

Kirche

Herrenhaus

Hauptstraße

Neustedterweg

地点 Münchehofe　时间 11/2006-02/2007　主办方 Gläserne Molkerei GmbH　参赛者 23 applicants; 7 participants
面积 about 1,700 sq m　竞赛费用 36,200 Euro　专业评奖委员会 Prof. Gisela Glass, Berlin/Munich; Doris Gruber, Berlin;
Tim Heide, Berlin; Prof. Claudia Lüling, Berlin/Frankfurt/Main　专家评奖委员会 Hubert Böhmann, head Gläserne Meierei GmbH, Upahl;
Dr. Peter Danckert, MdB, Berlin; Meinrad Schmitt; head and holder TERRA Naturkost Handelsgesellschaft, Berlin

In Münchehofe, in the Dahme-Heideseen nature reserve southeast of Berlin, the Gläserne Molkerei GmbH runs an ecofriendly dairy and cheese factory. Here, milk produced by environmental sound methods is packaged or processed into yoghurt, curd, cream and cheese. The name "Gläserne Molkerei" ("The Glass Dairy") reflects standards of transparency and authenticity in its dairy production. The basis of the competition was to extend the dairy and construct a new cheese factory. In keeping with the company's name, the plant conversion was indeed envisaged in terms of a "glass dairy" where visitors can actually witness the manufacturing process. The aim is to promote consumer interest in ecological sensitive approaches to food production and thereby promote customer loyalty. Extant buildings included, the spatial program comprises about 1,700 square metres of useable area, 900 square metres of which are to be new. It also includes building a small conference hall The aim of the competition was to come up with an approach that is both functional and aesthetically striking, one that does not imitate standard patterns for plants of this kind. Accordingly, the new building was to be harmoniously aligned with its environs, the village locale of Münchehofe, as well as add qualitative uplift.

Munchehofe在柏林东南的Dahme-Heideseen自然保护区里，Glaserne Molkerei GmbH开办一家生态环保的乳品和乳酪工厂，把用环保安全方法生产出来的牛奶包装或加工成酸奶、豆腐、冰淇淋和乳酪。"玻璃奶品厂"这个名字反映乳品生产过程的透明性和可靠性的标准。竞赛的目的是扩建奶品厂，建成一家新乳酪工厂。该项目被命名为参观者可以目睹生产过程的"玻璃奶品厂"，目的是提高消费者对食品生产的生态环保方法的兴趣和信任度。包括现有建筑，空间规划包括约1700平方米的使用面积，其中900平方米是新建的，也包括一个小会议厅。竞赛的目的是提出一个与这类工厂标准模式不同的在功能性和美学方面引人注意的方法。因此，新建筑不但会增强质量方面的提升，而且将与周围环境和谐统一。

Aerial view　鸟瞰图

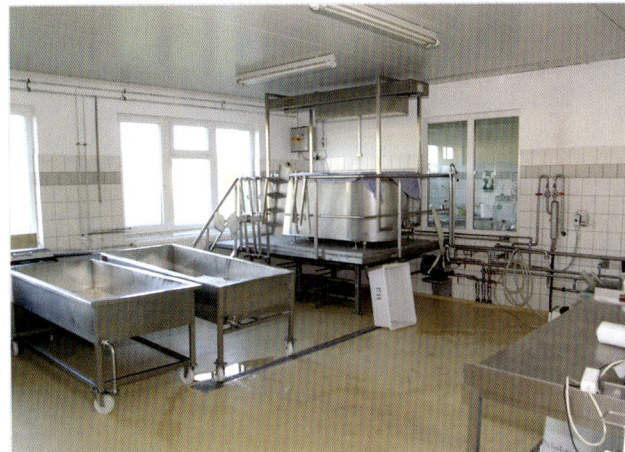

Cheese dairy　乳酪制品

Competition site　竞赛地点

View of the village　村庄景色

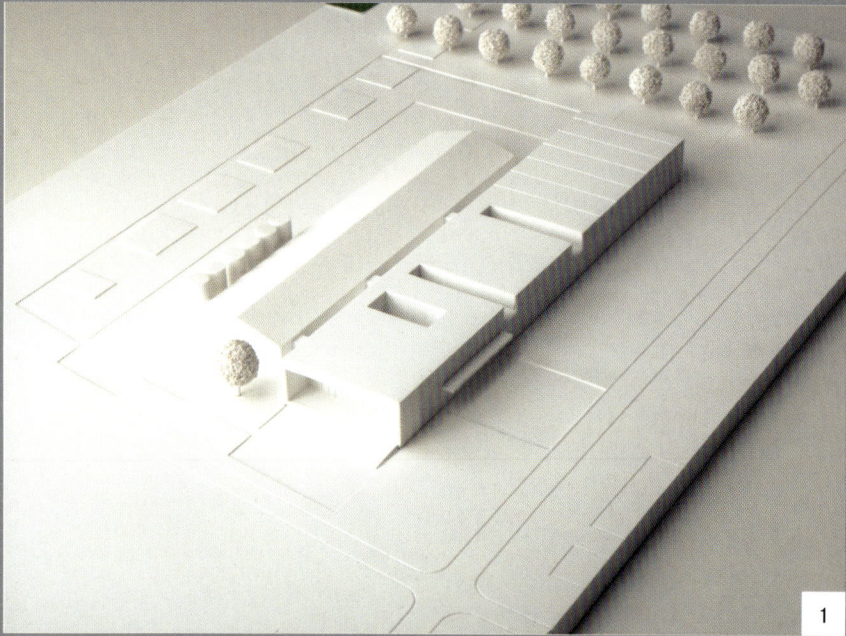

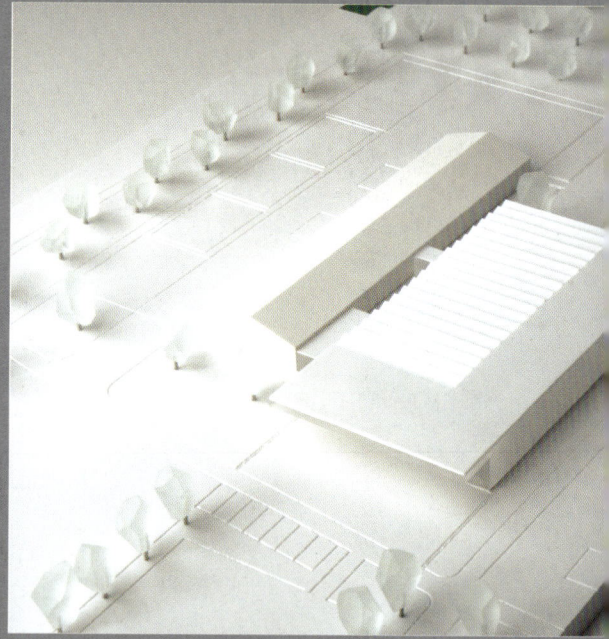

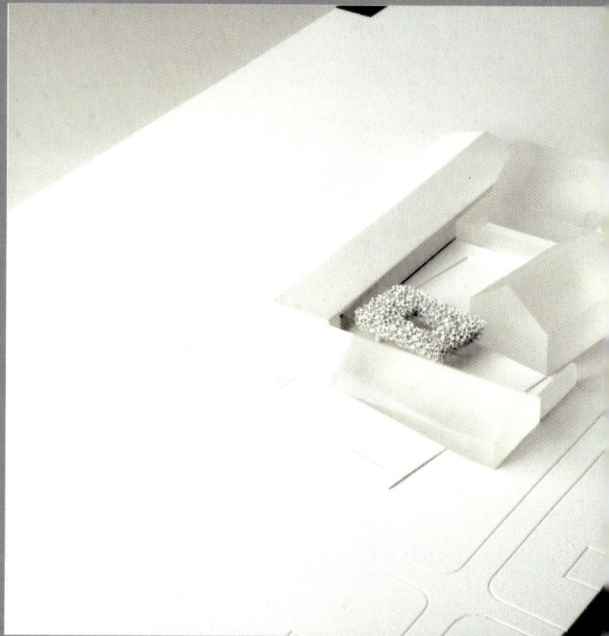

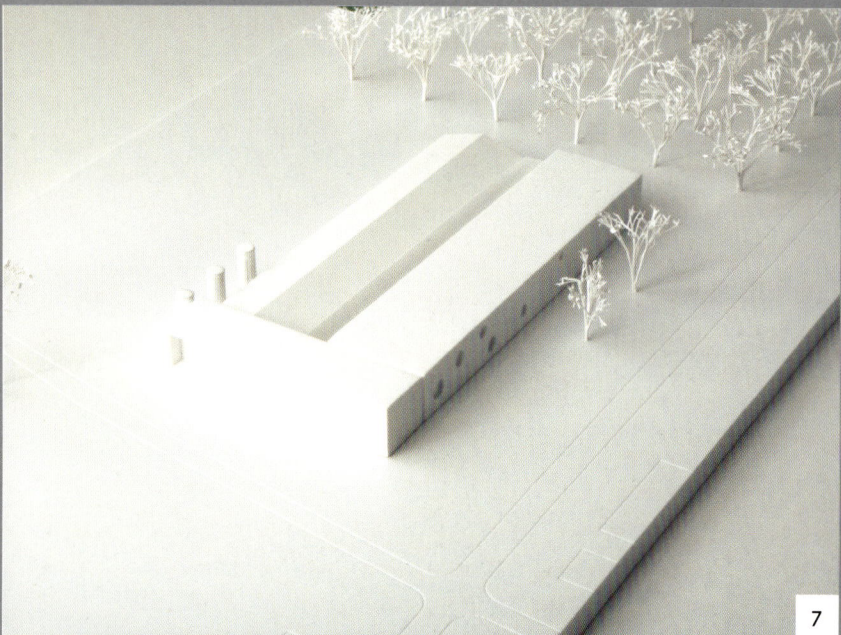

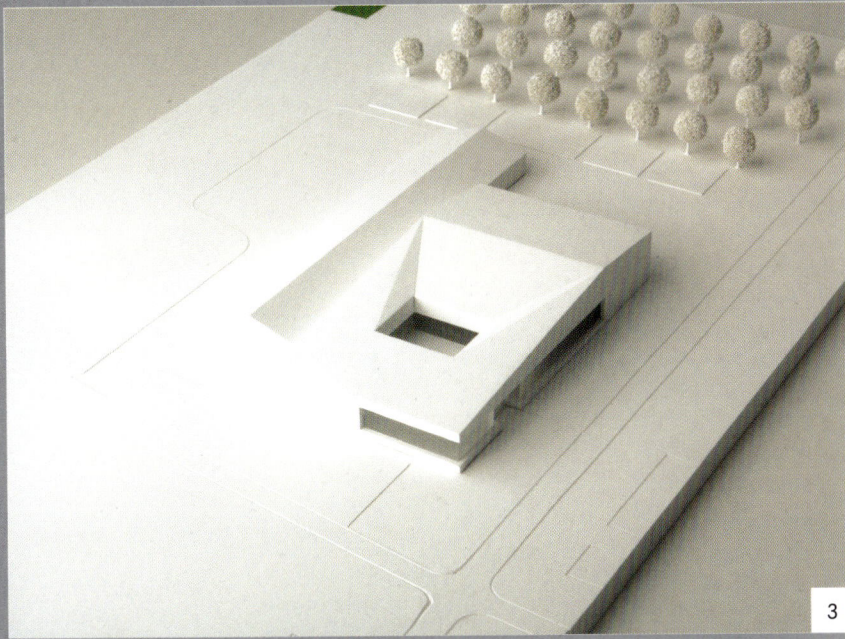

Qualified participants
合格的参赛者

1
1st prize 一等奖
Lehrecke Architekten, Berlin

2
2nd prize 二等奖
keller mayer wittig, Cottbus

3
3rd prize 三等奖
Modersohn & Freiesleben
Architekten, Berlin

4
Further participant 其他参赛者
Grüntuch Ernst Architekten, Berlin

5
Further participant 其他参赛者
Anderhalten Architekten, Berlin

6
Further participant 其他参赛者
Kretzschmar & Weber Architekten,
Oranienburg

7
Further participant 其他参赛者
Architekturbüro Kühn-
von Kaehne + Lange, Potsdam

Lehrecke Architekten Berlin

Lehrecke建筑事务所 柏林

"The structure is endowed with [...] an elegant air of sobriety that confidently inserts itself into the context of Münchehofe's agricultural scenery."

"给建筑结构赋予一种严肃的优雅气氛，自信地融入到Munchehofe的农业景色之中。"

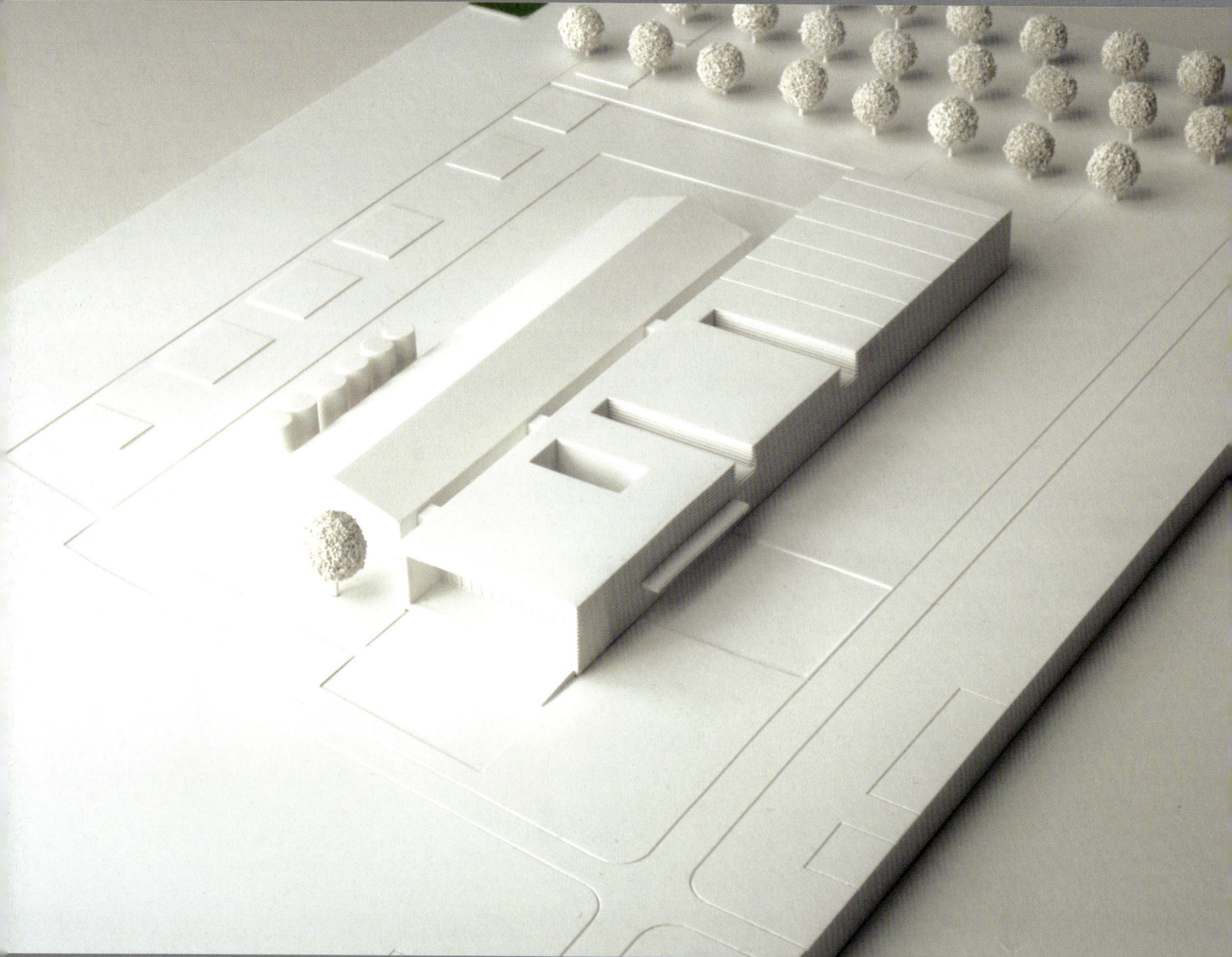

作者 Jakob Lehrecke   合作伙伴 Florian Kammerer, Ralf Tschöpe, Robert Witschurke
专家 structural analysis: Gerd-Walter Huske, Berlin; energy: Transsolar Energietechnik, Stuttgart

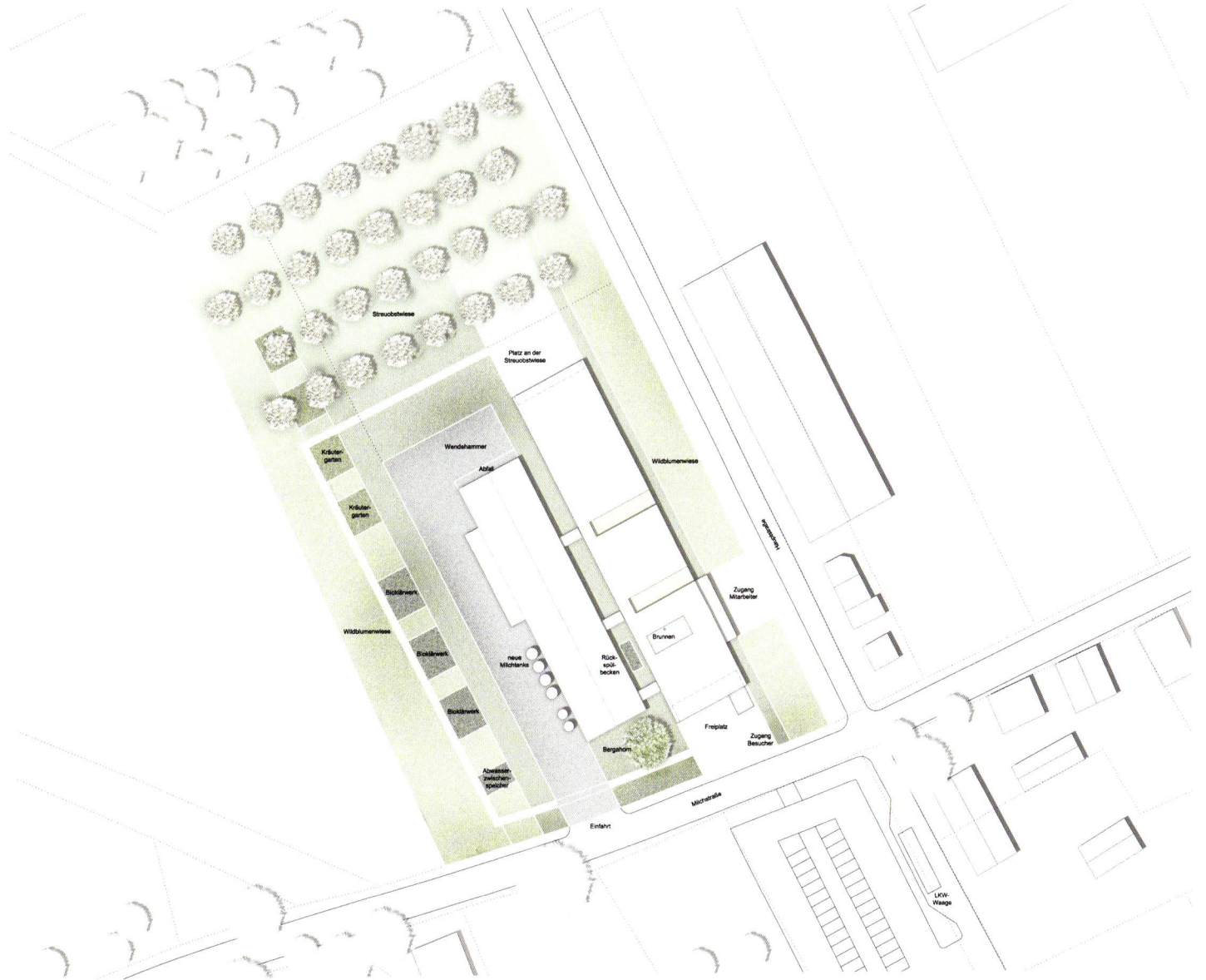

Streuobstwiese

Platz an der
Streuobstwiese

Kräuter-
garten

Wendehammer

Wildblumenwiese

Abfall

Kräuter-
garten

Bioklärwerk

Wildblumenwiese

Bioklärwerk

Zugang
Mitarbeiter

Bioklärwerk

neue
Milchtanks

Rück-
spül-
becken

Brunnen

Bioklärwerk

Freiplatz

Zugang
Besucher

Bergahorn

Abwasser-
zwischen-
speicher

Einfahrt

Milchstraße

LKW-
Waage

keller mayer wittig Cottbus
Keller mayer wittig建筑事务所 科特布斯

"Its location [...] makes it certain to be the first port of call for every visitor with an urban background, the show operation of the Glass Dairy acting as an attraction in its own right."

"它的地理位置使它成为邀请城里人参观的第一站，玻璃奶品厂的生产过程的展示操作成为一大亮

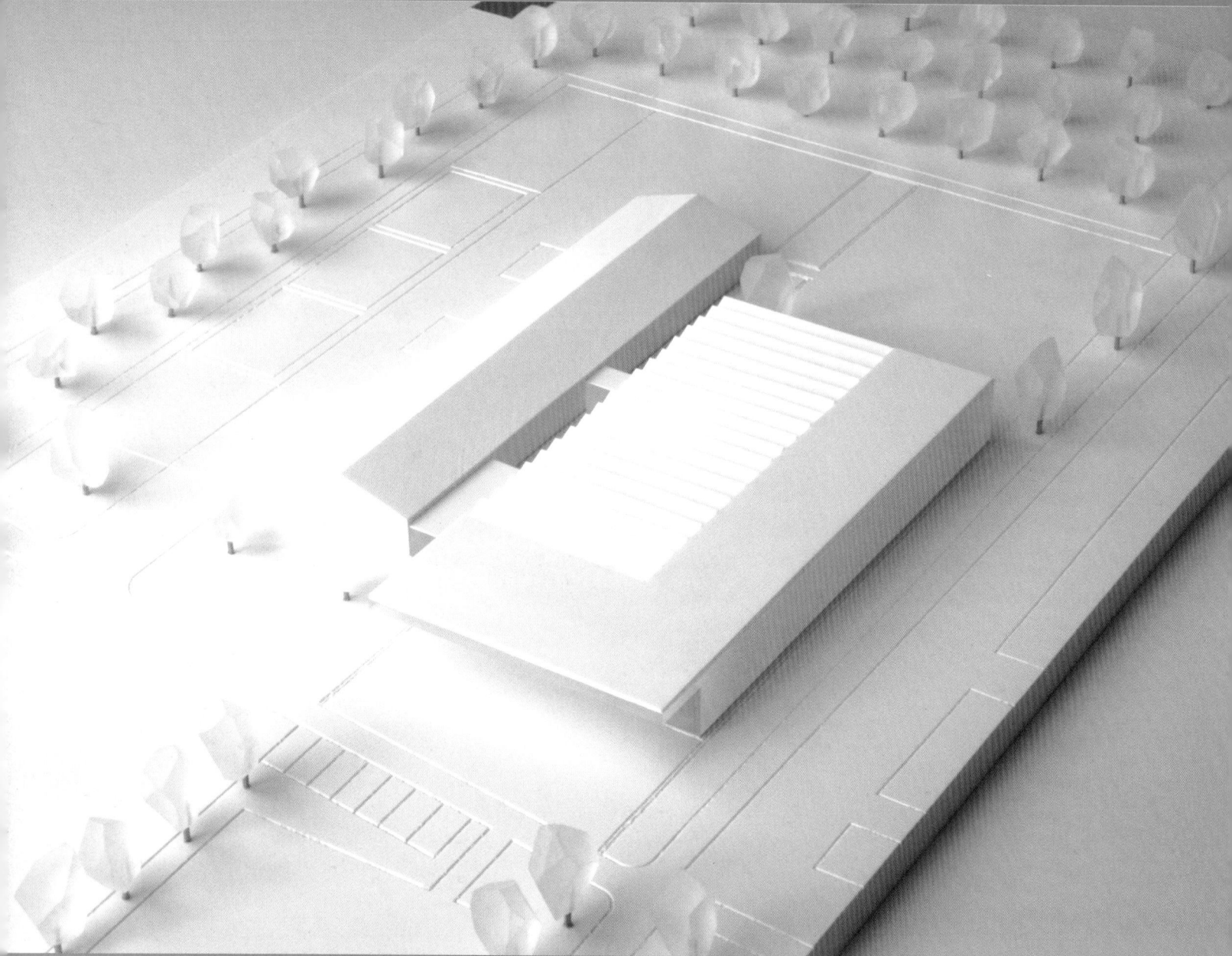

作者 Christian Keller   合作伙伴 Christoph Schulze, Martina Lehmann, André Krämer, Marco Laske   专家 structural analysis and energy: Hendrik Lindner, Prof. Pfeifer und Partner Ingenieure, Cottbus/Darmstadt; lighting: Zumtobel Staff, Berlin

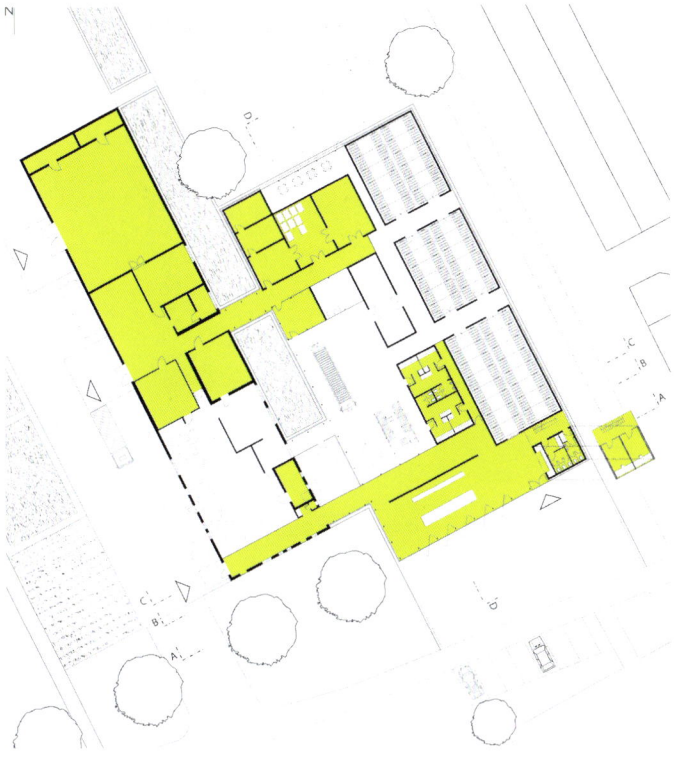
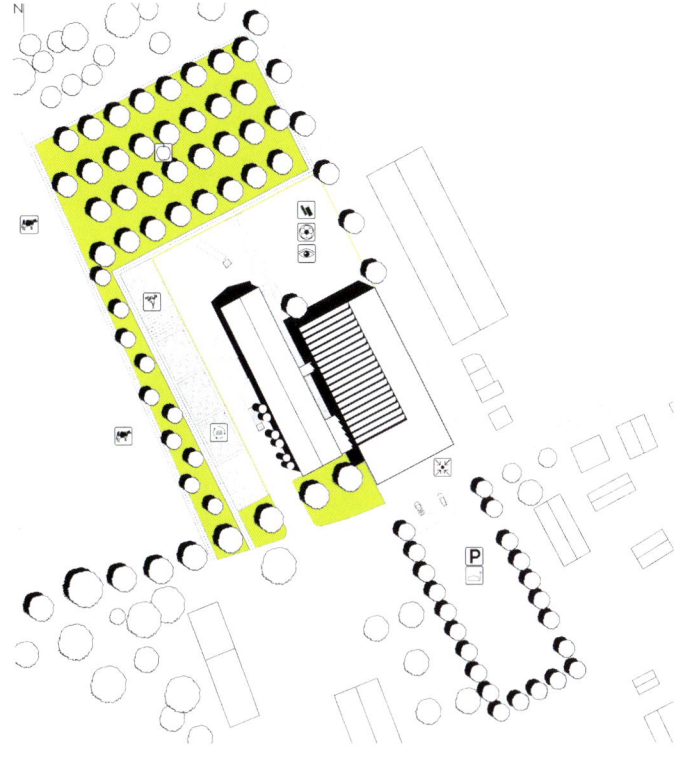

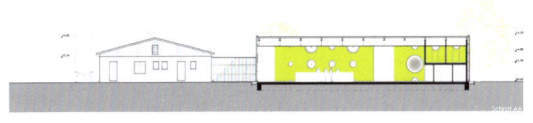

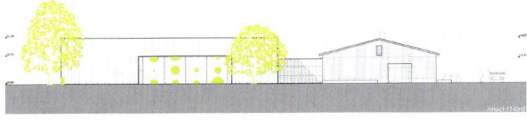
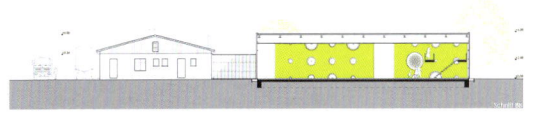

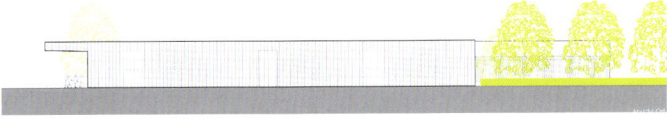
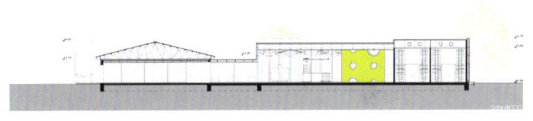

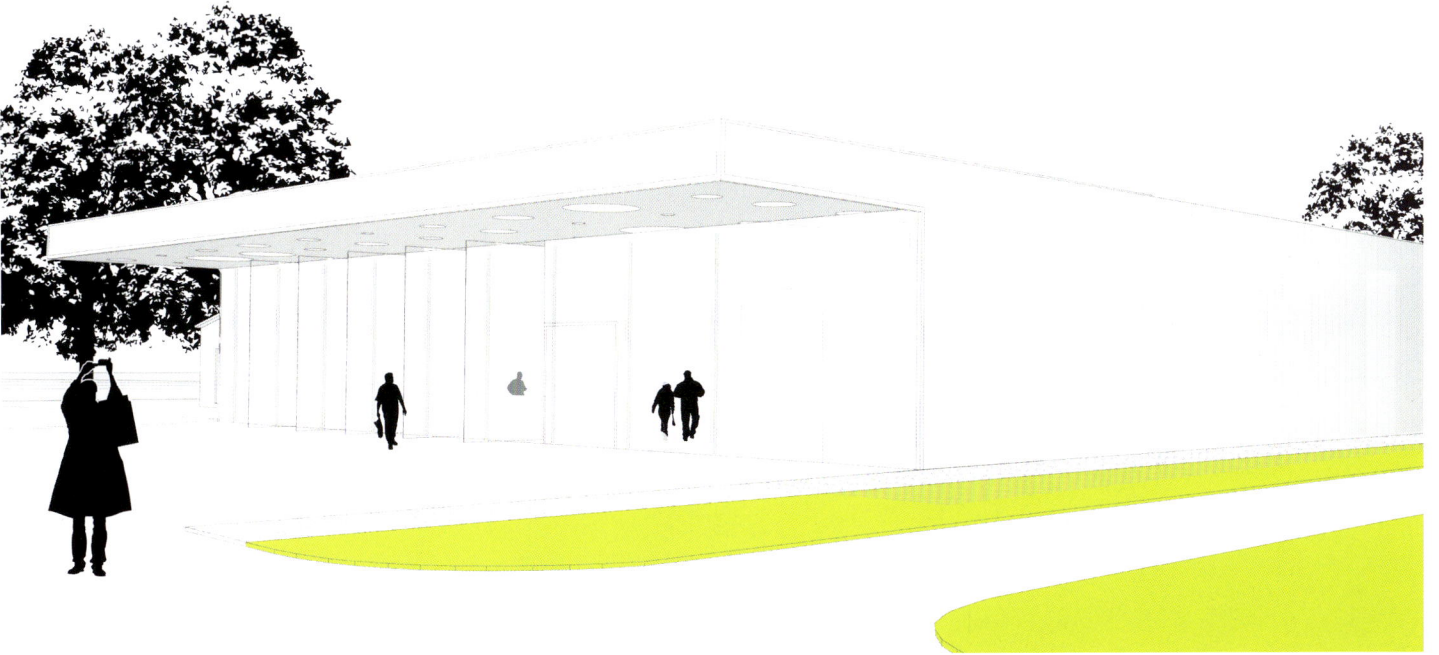

Modersohn & Freiesleben Architekten Berlin
Modersohn & Freiesleben建筑事务所 柏林

"The design presents a roofed building that opens up almost symbolically: A wood structure with wide inviting ribbon glazing, set on an inconspicuous plinth."

"设计一种带屋顶的建筑朝上，具有象征性。带有彩条玻璃窗的木质结构放置在一个不显眼的基座上。"

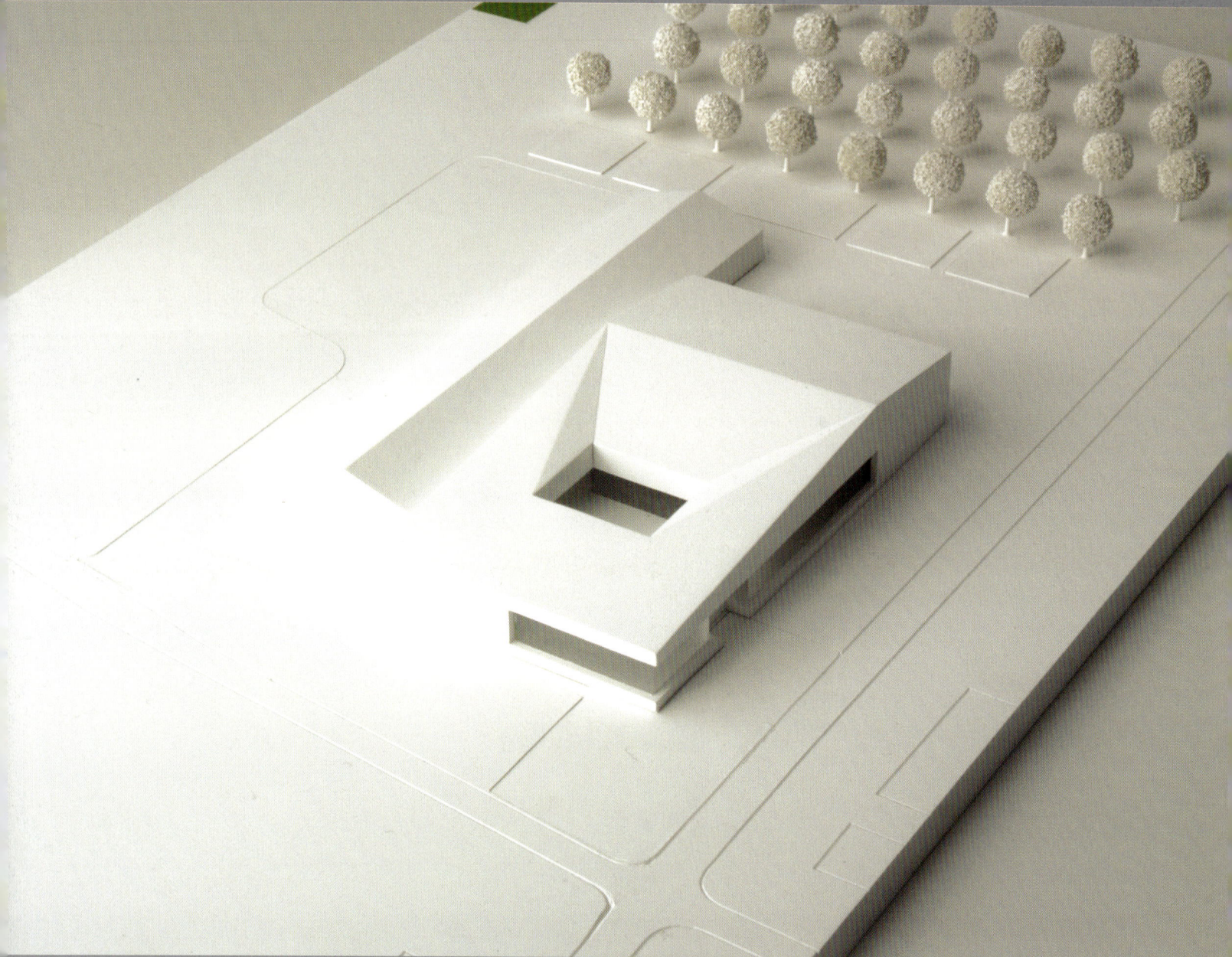

作者 Johannes Modersohn, Antje Freiesleben   合作伙伴 Christian Holthaus, Janine Ritz, Aimée Wolf
专家 structural analysis: Ingenieurbüro für Structural Analysis, Dr. Ing. C. Müller; technical equipment: Gebäudetechnik Dresden

Schwarzplan M 1:10.000

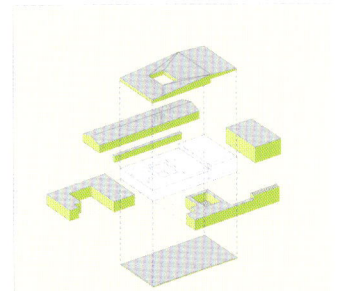

Explosionsskizze

Flächenschema Schwarz-, Grau-, Weißbereich + Besichtigungszone

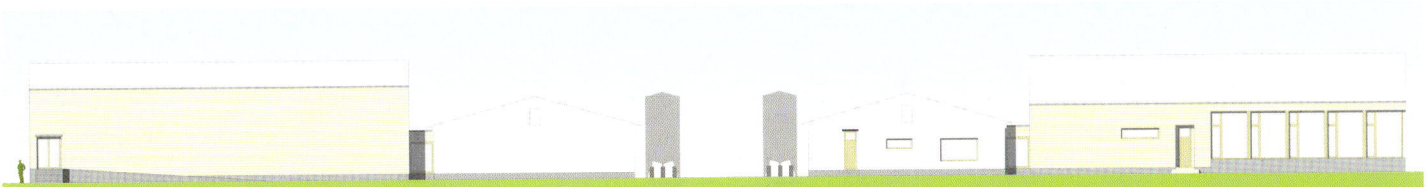

Lageplan mit Außenanlagen M 1:500

Ansicht Nord M 1:200

Ansicht Süd M 1:200

Grüntuch Ernst Architekten Berlin
Gruntuch Ernst建筑事务所 柏林

"The extension to the Glass Dairy in Münchehofe provides an opportunity to create a distinctive structure, embodying the idea of transparent and credible food processing within a setting of unique and unconventional architecture."

"Munchehofe里玻璃奶品厂的扩建产生一种有特色的建筑结构，用一种独特的非常规的建筑来体现食品加工透明可信的观点。"

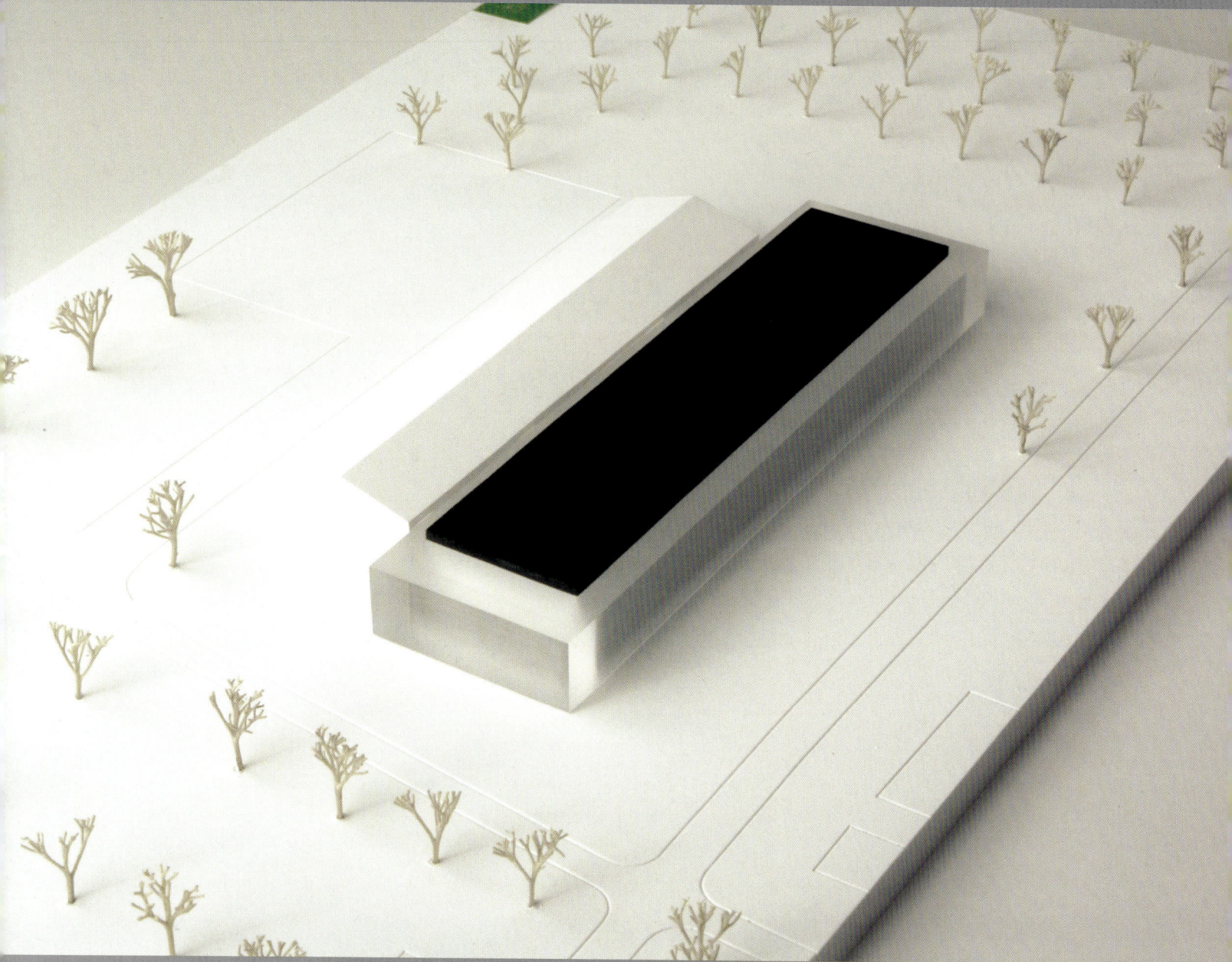

作者 Armand Grüntuch 合作伙伴 Arno Löbbecke, Pascale Busch, Stefan John, Allessio Fossati

Realisierungswettbewerb
Gläserne Molkerei in Müchehofe

1

Umgebungsplan M 1:2000

Gerber Architekten

Thomas M

behet bondzio lin arch

**Riedberg Campus Commons
in Frankfurt/Main**

Riedberg校园公用地
德国美因河畔法兰克福

üller Ivan Reimann

tekten

Riedberg Campus Commons Frankfurt/Main
Riedberg校园公用地 德国美因河畔法兰克福

Restricted interdisciplinatory project competition for architects and engineers preceded by an application procedure
限制严格的建筑师和工程师多科项目竞赛，开始前有申请程序。

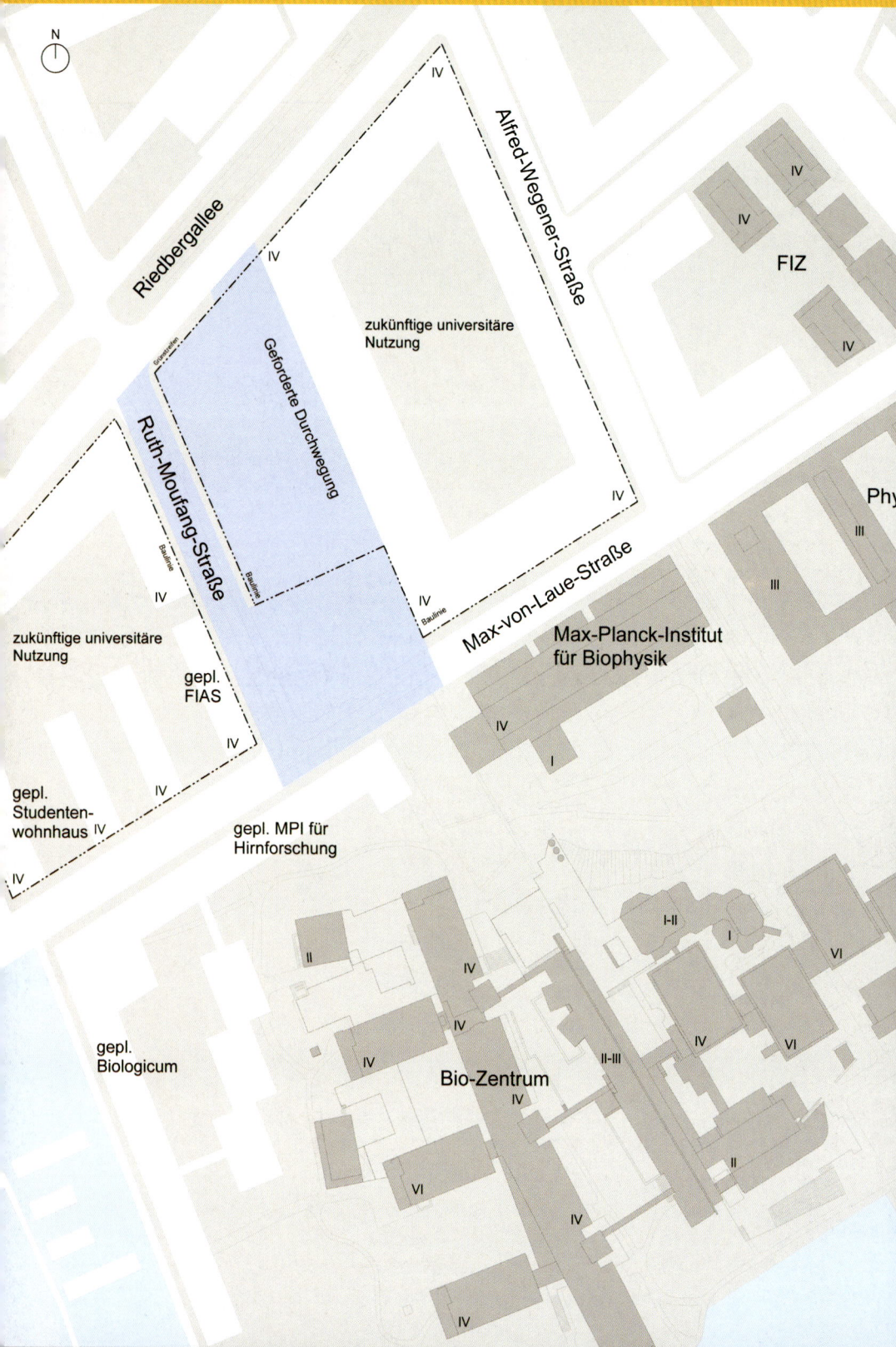

Map labels:

- Riedbergallee
- Alfred-Wegener-Straße
- zukünftige universitäre Nutzung
- Geforderte Durchwegung
- Ruth-Moufang-Straße
- Grünstreifen
- Bauline
- FIZ
- Max-von-Laue-Straße
- Max-Planck-Institut für Biophysik
- Phy...
- zukünftige universitäre Nutzung
- gepl. FIAS
- gepl. Studentenwohnhaus
- gepl. MPI für Hirnforschung
- gepl. Biologicum
- Bio-Zentrum
- N

地点 Frankfurt/Main
时间 06/2006–12/2006
主办方 Ministry of Higher Education, Research, and the Arts, Federal State of Hessen, represented by minister of state Udo Corts, Wiesbaden
参赛者 25 participants
面积 6,000 sq m
竞赛费用 113,000 Euro
专业评奖委员会
Harald Clausen, head of division, Ministry Finance, Federal State of Hessen, Wiesbaden;
Prof. Ansgar Lamott, Stuttgart/Darmstadt;
Prof. Ulrike Lauber, Munich/Berlin;
Prof. Uwe Rotermund, building services engineer, Braunschweig/Münster;
Prof. Kirsten Schemel, Berlin/Münster
专家评奖委员会
Irene Bauerfeind-Roßmann, head of division, Ministry of Arts and Sciences, Federal State of Hessen, Wiesbaden;
Günter Schmitteckert, leading head of division, Ministry of Arts and Sciences, Federal State of Hessen, Wiesbaden
Prof. Dr. Rudolf Steinberg, president, Johann Wolfgang Goethe-Universität Frankfurt/Main;
Prof. Dr. Horst Stöcker, designated vice president , Johann Wolfgang Goethe-Universität, Frankfurt/Main

The Riedberg Campus is situated in Frankfurt's north-west, and intended to house all the science institutions of the university. The spatial proximity of the university buildings to each other and the openness of the campus can be used to strengthen interdisciplinary ties. The task of this competition was the new built construction of the Central Commons. This is essential to the development plan for Riedberg Campus. The site under competition measures approx. 15,000 square metres and is located at the northern edge of the campus, next to downtown Riedberg: Its open-air spaces are to define the venue for entering the campus from there. A part of the task was to design a plaza environment on the southern portion of the site. With approx. 6,000 square metres of useful area, the Central Commons focuses essential supply functions in such a way as to provide the campus with an attractive, well functioning centerpiece. The program of spaces comprises a departmental library, a cafe-teria, lecture and seminar facilities plus preparation and storage areas and offices. The competition task took an interdisciplinary approach, including concepts for load-bearing structures and mechanical services as well as architectural design.

Riedberg校园坐落在法兰克福的西北部，大学的全部自然科学机构全部坐落于此。建筑物之间的空间上的接近和校园的开放性能够增加各学科的联系。竞赛任务是中心公用地的重建，该项目对Riedberg校园整个的开发规划非常重要。竞赛地点测量约占1.5万平方米，位于校园的北角，紧挨着Riedberg中心：露天空间作为进入校园的场所，任务之一是在竞赛地点的南部设计一个广场。中心公用地使用面积约6000平方米，强调基本的供给功能，成为校园里引人注意、功能良好的一个中心。空间规划包括一个院系图书馆、咖啡厅、讲座和研讨会设施及准备和储存区和办公区。竞赛任务采用跨学科方法，包括支承结构、力学服务及建筑设计的观念。

Aerial view　鸟瞰图

Competition site　竞赛地点

Competition site　竞赛地点

Skyline of Frankfurt/Main　美因河畔法兰克福的地平线

1

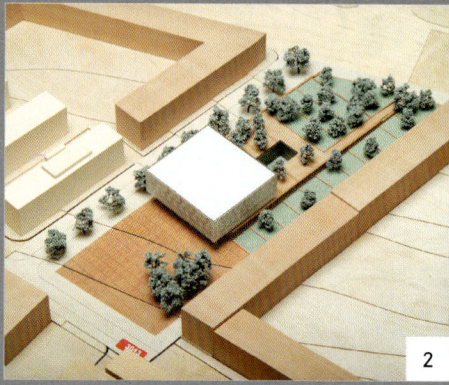

2

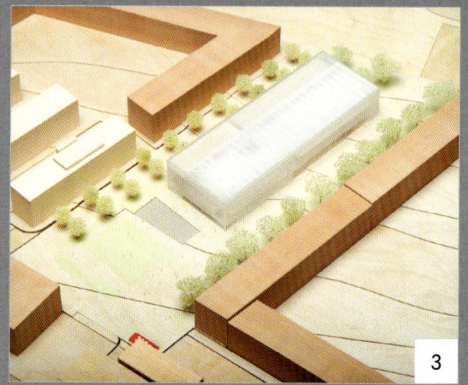

3

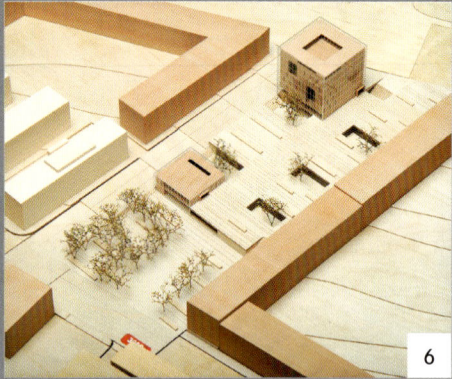

6

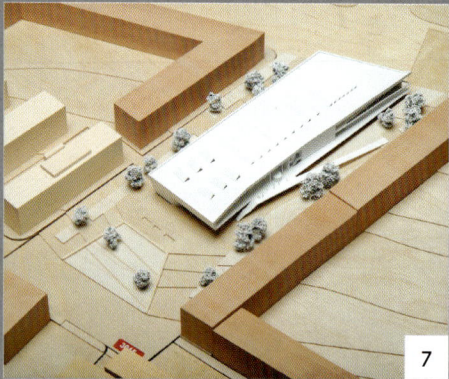

7

8

11

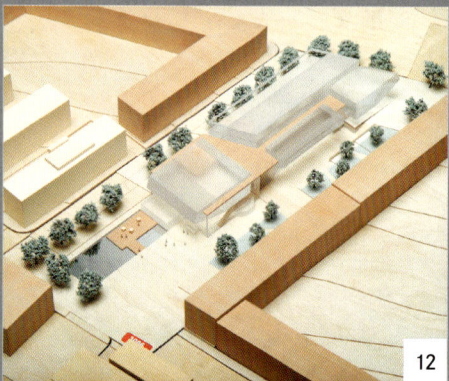

12

13

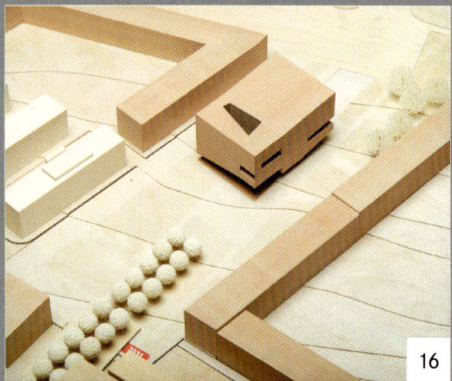

16

17

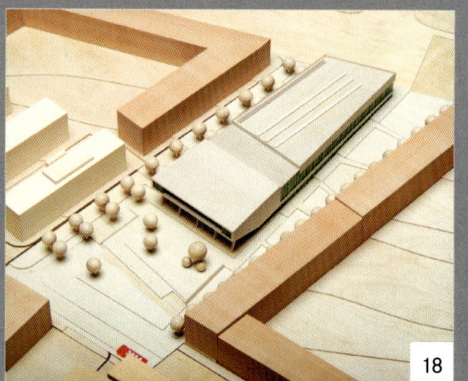

18

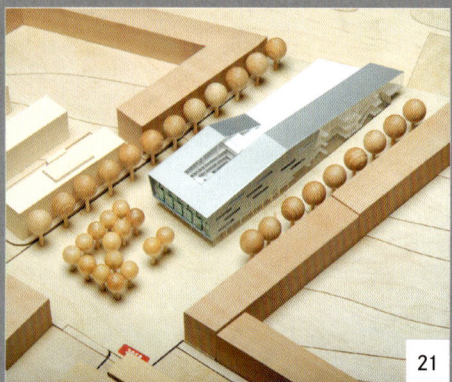

21

22

23

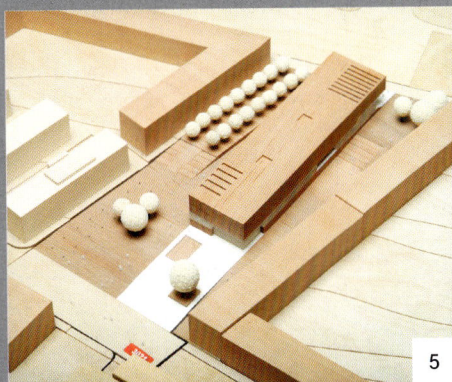

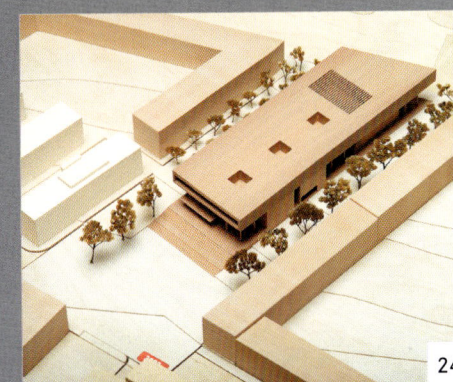
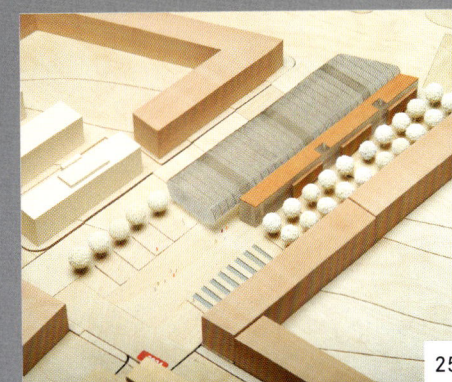

Qualified participants
合格的参赛者

1
1st prize　一等奖
Gerber Architekten, Dortmund
2
2nd prize　二等奖
Thomas Müller –
Ivan Reimann, Berlin
3
3rd prize　三等奖
Heinle, Wischer und Partner, Berlin
4
4th prize　四等奖
Ferdinand Heide, Frankfurt/Main
5
5th prize　五等奖
behet bondzio lin architekten, Münster
6
Acquisition　购买
Jockers Architekten, Stuttgart
7
Acquisition　购买
Kissler + Effgen, Wiesbaden
8
Acquisition　购买
kister scheithauer gross, Cologne
9
Prof. Dipl.-Ing. Architekt Benedict
Tonon, Berlin
10
Glass Kramer Löbbert Gesellschaft
von Architekten, Berlin
11
Chestnutt_Niess Architekten, Berlin
12
Kühnl & Schmidt, Karlsruhe
13
sauerbruch hutton, Berlin
14
Henn Architekten, Munich
15
Kleihues + Kleihues, Dülmen-Rorup
16
Wulf & Partner, Stuttgart
17
Nickl & Partner, Munich
18
hks hestermann rommel, Erfurt
19
HASCHER JEHLE, Berlin
20
ReimarHerbst.Architekten, Berlin
21
BMBW Architekten + Partner, Munich
22
b2Architekten Dittmann & Luft, Bonn
23
pfp Architekten, Hamburg
24
Klein & Sänger, Munich
25
KBK Architekten Belz Lutz
Guggenberger, Stuttgart

Gerber Architekten Dortmund
Gerber建筑事务所 多特蒙德

"The configuration and façade design derive from and are in keeping with the conceptual basis and functional assignment of the three built volumes – lecture halls, library, and foyer."

"建筑结构和外观源于并且与三大建筑物——讲座厅、图书馆和休息厅的概念的基础和功能解析协调一致。"

作者 Prof. Eckhard Gerber   合作伙伴 Nils Kummer, Alexandra Kranert, Stefan Lemke, René Albrecht, Martin Pellkofer, Matthias Deilke, Benjamin Sieber, Siegbert Hennecke   专家 building services: Energy Design, Braunschweig; structural analysis: Prof. Pfeifer und Partner, Darmstadt; landscape architecture: Gerber Architekten

Norden

Lageplan M 1/500

Riedbergallee

Ruth-Moufang-Straße

Altred-Wegener-Straße

FIZ

zukünftige universitäre Nutzung

zukünftige universitäre Nutzung

Max-Planck-Institut für Biophysik

gepl. Kita

gepl. FIAS

gepl. Studentenwohnhaus

Max-von-Laue-Straße

gepl. MPI für Hirnforschung

Campus Platz

IV

IV

IV

IV

IV

IV

IV

IV

I

Campus Platz

Gartenhof

Cafeteria

Foyer

Oberes Eingangsgeschoß M 1 / 200

Riedbergallee

Ruth-Moufang-Straße

zukünftige universitäre Nutzu

IV

IV

Längsschnitt Hörsäle M 1 / 200

Ansicht Ruth-Moufang-Straße M 1 / 200

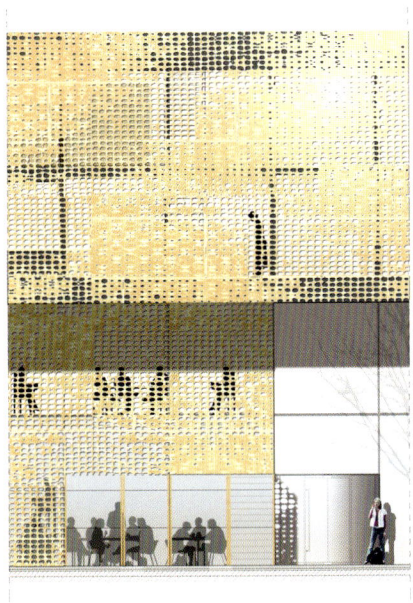

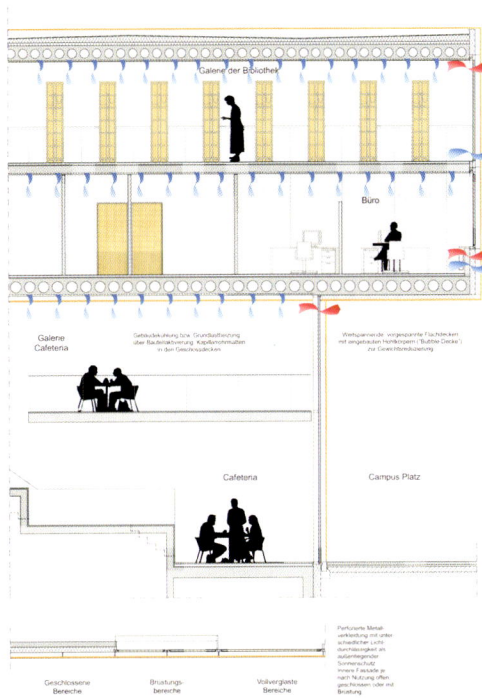

Detailansicht   M 1/50

Detailschnitt   M 1/50

Detailgrundriss   M 1/50

Tragwerk / Systemskizzen

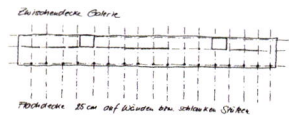

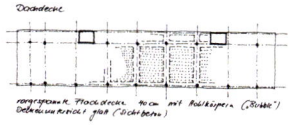

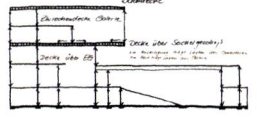

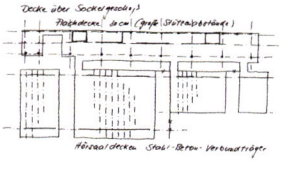

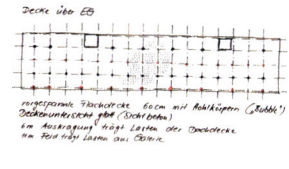

Tragwerkskonzept

Das räumliche Konzept des Entwurfs zeichnet sich durch Offenheit, Großzügigkeit und Transparenz aus.

Das Tragwerk für das Gebäude unterstützt diesen Ansatz durch leistungsfähige Konstruktionen mit großen Spannweiten auf Wänden und wenigen dünnen Stützen.

Energiekonzept / Systemskizzen

Energiekreislauf

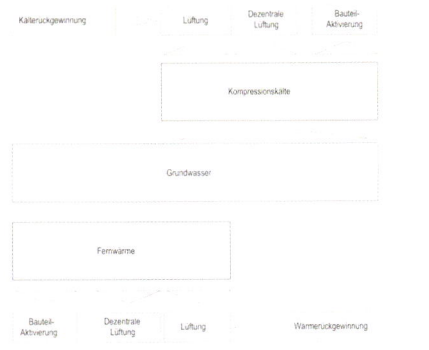

Energiekreislauf Kühlung

Energiekreislauf Heizung

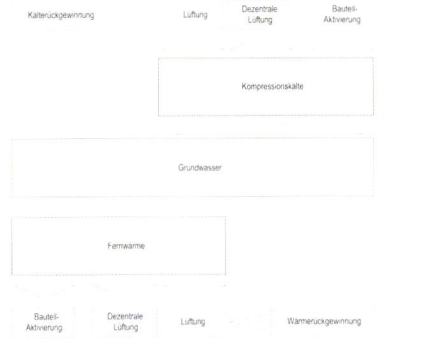

Zonierung

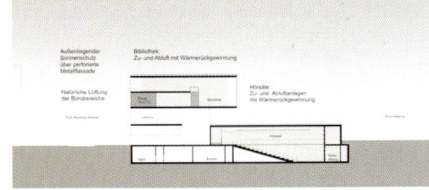

Sommer

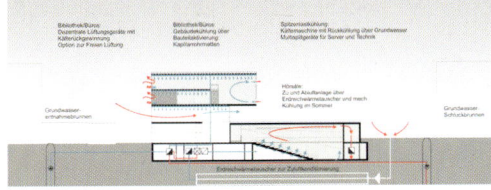

Winter

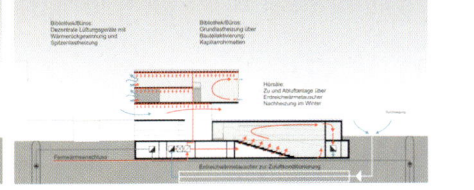

Thomas Müller Ivan Reimann Berlin

Thomas Muller Ivan Reimann建筑事务所 柏林

"We imagine the new Campus Commons not only as a building but also as a sequence of differing public spaces."

"我们认为新校园公用地不仅是一个建筑物，而且也是一连串的不同的公众空间。"

作者 Thomas Müller, Ivan Reimann   合作伙伴 Anna Lemme, Jens Wesche, Marius Förster, Kristina Knapp, Milos Linhard, Thomas Kautsch, Nils Noud, Jana Galicka   专家 building services: IC Ingenieurconsult GmbH, Frankfurt/Main; structural analysis: GSE Ingenieur-Gesellschaft mbH, Berlin; landscape architecture: Jürgen Weidinger Landschaftsarchitekten, Berlin; fire protection: Peter Stanek, Berlin

Perspektive Campusgarten

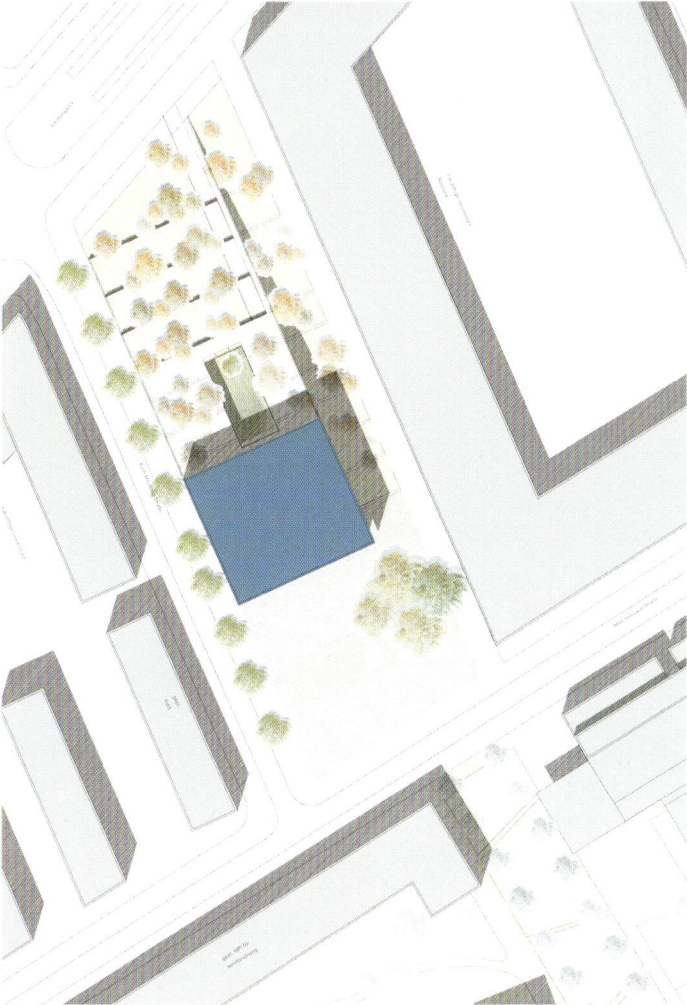

Perspektive Campusplatz

Lageplan M 1:500

Ansicht Süd M 1:200

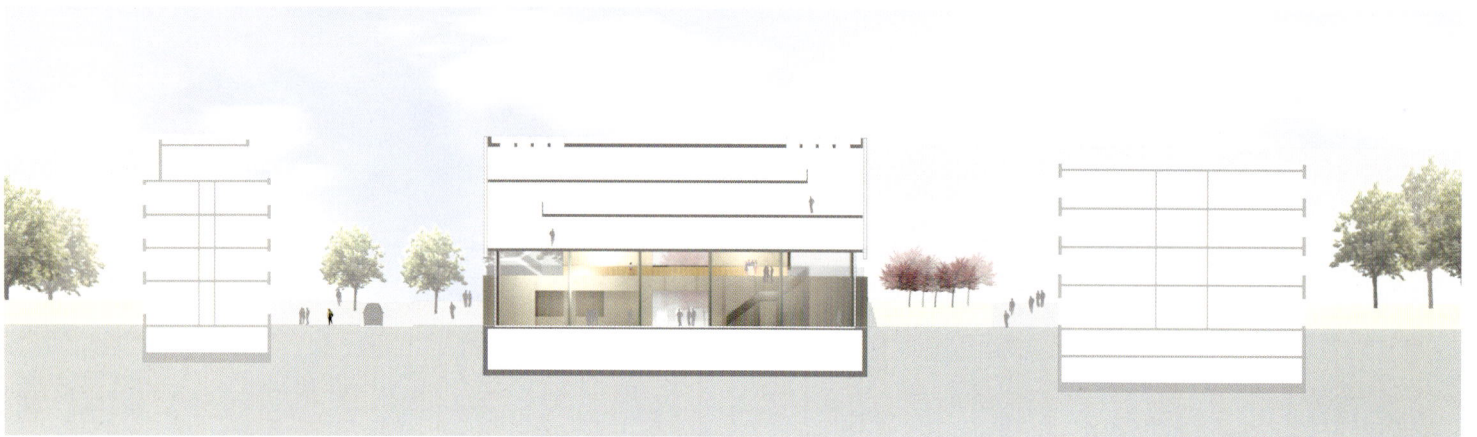

Querschnitt M 1:200

Ansicht Ost M 1:200

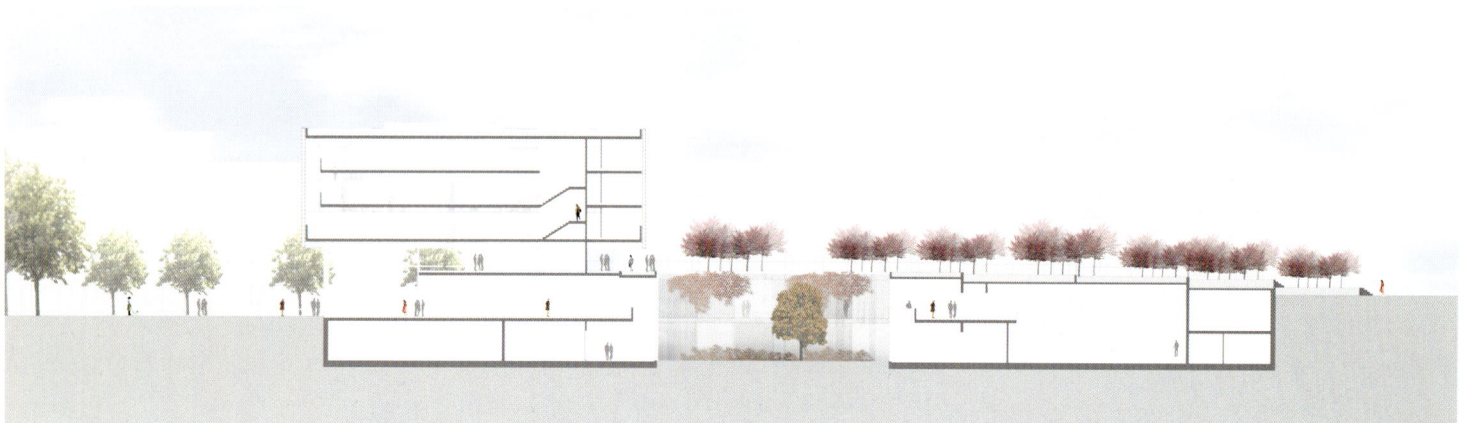

Längsschnitt M 1:200

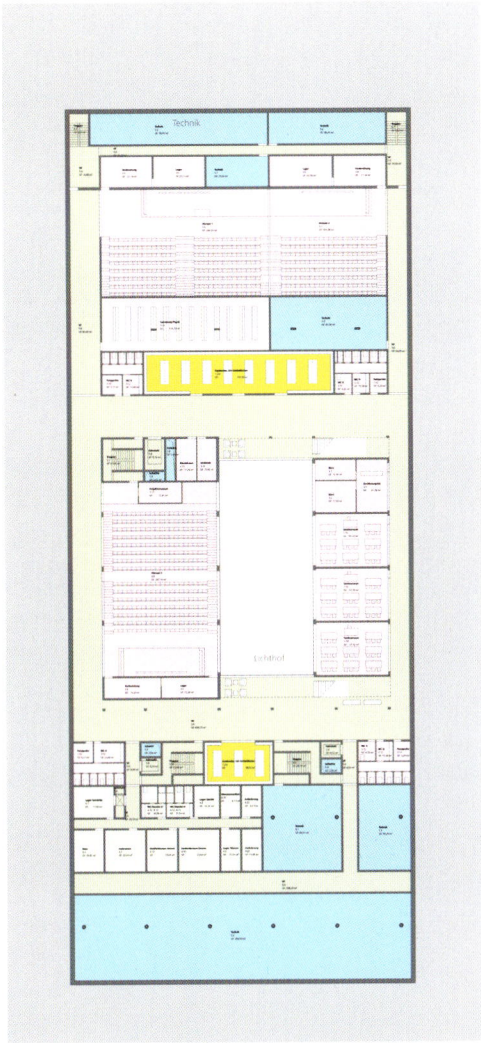

Hörsaalzentrum Sockel M 1:200

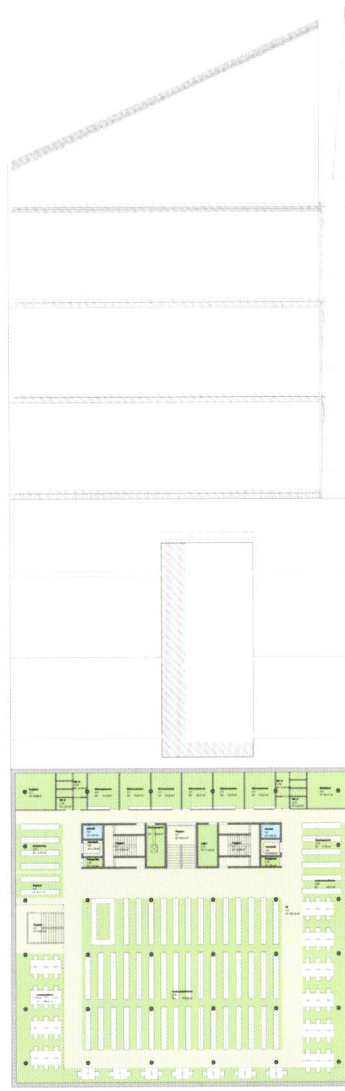

Bibliothek 2.OG M 1:200

Bibliothek 3.OG M 1:200

Bibliothek 4.OG M 1:200

Querschnitt M 1:200

behet bondzio lin architekten Münster
behet bondzio lin建筑事务所 蒙斯特

"It is the primary goal of this concept to find an overall topic of town planning for the whole campus."
"为整个校园找到一个全面的城市规划主题是这个观念的最重要的目标。"

作者 Martin Behet, Roland Bondzio, Yu-Han Michael Lin　合作伙伴 Ulf Düsterhöft, Malte Petersen, Sonja Strickmann, Britta Kasner, Paulo de Aranjo　专家 building services: Ingenieurbüro Nordhorn, Klaus Nordhorn, Münster; collaborator: Thilo Ihle; structural analysis: Ingenieur ARGE HJW, Dr. Jaenisch, Leipzig; collaborator: Mr. Krüger

Riedberg Campus Commons, Frankfurt/Main

S-Bahn H...

+2.00

Eingang Nord

+2.00

+4.00

Zentraler Campusplatz

0,00

Anlieferung

0,00

Passage

+2.00

Anlieferung Küche

+4,00

Foyer

0,00

Eingang Süd

Wasserbecken

0,00

Zentraler Campusplatz

-1,00

Erdgeschoss 2

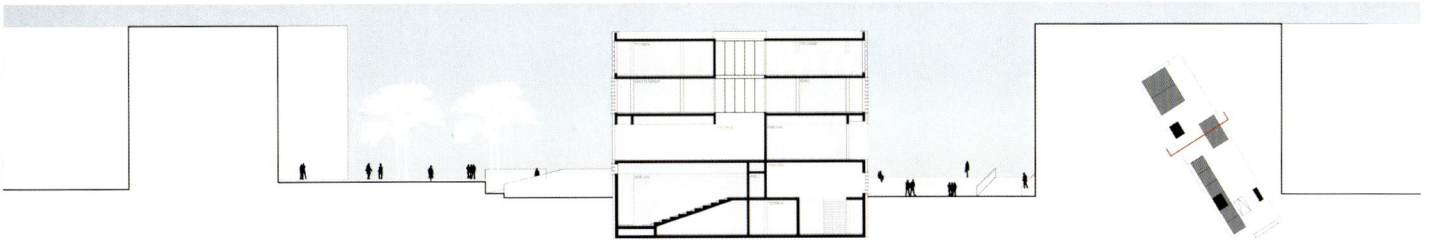

SCHNITT 1:200

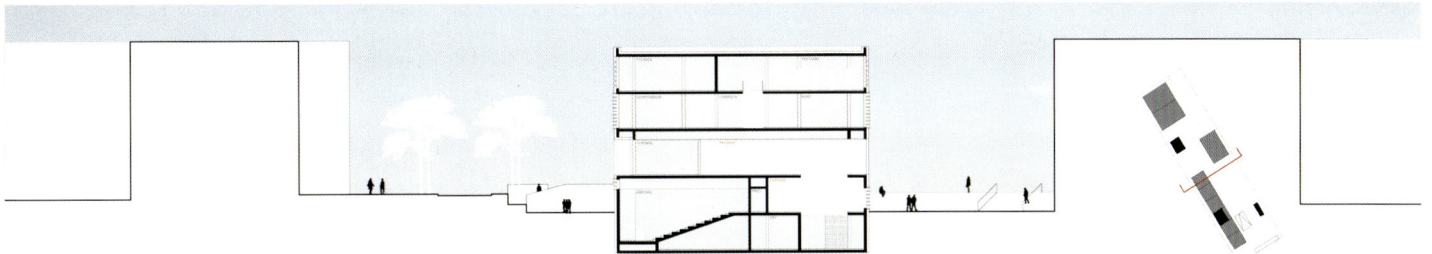

SCHNITT 1:200

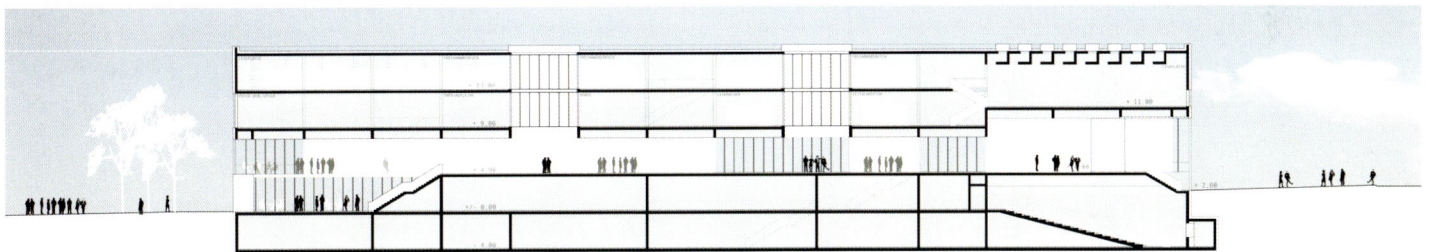

SCHNITT 1:200

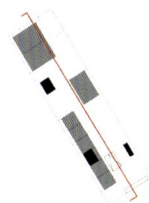

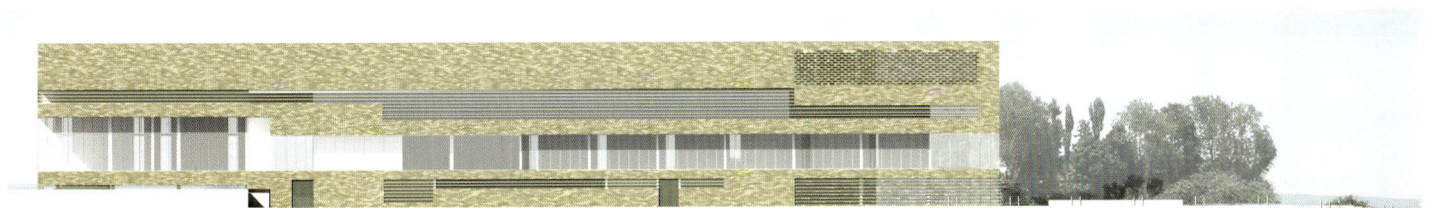

ANSICHT OST 1:200

behet bondzio lin architekten, Münster

Riedberg Campus Commons, Frankfurt/Main

HASCHER JEHLE

Aue.

KSP Engel und Zimme

New KfW Building at the Senckenberganlage in Frankfurt/Main

新KfW大楼
Senckenberganlage，德国美因河畔法兰克福

+Weber +Assoziierte

mann

New KfW Building at the Senckenberganlage Frankfurt/Main

新KfW大楼，Senckenberganlage
德国美因河畔法兰克福

Restricted project competition for general planners preceded by an application procedure
限制严格的一般规划人员的项目竞赛，开始前有申请程序。

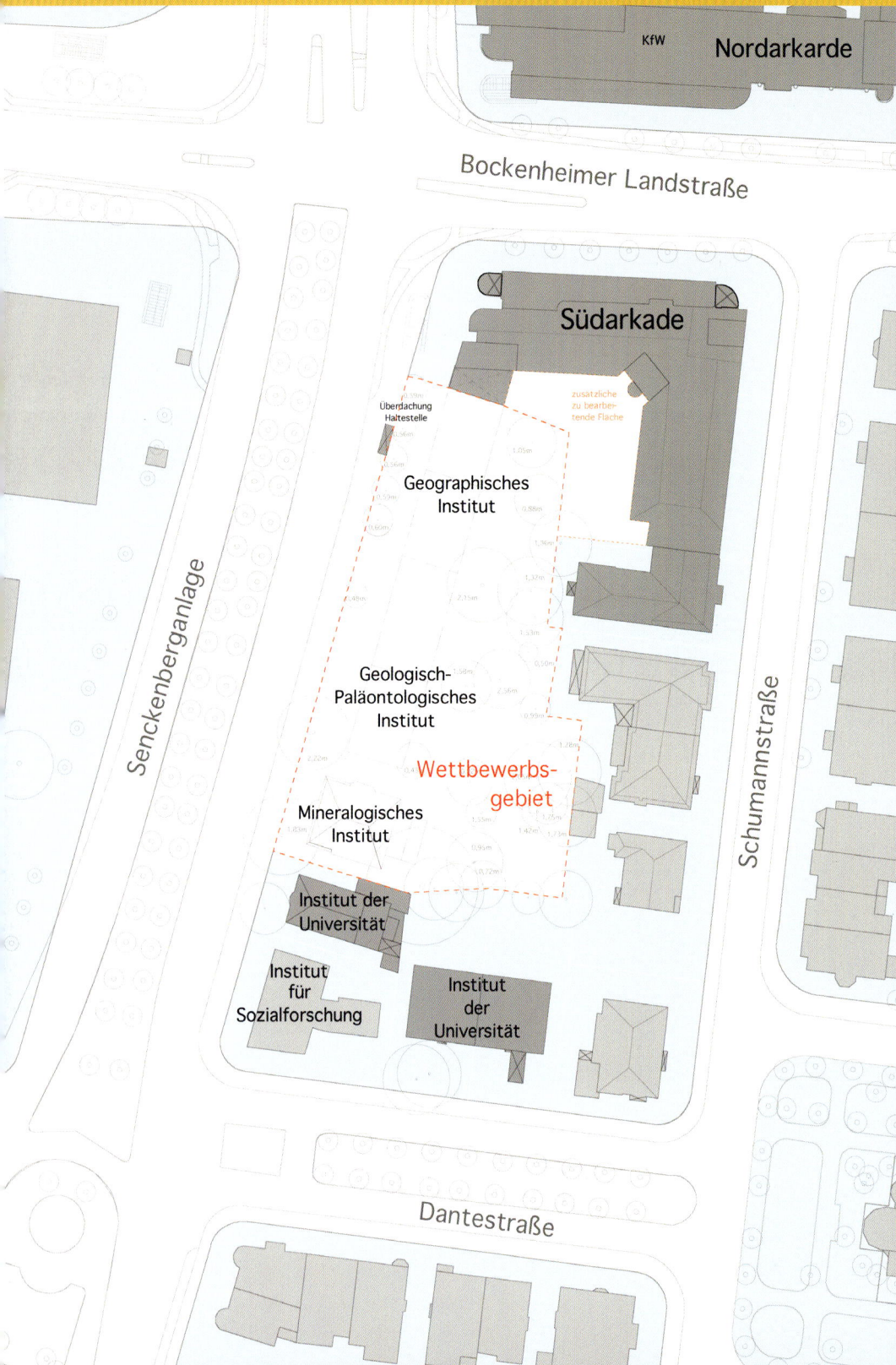

地点 Frankfurt/Main
时间 10/2005-06/2006
主办方 KfW
参赛者 10
面积 5,400 sq m
竞赛费用 127,000 Euro
专业评奖委员会
Prof. Ulrike Lauber,
Munich/Berlin;
Dieter von Lüpke,
head of city planning office,
Frankfurt/Main;
Prof. Matthias Sauerbruch,
Berlin/Stuttgart;
Alexander Theiss,
Frankfurt/Main
专家评奖委员会
Klaus J. Helms, KfW,
head of city planning office;
department new construction –
design and construction, Frankfurt/Main;
Detlef Leinberger, head of KfW,
Frankfurt/Main;
Hans W. Reich,
chairman of management board KfW,
Frankfurt/Main

The KfW (in full: Kreditanstalt für Wiederaufbau, roughly translatable as reconstruction loan institute) was set up in 1948. As a state-run promotional bank, the KfW made an important contribution to the reconstruction of Germany after World War Two. Subsequently, it was repeatedly assigned new tasks, mainly promoting small and medium-sized businesses and financing aid packages for developing countries. The lot (approx. 5,400 square metres) occupies prime inner-city space at Senckenberganlage, right next to the corporation's longterm headquarters in Frankfurt on the Main. Besides the Senckenberganlage project, the north and south arcade buildings were to be restored and a new structure was to be built at the west arcade. The aim was to increase performance at the site, facilitate workflow and provide more area for use without unduly burdening the neighborhood. The new building on the Senckenberganlage was to connect to the extant south arcade building, without looking like an extension. Rather it was intended to present itself as an independent structure. With a gross floor area of 10,000 square metres (plus two basement levels with space for 100 cars), the new building houses 350 workplaces. The KfW considers flexibility of use a precondition for meeting future needs. This was therefore a key element of the competition task, as well as providing engineering solutions for load-bearing structures and mechanical services.

KfW于1948年建立,为二战后德国重建作出重要贡献。作为一家国立发展促进银行,其工作任务不断增加新的内容,主要是促进中小型企业发展和为发展中国家提供经济援助一揽子计划。竞赛地点位于Senckenberganlage最重要的地点,紧挨着公司在法兰克福的长期总部。除了Senckenberganlage项目以外,还要恢复北面和南面拱形走道的大楼,而且在西面的拱形走道还要建一座新世界建筑物。
竞赛的目的是提高竞赛地点的性能,加速工作流程,在不给周围环境造成负担的前提下提供更多的使用面积。在Senckenberganlage上建的新大楼将连接到南面现有的建筑物,看上去不像一个延伸部分,而是一个独立的建筑。新大楼总建筑面积共有1万平方米,有350个工作室。KfW认为使用灵活是未来会议需要的一个前提。因此,除为支承结构和机械服务提供工程解决方案以外,使用的灵活性也是竞赛任务的一个关键要素。

Aerial view　鸟瞰图

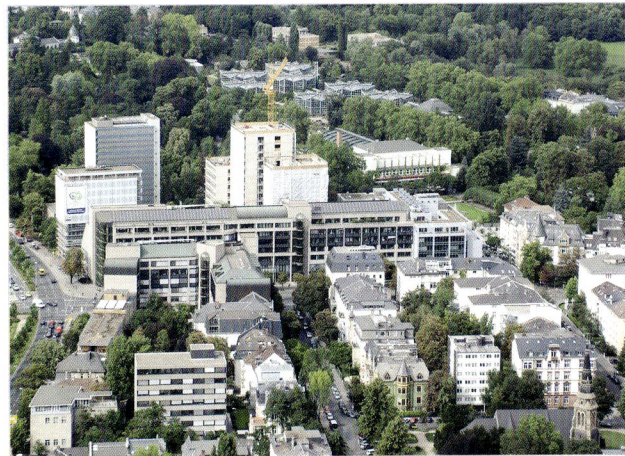

Competition site　竞赛地点

View of the City　城市景色

New KfW Building at the Senckenberganlage, Frankfurt/Main

1

2

6

7

Qualified participants
合格的参赛者

1
1st prize 一等奖
KSP Engel und Zimmermann
Architekten, Frankfurt/Main

2
2nd prize 二等奖
Auer+Weber+Assoziierte, Stuttgart/
Munich

3
3rd prize 三等奖
HASCHER JEHLE Architektur, Berlin

4
Acquisition 购买
Petzinka Pink Architekten,
Düsseldorf

5
Acquisition 购买
ASP Schweger Assoziierte
Gesamtplanung, Hamburg

6
Further participant 其他参赛者
Ingenhoven Architekten, Düsseldorf

7
Further participant 其他参赛者
BEHNISCH ARCHITEKTEN, Stuttgart

8
Further participant 其他参赛者
schneider + schumacher
Architekturgesellschaft,
Frankfurt/Main

9
Further participant 其他参赛者
struhk architekten
Planungsgesellschaft,
Braunschweig

10
Further participant 其他参赛者
Rhode Kellermann Wawrowsky,
Düsseldorf

KSP Engel und Zimmermann Architekten Frankfurt/Main
KSP Engel und Zimmermann建筑事务所 法兰克福/美因

"The volume of the extant villa is adapted to the volume metrics of the building."

"现有别墅的体积与大楼的体积韵律相适应。"

作者 Jürgen Engel   合作伙伴 Gregor Gutscher, Özgür Ilter, Antonio Vultaggio, Thomas von Girsewald, Anna Stoyanova, Ramona Becker, Silvia Grüning   专家 structural analysis: Ruffert & Partner Ing. GmbH, Limburg; Heinz-Georg Ruffert, Meinhard Rompel, Kay-Uwe Thorn; building services: HTW Hetzel, Tor-Westen + Partner, Düsseldorf; Ralf Tosetto, Sabine Hanel, Olaf Hasse   其他专家 Hegelmann, Dutt + Kist GmbH, Hanno Dutt, Luca Kist; IFFT Institut für Fassadentechnik, Herr Böhm, Karl Otto Schott

[lageplan] 1:500

[körnung] [motiv villa] [kommunikative mitte]

Architektonische Intension

Der Ausbau des Hauptsitzes der KfW Bankengruppe entlang der Senckenberganlage nach Süden bietet die Chance ein Ensemble einzigartiger Gebäude zu schaffen. Der vorliegende Entwurf trägt dem Rechnung und verfolgt folgende übergeordnete Ziele:

- Ein Signet der KfW an der Senkenberganlage
- Neue Interpretation des Themas „Villenviertel Westend"
- Flexibel nutzbare, wirtschaftliche Bürobereiche

Historischer Bezug

Interpretation der Villenstruktur
Das Frankfurter Westend wird geprägt durch eine Struktur einzelner Villen und Gebäude. Ein hohes Maß von Grünflächen und ein dichter Baumbestand charakterisieren weiterhin das Areal.
Im laufe der letzten Jahrzehnte ist diese Struktur durch Großbauten verdrängt worden. Entlang der Bockenheimer Landstraße entstanden eine Reihe Hochhäuser.
Der Mikrostandort wird auch heute noch von der ursprünglichen Idee des Westends geprägt. Der Entwurf nimmt diese Spuren auf und setzt Sie in einer modernen Interpretation der historischen Struktur um.

PLA N01

[perspekive senckenberganlage]

179236

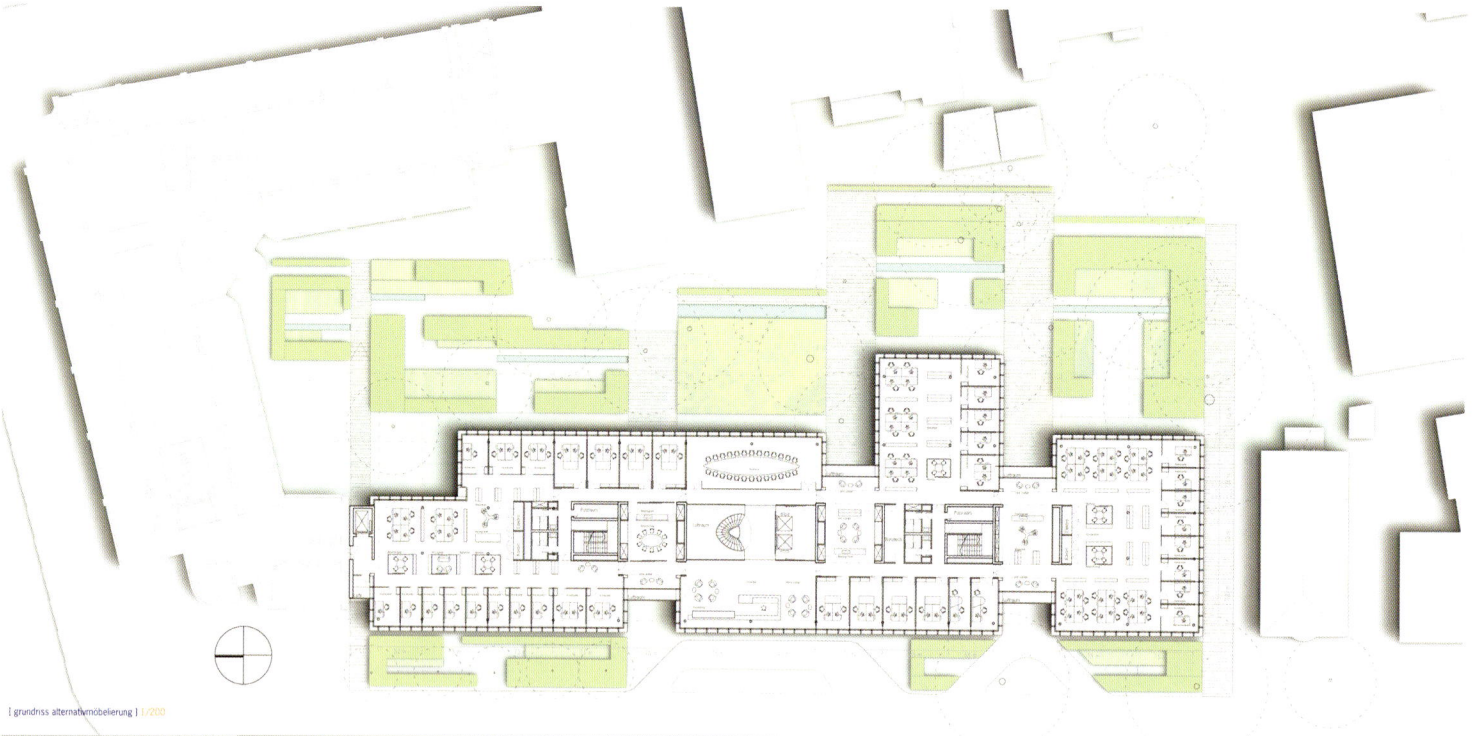

[grundriss alternativmöblierung] 1/200

PLA
NO5

[demokratische fassade]

fassadendetail 1/5

[systemschnitt · luftraum] 1/50

[lüftungskonzept]

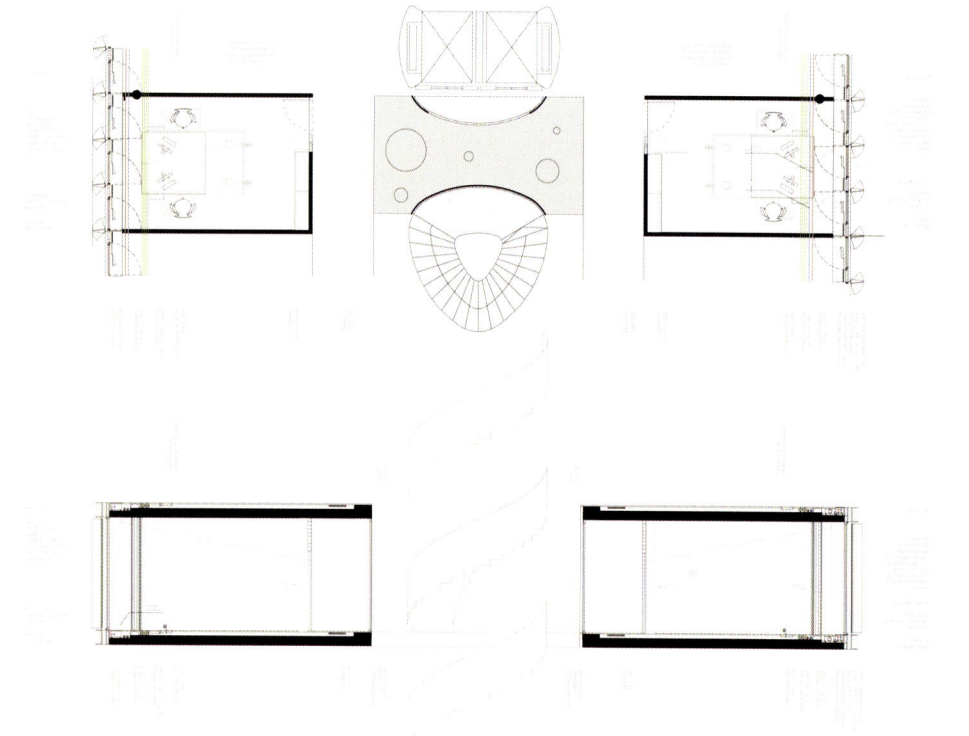

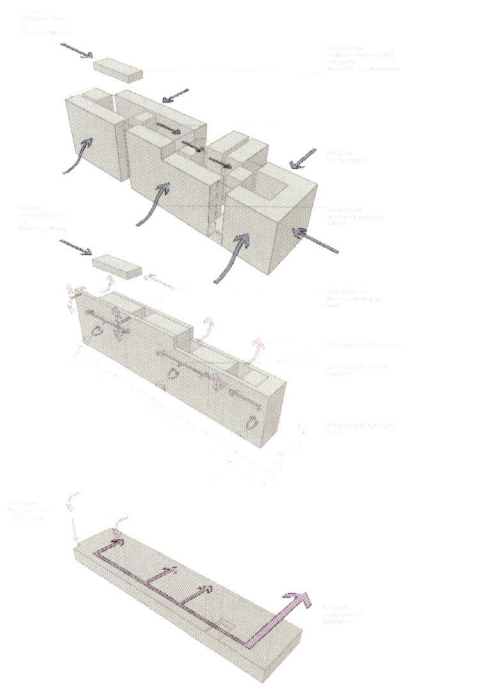

Fassade

Natürlicher Sonnenschutz mit Ausblick
Die Fassaden werden geprägt durch ihre Ost West Ausrichtung. Der außenliegende Sonnenschutz wird durch große, bewegliche Vertikallamellen sichergestellt.

Diese Lamellen erzeugen eine Grundverschattung, ohne die Aussicht maßgeblich zu behindern. Eine zentrale, sonnenstandsabhängige Steuerung kann so durchgeführt werden, ohne dass der einzelne Mitarbeiter gestört wird. Der Anteil der Fassade, der dem direkten Sonnenlicht ausgesetzt ist wird so auf ein Minimum reduziert (Restsonneneinstrahlung im Grenzfall Sommersonnenwende gegen 18.00Uhr).
Die nun noch entfallende Reststrahlung wird durch eine Sonnenschutzverglasung plus innenliegenden Blendschutz absorbiert. Der Blendschutz wird als Schiebeelement konzipiert und kann individuell der Sonne und dem Sonnenschutz nachgeführt werden (auch hier wäre eine Automation denkbar).
Ein hohes Maß an indirektem Tageslicht gelangt so noch in die Bürobereiche und reduziert den Kunstlichtanteil bei der Beleuchtung.

Materialität

Hochwertige Oberflächen entsprechen dem Anspruch der KfW
Die Struktur des Sonnenschutzes prägt das Erscheinungsbild des Gebäudes. Die Lamellen werden entsprechend der Ausrichtung zur Sonne mit zwei unterschiedlichen Oberflächen ausgestattet.
Zur Sonne, bzw. Außenseite hin, wird eine etwa 4cm starke Naturstenschicht eingelassen. Entsprechend der Gliederung des Gebäudes werden leicht differierende Oberflächen verwendet.

Die zum Gebäude zeigende Seite wird, wie die gesamte Lamellenkonstruktion in Aluminium ausgeführt. Die Oberfläche in Eloxal erzeugt eine helle Fläche mit hohem Reflexionsgrad für Tageslicht an den Arbeitsplätzen.

Je nach Öffnungsgrad der Lamelle erhält das Gebäude so eine eher steinerne oder metallische Erscheinung. Die horizontalen Halteprofile der Lamelle werden in Chromstahl ausgeführt. Die warm glänzende Oberfläche schafft einen dezenten Kontrast zu den Materialien der Lamelle.
Die gewählten Oberflächen geben dem Neubau der KfW ein zeitloses Design, dass sich sowohl in die städtebauliche Umgebung, als auch in das Ensemble der KfW einfügt.

Arbeitswelten

Flexibel nutzbarer, transparenter Dreibund
Die Arbeitsbereiche liegen, mit gleichberechtigten Ausblickmöglichkeiten, entlang der Fassade. Die Kontur des Gebäudes ist auf eine möglichst große Abwicklungslänge hin konzipiert, um eine größtmögliche belichtete Fläche zu erzeugen.
Die Innenzone, der kommunikativen Mitte, des Gebäudes liegen, neben der Infrastruktur, den Sanitärbereichen und der Erschließung auch Besprechungsräume mit Teeküche, Meetingpoints, sowie flexibel nutzbare Bereiche. Diese Zonen sind den Einschnitten zugeordnet, um eine der Nutzung entsprechende natürliche Belichtung zu gewährleisten.
Über die Mittelzone hinweg sind offene Wegbeziehungen zwischen den Abteilungen und offene Möblierungen möglich.

[lüftungskonzept]

[querschnitt]

PLA N06

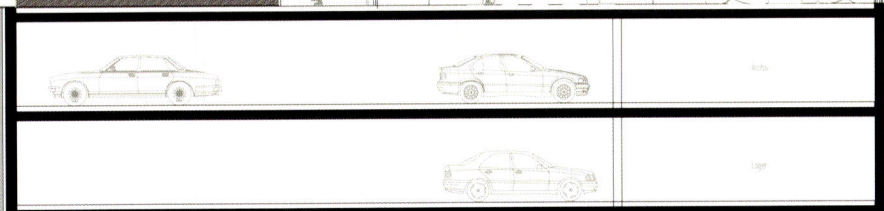

[innenraumperspektive]

[lüftungskonzept]

Lüftungstechnische Konzeption

[energiekonzept]

Geothermiezentrale

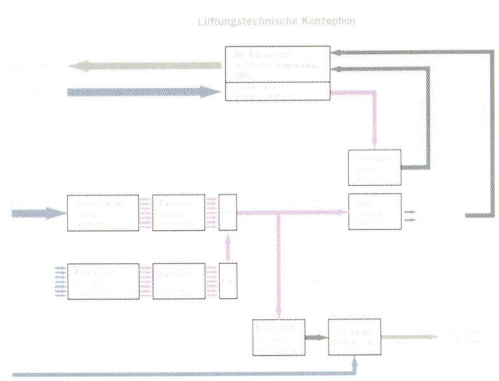

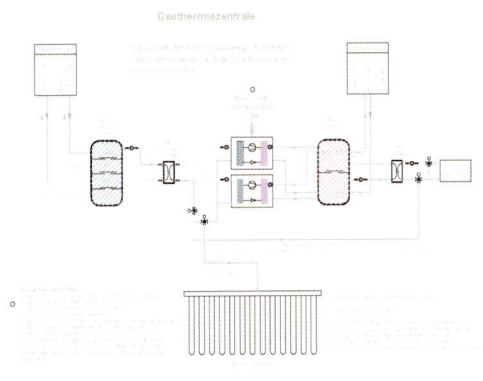

Auer + Weber + Assoziierte Stuttgart/Munich
Auer+ Weber+Assoziierte建筑事务所 斯图加特/慕尼黑

"What is intended is an integrative, while autonomous building that responds in a pliant, easy-to-get-along-with manner to the various requirements."

"我们的目的是综合性的、然而也是自治的大楼，柔韧地、随和地对不同的要求作出反应。"

作者 Auer+Weber+Assoziierte, Stuttgart/Munich   合作伙伴 Achim Söding (associate), Henrike Schlinke, Karsten Schuch, Rainer Oertelt, Daniel Hänelt, Jan Huettel, Marianne Strauss   专家 structural analysis: Pfefferkorn Ingenieure, Stuttgart; building services: Zibell, Willner + Partner, Munich   其他专家 Jörg Stötzer, landscape architect, Waldkirch

New KfW Building at the Senckenberganlage, Frankfurt/Main

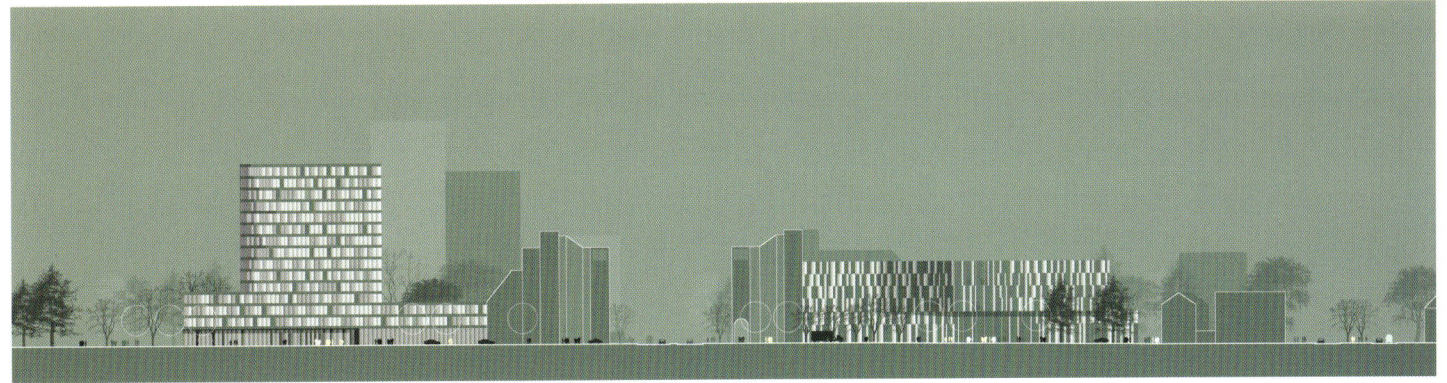

HASCHER JEHLE Architektur Berlin
HASCHER JEHLE建筑事务所 柏林

"Individual 'villas' with a depth effect blend into the texture of the surroundings. Glazed halls set back from the building line connect these 'villas' and turn them into a functional unit."

"有深邃效果的别墅个体融入周围环境的质地中，从建筑线中后移的镶有玻璃的大厅把这些'别墅'连接起来组成一个功能体。"

作者 Prof. R. Hascher, Prof. S. Jehle   合作伙伴 Fleur Keller, Michael Meier, Moritz Müller-Werther, Florian Sell   专家 structural analysis: RPB Rückert GmbH Planer + Berater, Berlin; building services: SCHOLZE Ingenieurgesellschaft mbH, Berlin
Weitere Fachberater hutterreimann Landschaftsarchitektur, Berlin

Perspektive Senckenberganlage

Bär, Stadelmann, Stöc

netzwerl

Burger Rudacs Archite

Thoma

SCHULTl

Gerber Architekten

agps

University and State Library in Darmstadt

达姆施塔特大学和州立图书馆
德国

architekten

Müller Ivan Reimann

S FRANK ARCHITEKTEN

University and State Library Darmstadt
达姆施塔特大学和州立图书馆 德国

Restricted two-stage project competition preceded by an application procedure
限制严格的项目竞赛，分为两个阶段，开始前有申请程序。

地点 Darmstadt   时间 06/2005-12/2005   主办方 TU Darmstadt   参赛者 1st stage: 55; 2nd stage: 14   面积 19,000 sq m
竞赛费用 172,000 Euro   专业评奖委员会 Prof. Hilde Barz-Malfatti, Weimar; Prof. Rebecca Chestnutt, Berlin; Prof. Markus Gasser, Darmstadt/Zurich; Prof. Wolfgang Lorch, Darmstadt/Saarbrücken; Prof. Matthias Sauerbruch, Berlin/Stuttgart   专家评奖委员会 Dr. Hans-Georg Nolte-Fischer, director ULB Darmstadt; Günter Schmitteckert, ministry of higher education, research and the arts, Federal State of Hessen, Wiesbaden; Dieter Wenzel, local council, City of Darmstadt; Prof. Dr.-Ing. Johann-Dietrich Wörner, president TU Darmstadt

It sees itself as a modern hybrid library: As its first function, it will be an academic service center for information retrieval, a place for study and work, for communication and human encounter. The second, more public function is that of a state library whose importance, in view of its highly valued historic collection, surpasses regional boundaries. The project site is located at TUD's long-established premises, adjacent to the palace and Darmstadt's city center. It is framed around an inner courtyard which is listed as a heritage site and borders on the "old suburb", likewise a listed site. The area lies in between the TUD's historic main building and the university commons which dates from the 1970s. The main task set to the architects was to find sensitive and original ways to integrate the new structure into the context of a highly eclectic group of buildings, some of them are listed, and thus create a new overall architectural scenario. The program of spaces includes various reading rooms, administrative areas, event venues, magazines, and workshops.

规划兴建的新大学和州立图书馆是一个现代的混合图书馆。首先，它是一个学术服务中心，可以检索信息、工作学习、交往沟通。其次，州立图书馆由于其历史收藏珍贵，其重要性已远远超出地区的界限。项目地点位于达姆施塔特的技术大学历史悠久的校园，紧临宫殿和市中心。项目在邻近"老郊区"的一个被列为历史地段的一个内庭旁边进行，位于达姆施塔特的技术大学的具有历史意义的主楼和始建于70年代的大学公用地之间。竞赛的主要任务是找到敏感的、新颖的方式使新建筑物与一些非常折中的建筑群浑然一体。空间规划包括各种阅览室、管理区、活动场馆、杂志室和工作室。

Aerial view　鸟瞰图

Competition site　竞赛地点

Competition site　竞赛地点

View of the City　城市景色

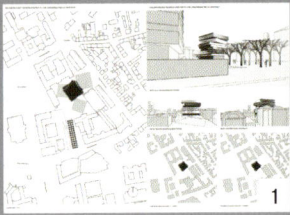

1

2

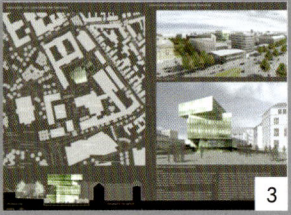

3

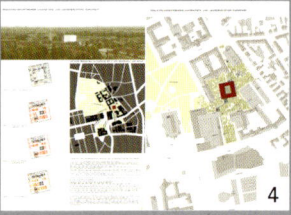

4

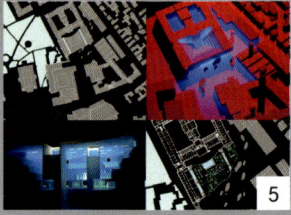

5

8

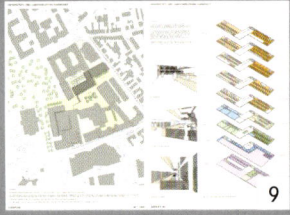

9

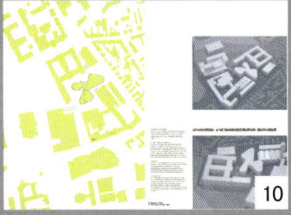

10

11

12

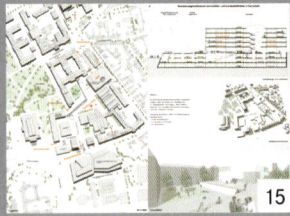

15

16

17

18

21

22

23

24

25

28

29

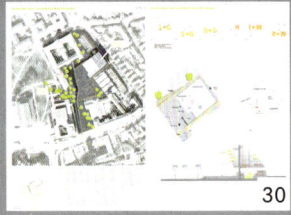

30

31

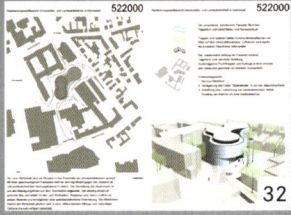

32

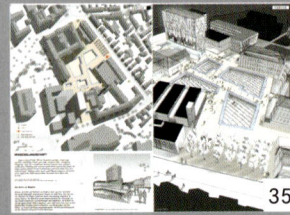

35

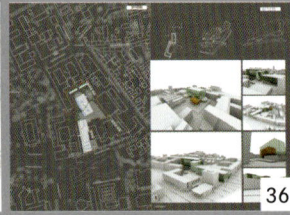

36

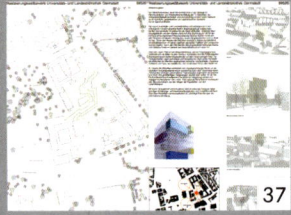

37

38

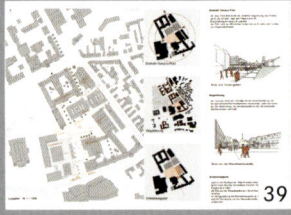

39

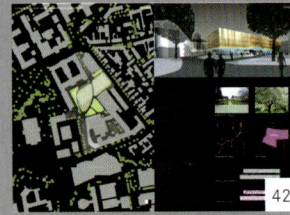

42

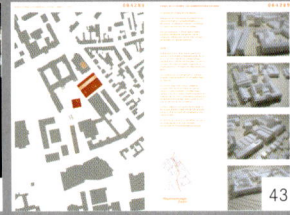

43

44

45

46

49

50

51

52

53

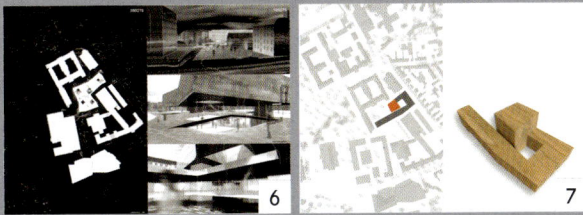

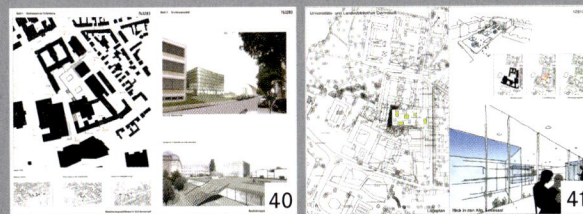

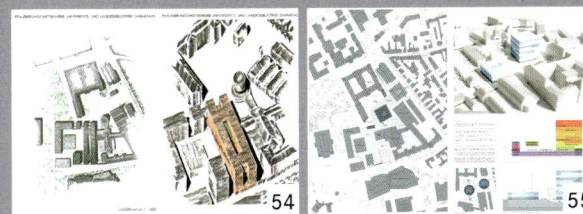

Qualified participants 1st stage 第一阶段合格参赛者

1 Peter Kulka Architektur, Cologne **2** ff-Architekten, Berlin **3** Auer+Weber+Assoziierte, Stuttgart/Munich **4** Gerber Architekten, Dortmund **5** SCHULTES FRANK ARCHITEKTEN, Berlin **6** Kirsten Schemel Architekten BDA, Berlin **7** Bär, Stadelmann, Stöcker Architekten BDA, Nuremberg **8** KSP Engel und Zimmermann, Frankfur/Main **9** Thomas Müller Ivan Reimann, Berlin **10** Burger Rudacs Architekten, Munich **11** netzwerkarchitekten, Darmstadt **12** ASTOC Architects & Planners, Cologne **13** agps architecture, Zurich **14** Plasma Studio, London

Not qualified participants 1st stage 第一阶段不合格参赛者

15 Hoechstetter und Partner, Darmstadt **16** Baumschlager-Eberle Ziviltechniker GmbH, Lochau **17** Ferdinand Heide Architekt, Frankfurt/Main **18** Ortner + Ortner Baukunst, Berlin **19** Goldfinger A. Roloff Ruffing Sill, Hamburg **20** Pahl + Weber-Pahl Architekten, Darmstadt **21** Du Besset-Lyon Architectes, Paris **22** Heckmann - Jung Freie Architekten, Stuttgart **23** Abelmann Vielain Pock Architekten, Berlin **24** HASCHER JEHLE Architektur, Berlin **25** Prof. Jörg Friedrich, PFP Architekten, Braunschweig **26** Schneider + Sendelbach Architektengesellschaft, Braunschweig **27** Gatermann + Schossig Architekten Generalplaner, Cologne **28** gmp - Architekten von Gerkan, Marg und Partner, Hamburg **29** Blauraum Architekten, Hamburg **30** HG Merz Architekten, Berlin **31** Henning Larsen Architects, Kopenhagen **32** K+P Architekten und Stadtplaner GmbH – Koch Drohn Schneider Voigt, Munich **33** Van Den Valentyn - Architektur, Cologne **34** Scheuring und Partner Architekten, Cologne **35** Architectenbureau Micha de Haas, Amsterdam **36** Keith Williams Architects, London **37** ASP Schweger Assoziierte Gesamtplanung GmbH, Hamburg **38** AssmannSalomon, Berlin **39** Gössler Architekten, Berlin **40** Bernhardt + Partner, Darmstadt **41** Schuster Architekten, Düsseldorf **42** Benthem Crouwel GmbH, Aachen **43** Herzog + Partner, Munich **44** Braunfels Architekten, Berlin **45** AS&P - Albert Speer und Partner GmbH, Frankfurt/Main **46** waechter+waechter architekten, Darmstadt **47** Max Dudler, Berlin **48** feuerstein + gerken, Munich **49** Anin·Jeromin·Fitilidis & Partner Architekten & Ingenieure, Düsseldorf **50** Lederer+Ragnarsdóttir+Oei, Stuttgart **51** Univ.Prof.Arch.DI Klaus Kada, Graz **52** Opus Architekten, Darmstadt **53** Bez+Kock Architekten, Stuttgart **54** Architekturbüro Böhm **55** Knoche Architekten, Stuttgart

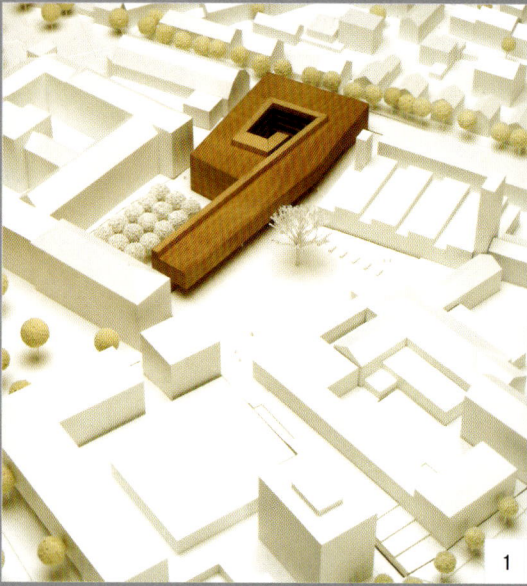

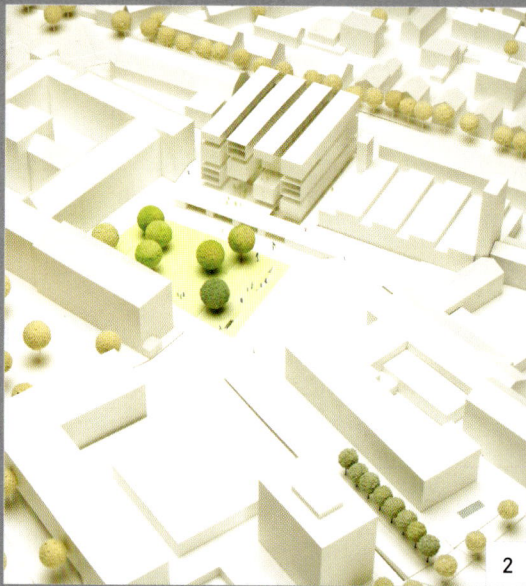

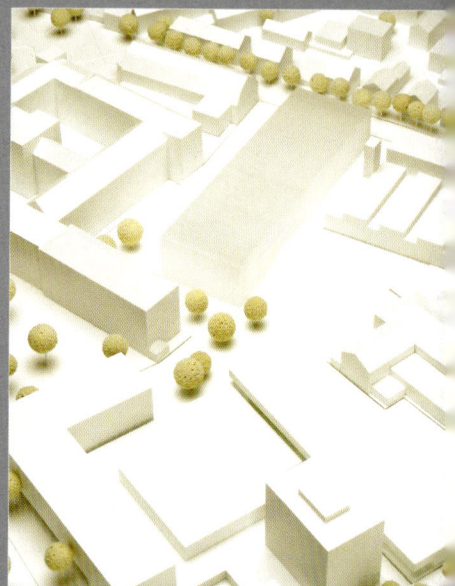

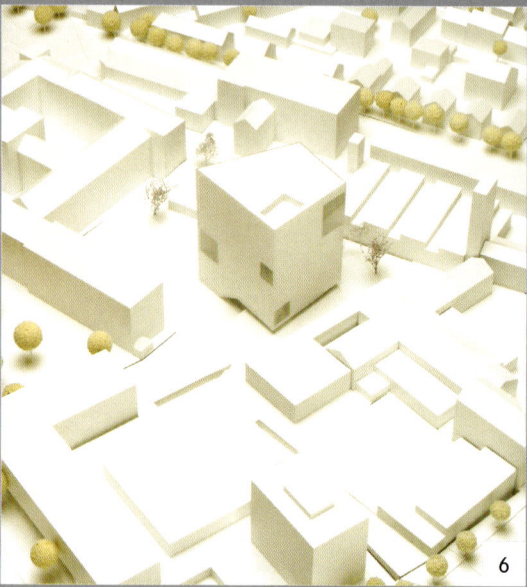

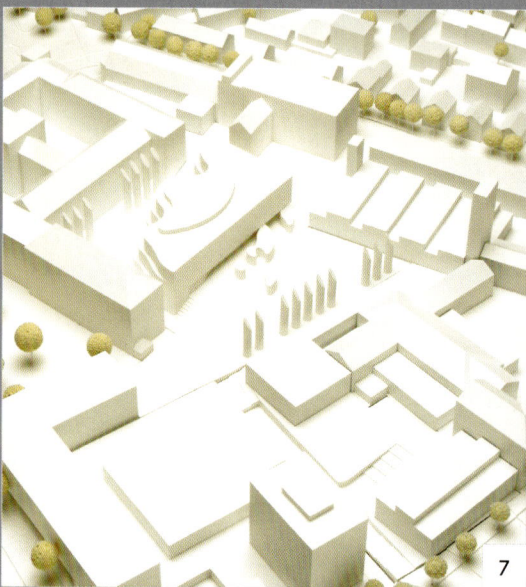

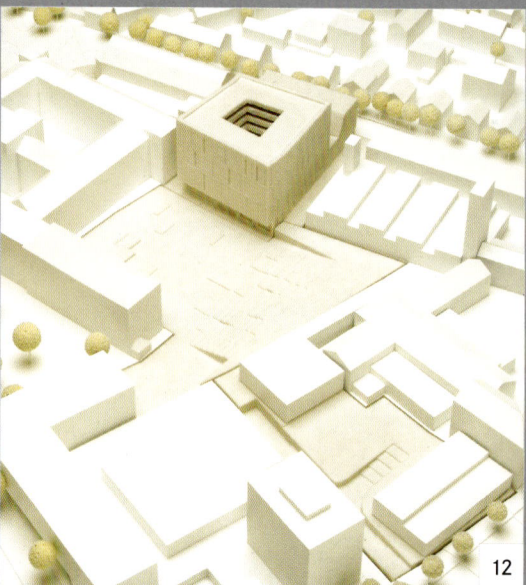

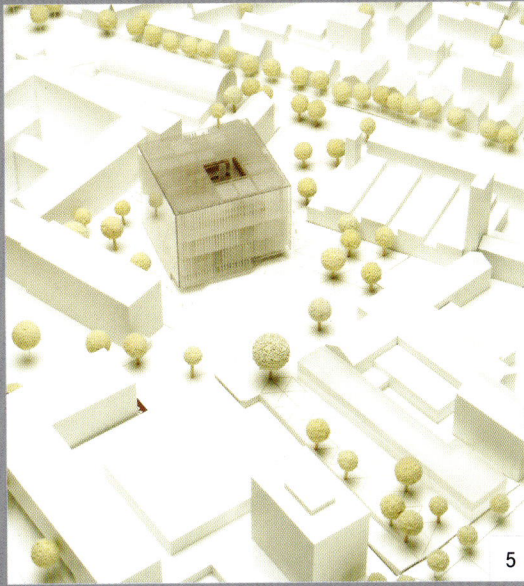

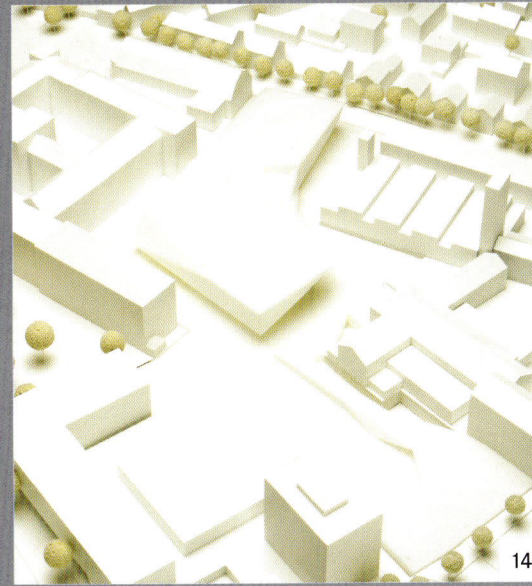

Participants 2nd stage
第二阶段参赛者

1
1st prize   一等奖
Bär, Stadelmann, Stöcker
Architekten, Nuremberg

2
2nd prize   二等奖
netzwerkarchitekten, Darmstadt

3
3rd prize   三等奖
Burger Rudacs Architekten, Munich

4
3rd prize   三等奖
Thomas Müller Ivan Reimann, Berlin

5
5th prize   五等奖
Gerber Architekten, Dortmund

6
1st acquisition   购买一等奖
agps, Zurich

7
Acquisition   购买
SCHULTES FRANK ARCHITEKTEN,
Berlin

8
Acquisition   购买
Plasma Studio, London

9
2nd round   第二轮
Peter Kulka Architektur, Cologne

10
2nd round   第二轮
ASTOC Architects & Planners,
Cologne

11
2nd round   第二轮
Kirsten Schemel Architekten, Berlin

12
1st round   第一轮
Auer+Weber+Assoziierte, Stuttgart/
Munich

13
1st round   第一轮
ff-Architekten, Berlin

14
1st round   第一轮
KSP Engel und Zimmermann
Architekten, Frankfurt/Main

357

Bär, Stadelmann, Stöcker Architekten Nuremberg
Bar, Stadelmann, Stocker建筑事务所 纽伦堡

"The design is marked by the idea of space creation, interweaving, and identification, and by the respect for the university's heterogeneous environs."

"该设计以其空间创造、相互交织和等同的思想，也以其对大学周围各异环境的尊重而引人注目。"

作者 Friedrich Bär   合作伙伴 Anja Vogl   专家 building services: Ingenieurbüro Hausladen GmbH, Josef Bauer, Kirchheim

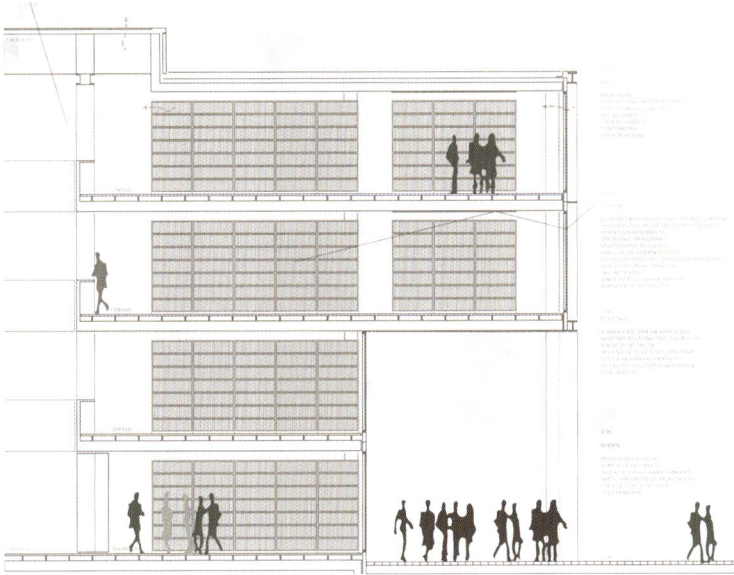

Detail 1:50

Schnitt + Klimakonzept

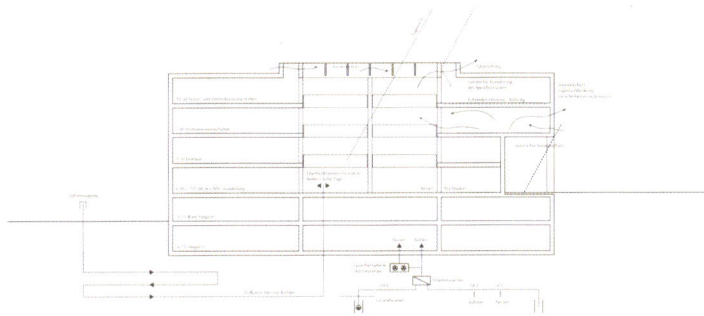

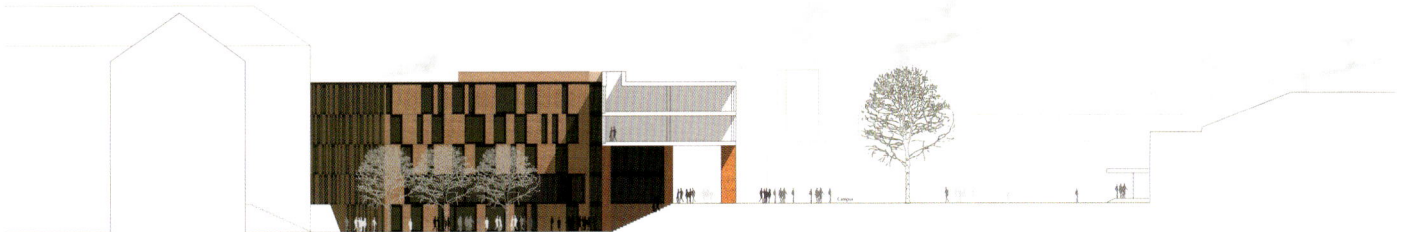

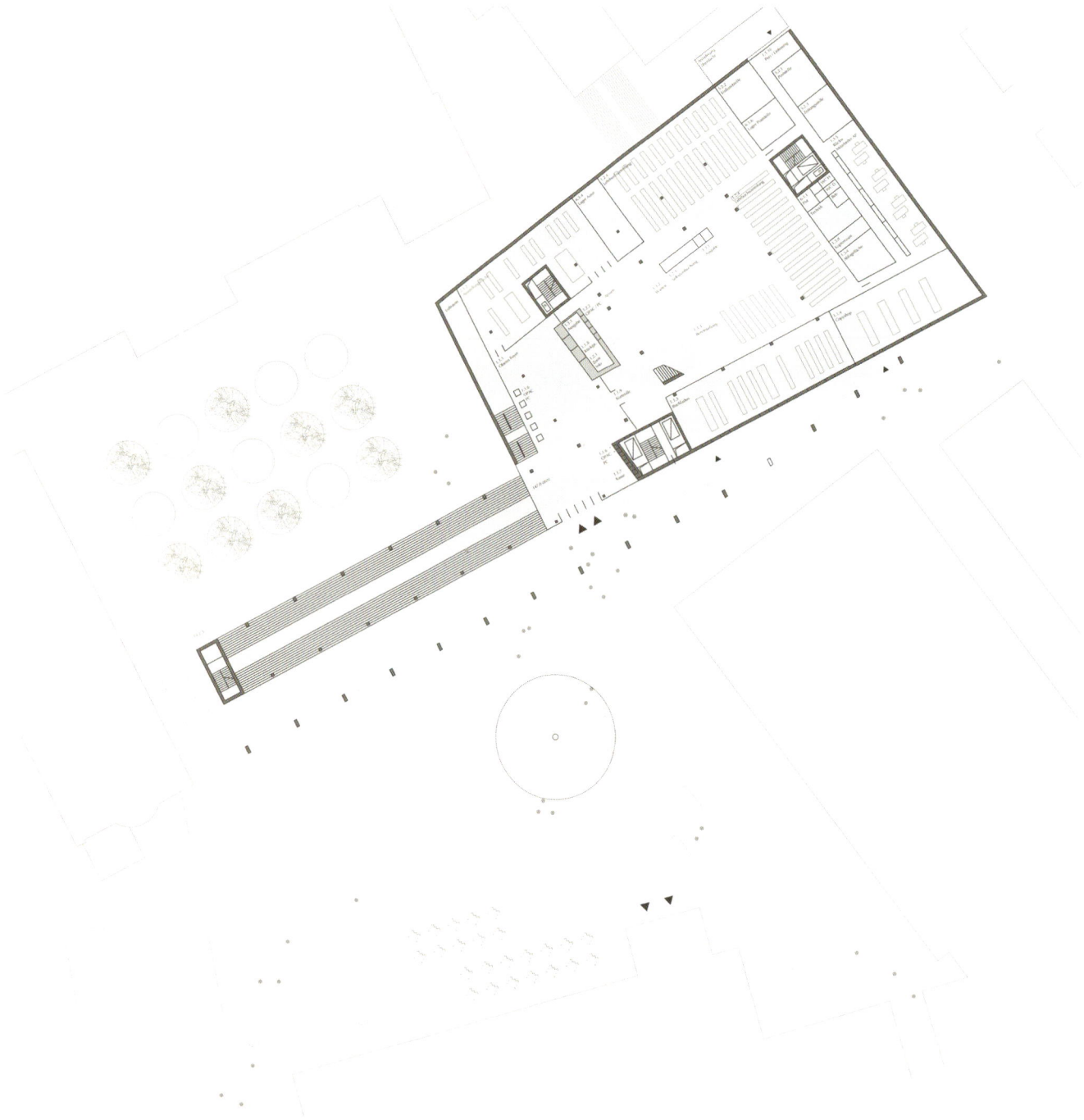

Erdgschoss 1:200

3

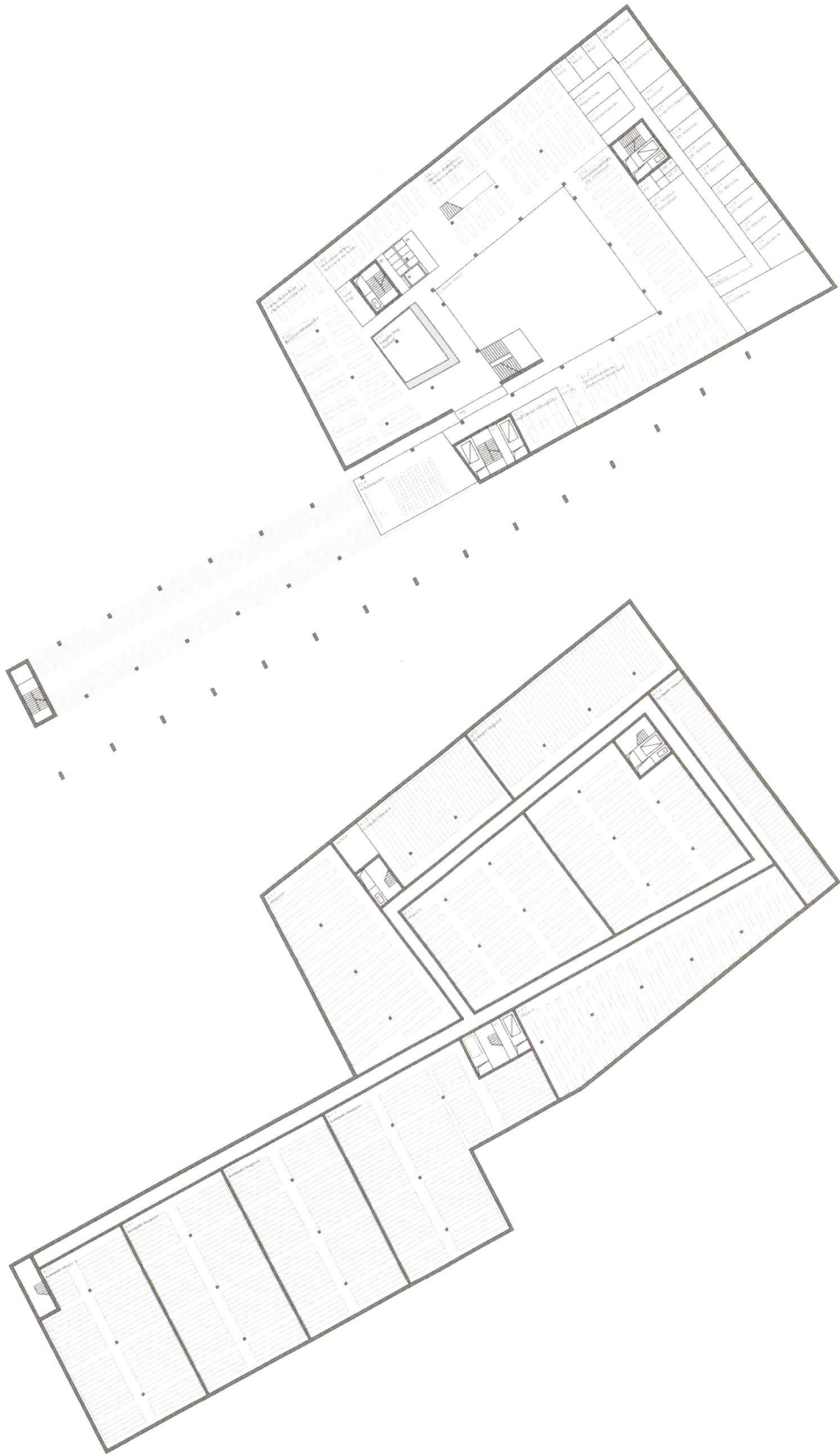

1. Obergeschoss 1:200

UG 1:200

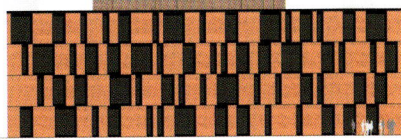

netzwerkarchitekten Darmstadt
Netzwerk建筑事务所 达姆施塔特

"The structure of the new university and state library will be located between Magdalenen Straße and the block's interior, thus freeing space for an ample campus."

"新大学和州立图书馆建筑将位于Magdalenen StraBe和区域内部之间，从而为校园释放了更多的空间。"

作者Thilo Höhne, Karim Scharabi, Philipp Schiffer, Jochen Schuh, Marcus Schwieger, Oliver Witan 合作伙伴 Petra Lenschow, Jeremias Lorch , Marvin King 专家 structural analysis: Bollinger und Grohmann, Frankfurt/Main; building services: Platzer Ingenieure, Bad Nauheim; landscape architects: Club L94, Cologne

Detailschnitt

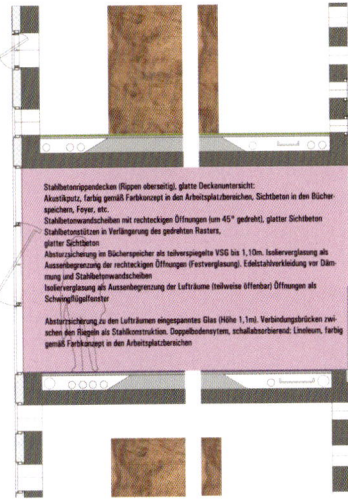

Stahlbetonrippendecken (Rippen oberseitig), glatte Deckenuntersicht:
Akustikputz, farbig gemäß Farbkonzept in den Arbeitsplatzbereichen, Sichtbeton in den Bücher-speichern, Foyer, etc.
Stahlbetonwandscheiben mit rechteckigen Öffnungen (um 45° gedreht), glatter Sichtbeton
Stahlbetonstützen in Verlängerung des gedrehten Rasters,
glatter Sichtbeton
Absturzsicherung in Bücherspeicher als teilverspiegeltes VSG bis 1,10m. Isolierverglasung als
Aussenbegrenzung der rechteckigen Öffnungen (Festverglasung), Edelstahlverkleidung von Däm-mung und Stahlbetonwandscheiben
Isolierverglasung als Aussenbegrenzung der Lufträume (teilweise öffenbar) Öffnungen als
Schwingflügelfenster

Absturzsicherung zu den Lufträumen eingespanntes Glas (Höhe 1,1m). Verbindungsbrücken zwi-schen den Riegeln als Stahlkonstruktion. Doppelbodensystem, schallabsorbierend: Linoleum, farbig
Farbkonzept in den Arbeitsplatzbereichen

Technik

Allgemein

Die kompakte Gebäudestruktur hat ein A/V-Verhältnis von ca. 0,3. Sie öffnet
sich mit seinen Lesebereichen im Wesentlichen gegen Nordwest und schließt
sich zu den wärmebelasteten Himmelrichtungen Nordost bis Südwest. Bei
Außentemperaturen oberhalb +17 °C werden Raumlufttechnische Anlagen
für den Außenzonenbereich (bis ca. 8 m Tiefe) abgeschaltet um diese natürlich
zu belüften und unterstützend durch wasser-basierte Kühldecken zu kühlen
um angenehme Raumzustände zu erreichen. Bei Außentemperaturen, die ein
Beheizen des Gebäudes nach sich ziehen, soll das Gebäude komplett geschlossen
werden.
Durch den Eintrag von Luft über eine Lüftungsanlage kann in hohem Maße eine
Wärmerückgewinnung erfolgen. Die Magazinbereiche werden im Wesentlichen
hermetisch abgeschlossen und mit einem ca. 2fachen Luftwechsel klimatisiert,
um sowohl Temperaturschwankungen wie auch Feuchteschwankungen in sehr
geringen Bandbreiten zu erreichen.

Lüftungskonzept:

Die Zuluftzuführung erfolgt über geschlitzte Doppelbodenstrukturen
(Druckboden), so dass sich eine Quelllüftung einstellt und mit minimalen
Luftmengen gearbeitet werden kann. In den Randbereichen des Gebäudes
werden Unterflurkonvektoren installiert, die die Glasflächen hinsichtlich ihrer
Oberflächentemperaturen abdecken und den Wärmeverlust ausgleichen.

Durch die besondere Struktur des Gebäudes und die offenen Raumbereiche
kann die Abluft und die Entrauchung über das gleiche Abluftsystem aus dem
Gebäude abgeführt werden das mit einem regenerativen Wärmerückgewinnungs-
Systemen ausgestattet ist. Zur Reduzierung des Kälteenergiebedarfs sind sie als
Desorptionsanlagen ausgebildet (Nutzung adiabat abgekühlter Abluftströme).

Leselandschaft

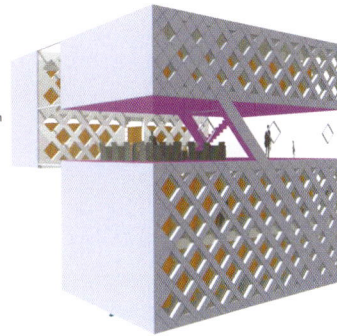

Buchbereiche

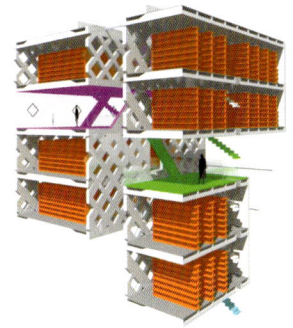

städtebauliche Ergänzung

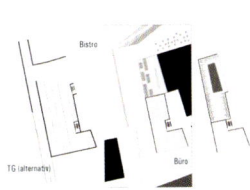

Bistro
Büro
TG (alternativ)

Lüftung /Entrauchung

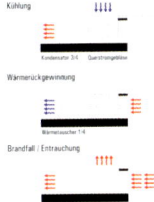

Kühlung
Wärmerückgewinnung
Brandfall / Entrauchung

Licht

Schall

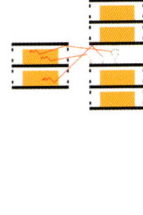

Erschließung

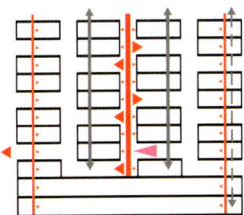

Buchförderanlage

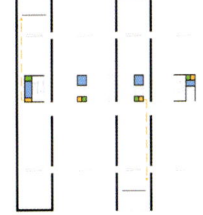

Westansicht

Erläuerungstext

Städtebauliche Einbindung

Der Baukörper für die neue Universitäts- und Landesbibliothek Darmstadt wird im Baufeld zwischen Magdalenenstraße und Blockinnenbereich platziert, so dass Raum für einen großzügigen Campus im Inneren entsteht. Zur Magdalenenstraße hin wird die Bibliothek aus den Gebäudefluchten der angrenzenden Bebauung über eine platzartigen Aufweitung zurückgesetzt. Hierdurch wird einerseits der Höhenentwicklung des Gebäudes im Verhältnis zum angrenzenden Quartier Rechnung getragen und andererseits die Durchwegung und Anbindung des Campus für den Fuss- und Radverkehr gestärkt. Die historische Maschinenhalle wird in ihrer Solitärcharakter thematisiert und erhält eine neue Kopfsituation zur Bibliothek hin. Das Gebäude stapelt sich mit seinen Bücherboxen und offenen Lesedecks in die Höhe und gibt dem neuen Universitätscampus sein Gesicht.

städtebauliche Einbindung

Der Campus wird als Kreuzungs- und Treffpunkt entwickelt, an dem sternförmig alle wesentlichen Fußwegeverbindungen der Hochschulumgebung anknüpfen, bzw. an dem die beiden wesentlichen Niveaus - Herrengarten und Plattform/ Uni- miteinander verflochten werden:
Von der Stadtseite her wird die Bestandsebene des Universitätszentrums großzügig über Sitzstufen an das obere Niveau von Mensa, Maschinenbaugebäude, Haupteingang Bibliothek und ehemaligem Hauptgebäude der Uni angeknüpft.
Vom Kongreßzentrum her kommend bereitet ein kleiner Platz das Entrée zur baumbegleiteten Wegeverbindung zum Campus. Die gegenwärtigen Gebäudeteile von Cafeteria und anschließenden Büroräumen werden durch einen neuen, gestreckten Gebäudeflügel entlang dieser Durchwegung ersetzt. Da in dieser Konfiguration nun die Anlieferung der Mensa und Cafeteria ostseitig des neuen Gebäudes und somit getrennt von der Rampe zur TG der Hochschule organisiert ist, kann die TG-Zufahrt in weiten Teilen abgedeckt werden. Die Hörsäle erhalten großzügige Oberlichter.
Das Foyer des Audimax, wie auch der Herrengarten bzw. die Hochschulstraße, sind nach wie vor an die untere Ebene des Campus angeschlossen.
Über die neue Mitte mit baumbestandener 'Leselounge' erreicht man das Sockelgeschoss des ehemaligen Hauptgebäudes und den unteren Eingangsbereich der Bibliothek an dem sich die öffentlichen Funktionen wie Cafeteria, Buchhandlung und Copyshop bündeln. Eine großzügige Treppenanlage mit diagonaler Rampe führt barrierefrei auf das obere Niveau zur Mensa hinauf.
Vom Kantplatz her führt eine gestreckte Rampe zwischen ehemaligem Hauptgebäude und Maschinenhalle zum Campus. An der Südwestecke des ehemaligen Hauptgebäudes, also in unmittelbarer Nähe des Haupteingangs der Bibliothek, wird an das 1. Obergeschoss angeschlossen, so dass die Bestandsbrücke entfallen kann und das Hauptgebäude in seiner gegenwärtigen Organisation kaum angepasst werden muss.

Strukturprinzip

Speicher

Ordnung

Schnittstelle

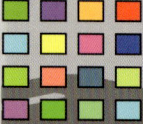

Verknüpfung

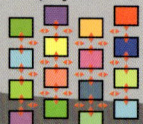

145,52
144,80
142,50
hist. Maschinenhalle
146,10
146,76
IIX
Hauptgebäude
143,50
Magdalenenstraße
146,87
Campus
144,08
147,22
Maschinenbauhalle
146,80
Kongreßhotel
Mensa
149,33
147,51
147,87
TU-Audimax
ehem. Kaserne
147,51
Städtebauliche Ergänzung
Universitätszentrum
149,29
Alexanderstraße
145,96
Kongress- und Wissenschaftszentrum
N

Lageplan M. 1:500

Ansicht Süd M. 1:500

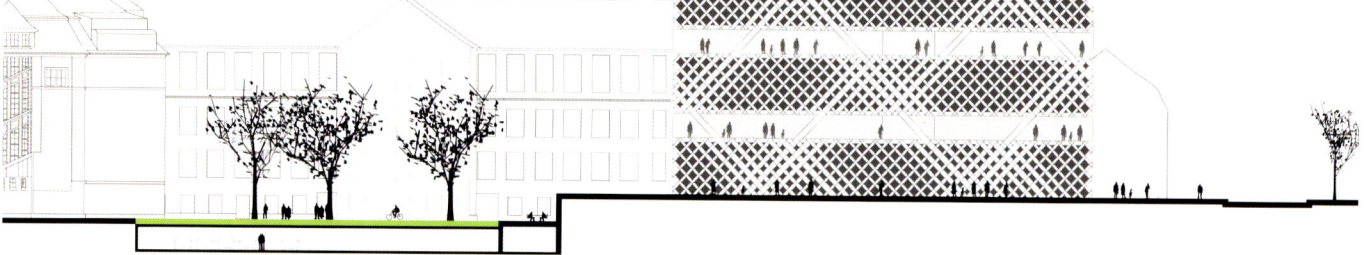

Blatt 1

Grundrisse M. 1:200

OG 5

OG 6

OG 7

OG 5

OG 6

OG 7

Querschnitt M. 1:200

Längsschnitt M. 1:200

Lesebereich 3. OG

Burger Rudacs Architekten Munich
Burger Rudacs建筑事务所 慕尼黑

"The new building for Darmstadt's university and state library is conceived as a large, four-storey, east-west-oriented bar."

"达姆施塔特的大学和州立图书馆新大楼被看成是一个大的、四层楼的、东西朝向的栅栏。"

作者 Stefan Burger, Birgit Rudacs   专家 landscape architects: Lohrer Hochrein landscape architects BDLA, Munich; Team Pawlowski, Ingenieurbüro im Bauwesen, Munich; building services: Schreiber Ingenieure GmbH, Ulm

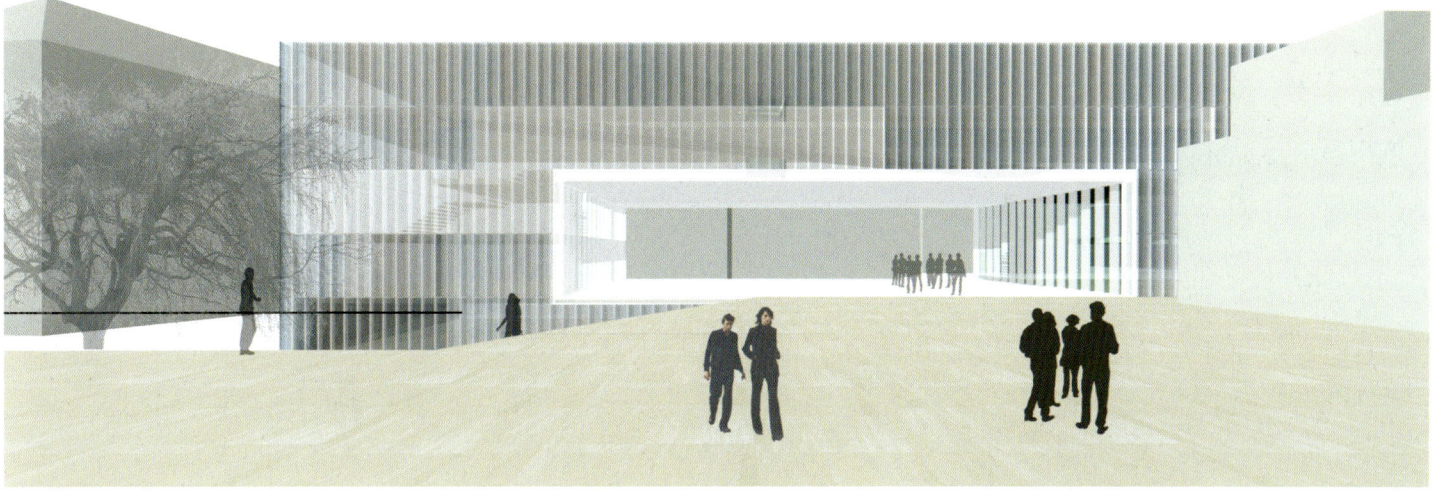

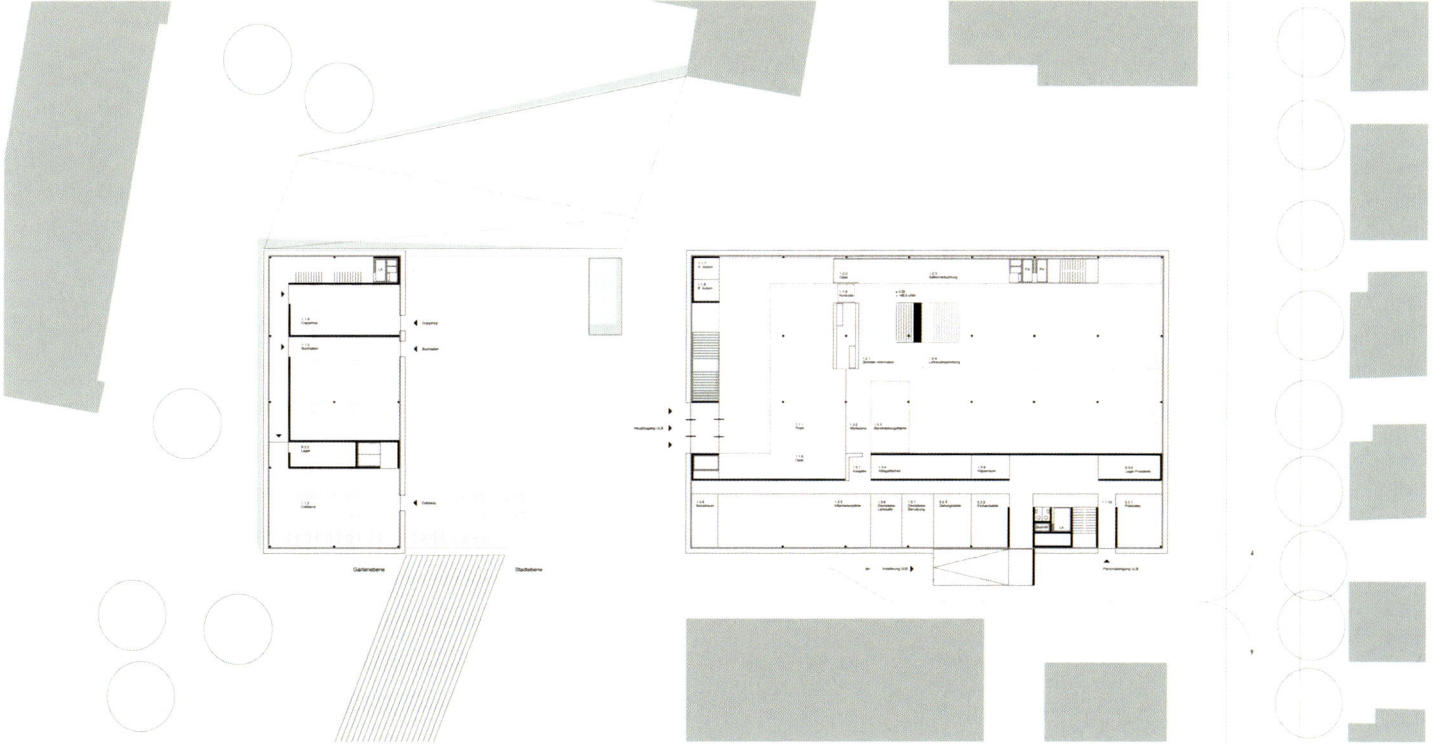

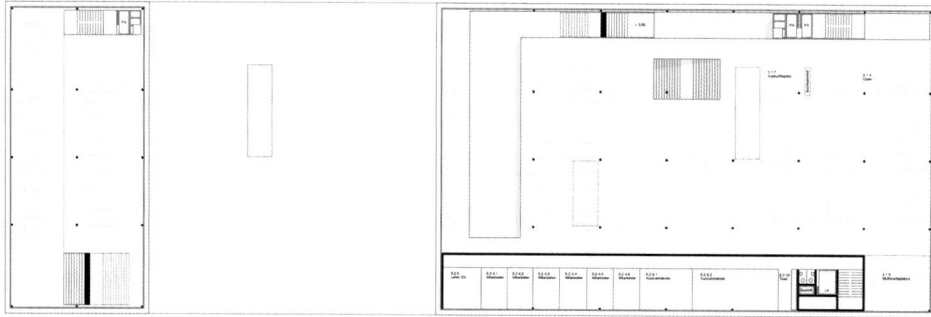

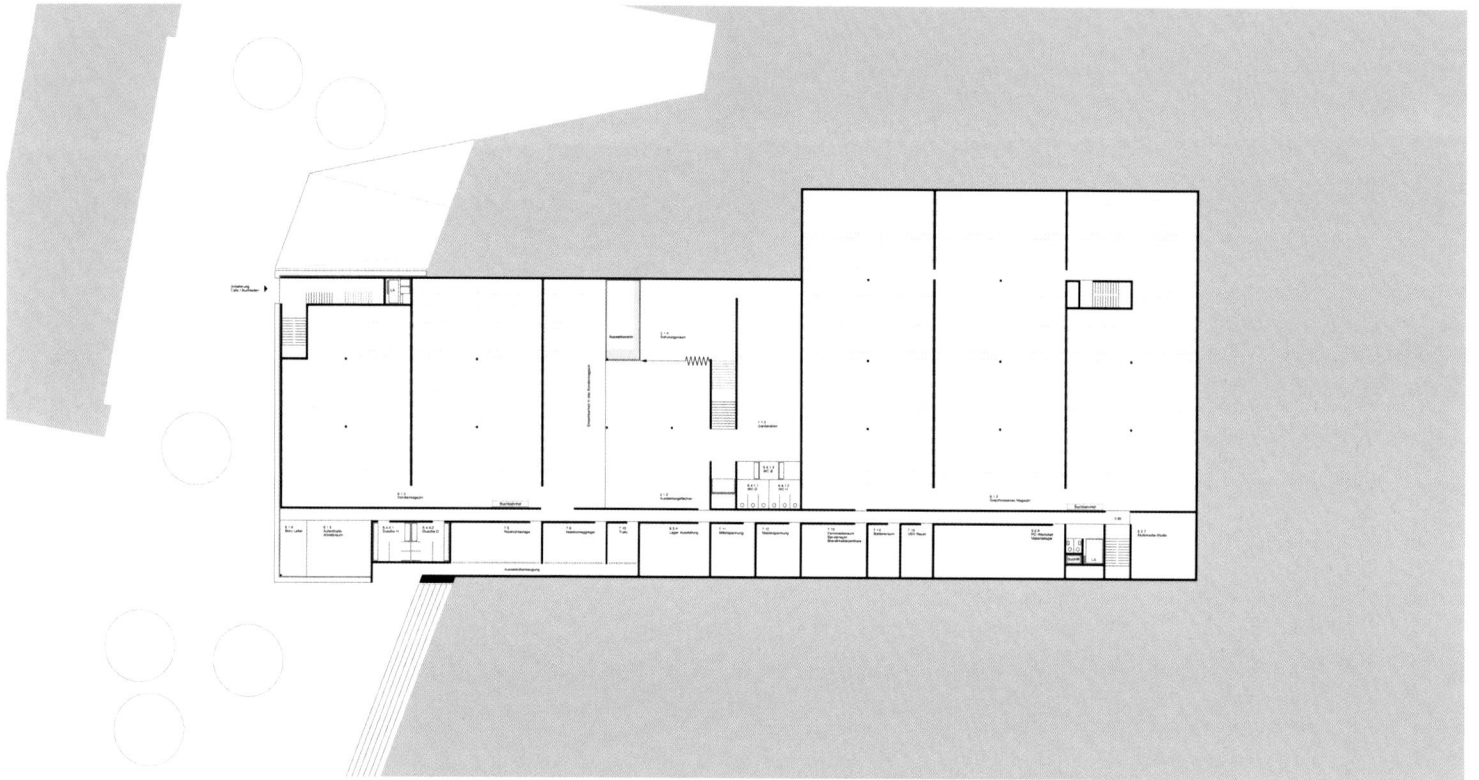

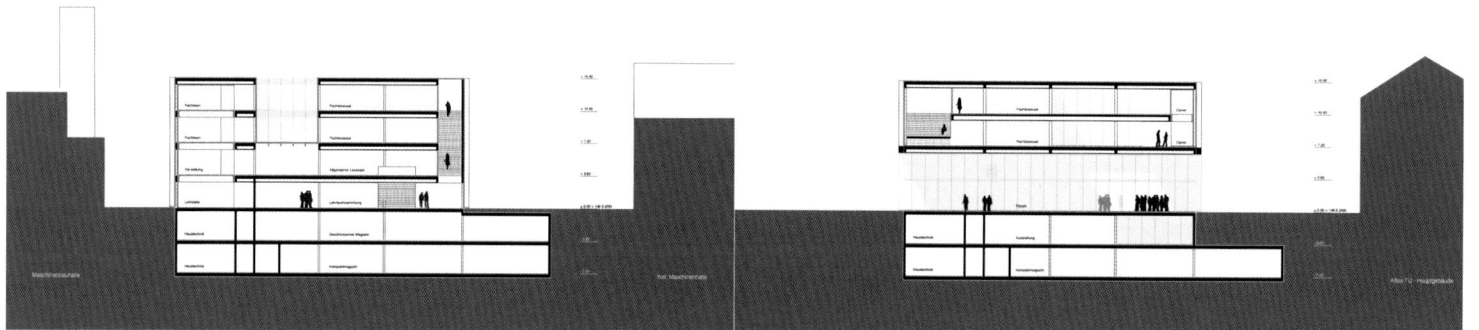

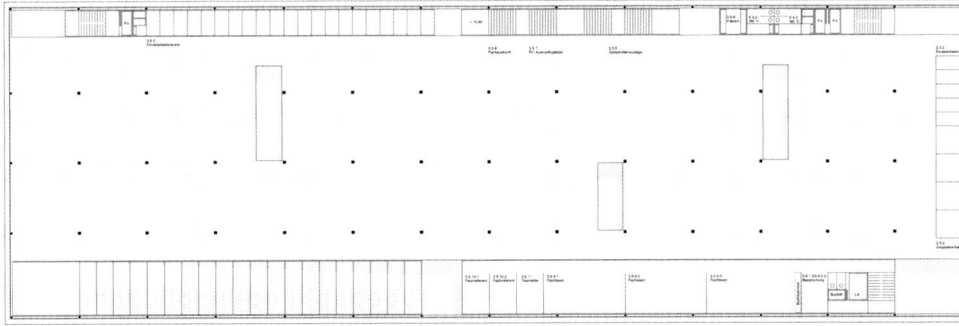

3. obergeschoss 1/200

detailansicht 1/40

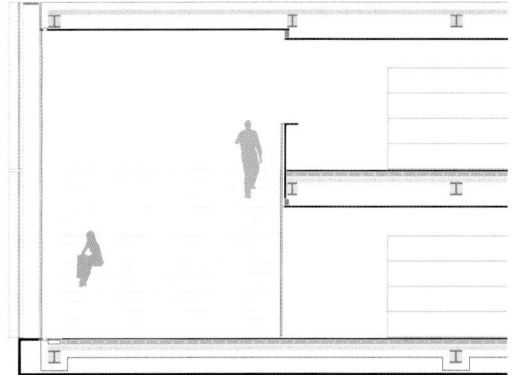

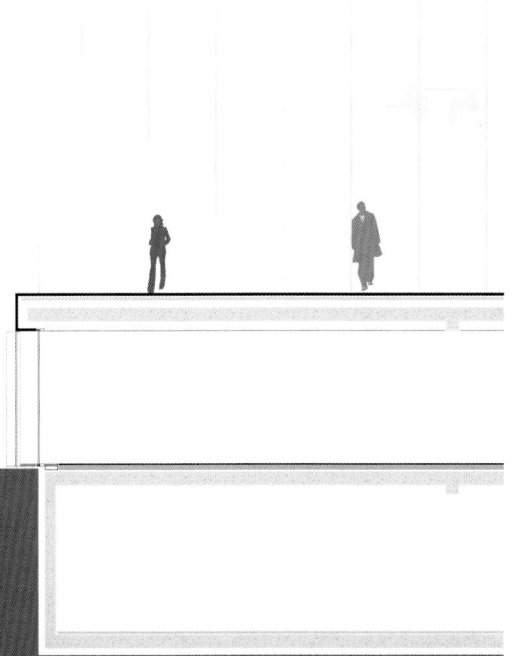

Fassade:
In den Lesesaalbereichen geschosshohe
Wärmeschutzverglasung auf Alu-Unterkonstruktion.
Aussenliegender beweglicher Sonnenschutz aus je
nach Anforderung unterschiedlich dicht gesetzten
bzw. bedruckten vertikalen Glaslamellen. Die
Oberflächen der Glaslamellen weisen einen leichten
Rauhigkeitsgrad auf, um einen eher absorbierenden
Charakter der Aussenhaut zu gewährleisten. Das
sich permanent verändernde Spiel der Aussenhaut
entsteht durch die unterschiedliche Bewegung der
einzelnen Lamellengruppen wie auch durch das
changieren der Oberfläche.

Thomas Müller Ivan Reimann Berlin
Thomas Muller Ivan Reimann建筑事务所 柏林

"The new library integrates smoothly with the extant built environment and complements latent spaces and circulation routes."

"新图书馆流畅地与现有建筑环境融合一起，补充潜在的空间和区内交通线。"

作者 Ivan Reimann   合作伙伴 Erik Frenzel, Ferdinand Oswald, Edna Lührs, Thomas Möckel, Jens Böttcher   专家
landscape architecture: Jürgen Weidinger; fire protection: Büro Stanek; further: Alhäuser + König, H. Dunschmann; energy:
Transsolar GmbH, H. Auer

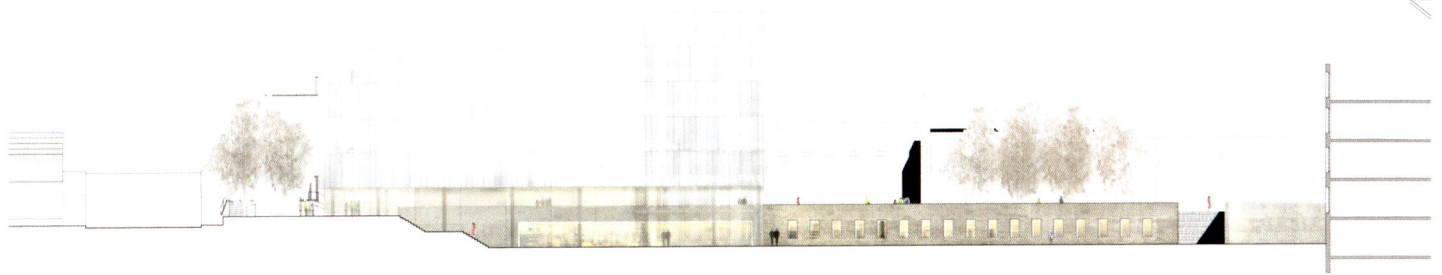

Ansicht Nord   M 1:200

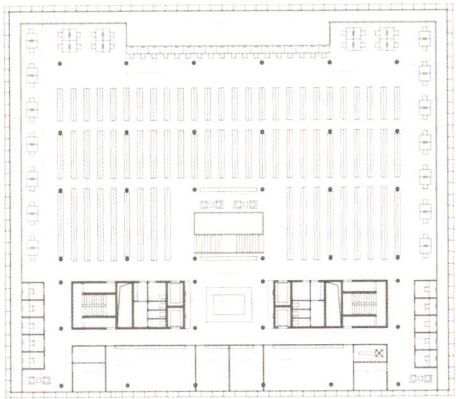

Technikwissenschaften   4.OG   M 1:200

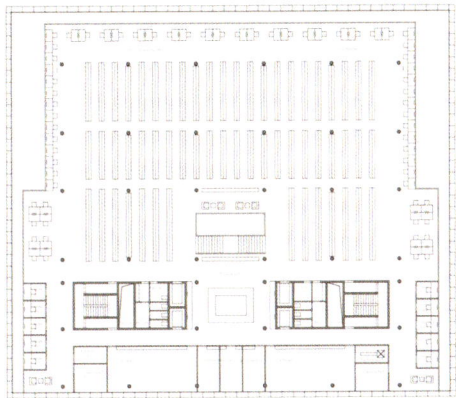

Technikwissenschaften   5.OG   M 1:200

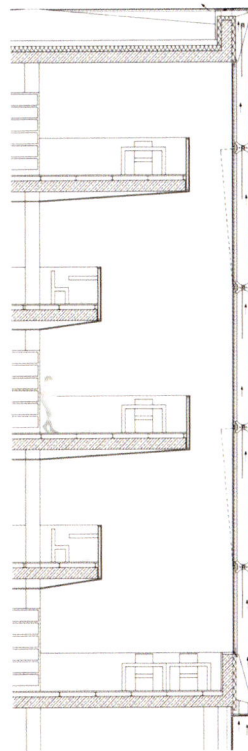

Fassadenausschnitt   M 1:50

Gerber Architekten Dortmund

Gerber建筑事务所 多特蒙德

"From a built volume reduced to the most basic geometric form, the cube, the energy of a central building would radiate into the heterogeneous environs."

"从一个被缩减为最基本的立体形式——立方体的建筑物中，中心大楼的能量向四周多样化的环境中辐射。"

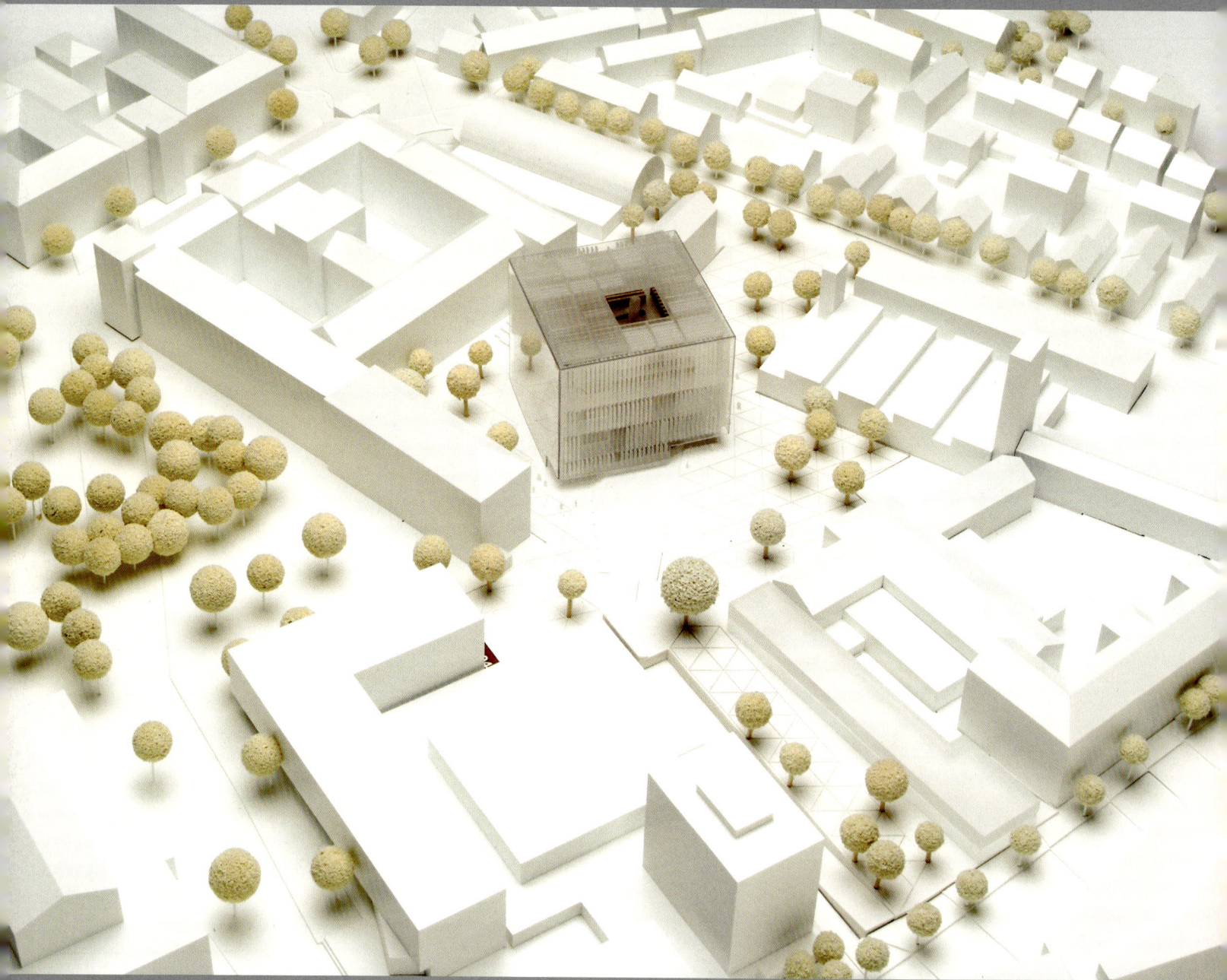

作者 Prof. Eckhard Gerber   合作伙伴及建筑系在校生 Sandra Kroll, Marius Puppendahl, Siegbert Hennecke, Van Hei Nyguen, Manuela Perz, Karsten Liebner, Benjamin Siebner, Lilian Panek, Vanda Govedarica; modelmaking: Henrik Hilsbos, Alexandra Kranert
专家 building services: DS-Plan AG, Hr. Moesle, Stuttgart; structural analysis: OSD, Prof. Kloft, Darmstadt; open space planning: Kienle Planungsgesellschaft Freiraum und Städtebau mbH, Stuttgart

Gerber Architekten, Dortmund

University and State Library, Darmstadt

Herrengarten

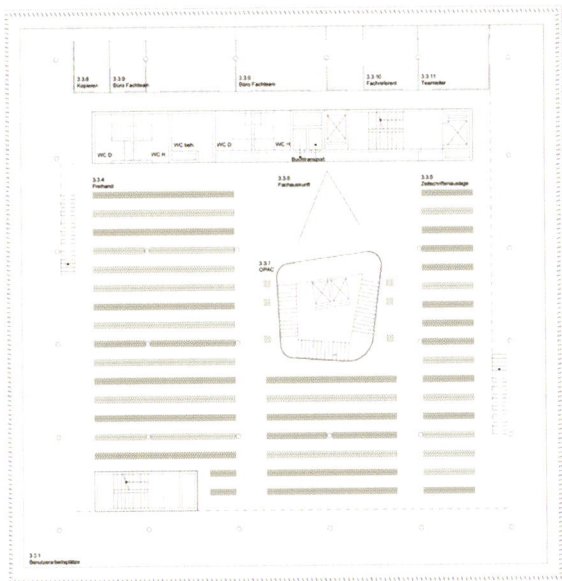

GRUNDRISS EBENE 03 1 _ 200

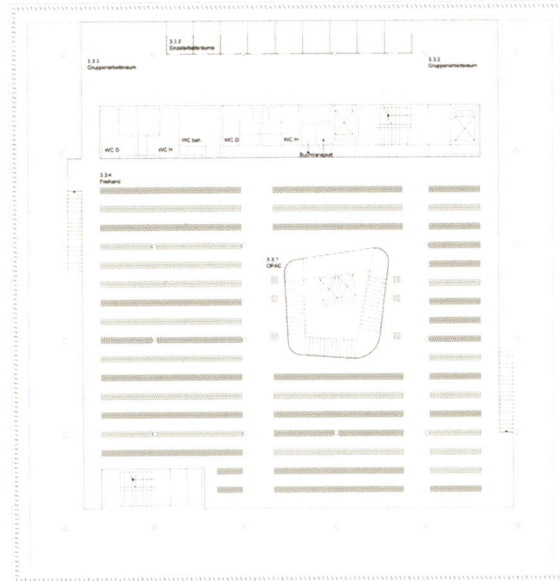

GRUNDRISS EBENE 04 1 _ 200

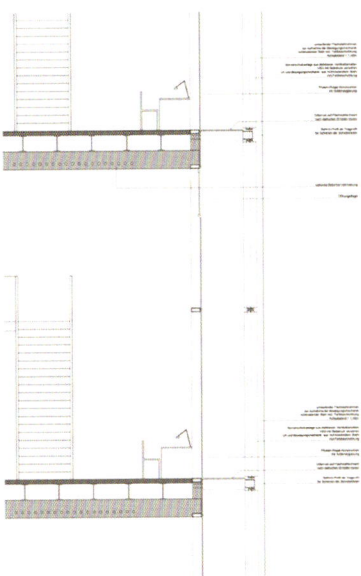

FASSADENDETAIL 1_50

LESESAALBEREICH

ANSICHT NORD 1_200

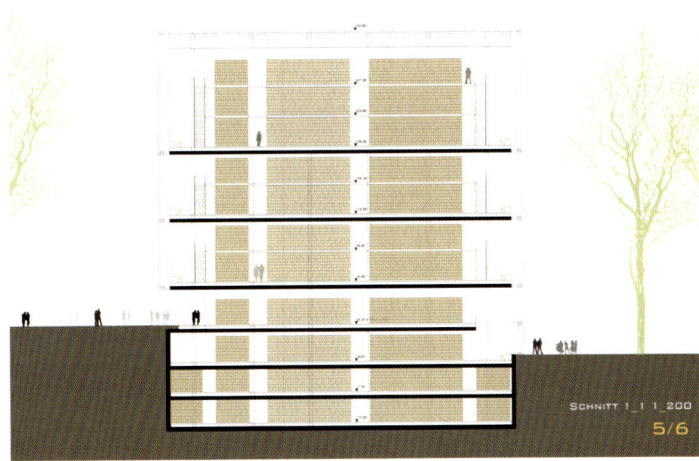

SCHNITT 1_1 1_200

5/6

agps Zurich
agps建筑事务所 慕尼黑

"To see and to be visible."

"看见，也可见。"

作者 Hanspeter Oester, Dr. Marc Angélil, Reto Pfenninger　合作伙伴 Denise Ulrich, Phil Steffen, Roger Naegeli, Thomas Summermatter, Andreas Kopp, Nelson Tam　专家 APT Ingenieure GmbH, Andreas Lutz, Zurich; Amstein+Walthert AG, Adrian Altenburger, Zurich

02

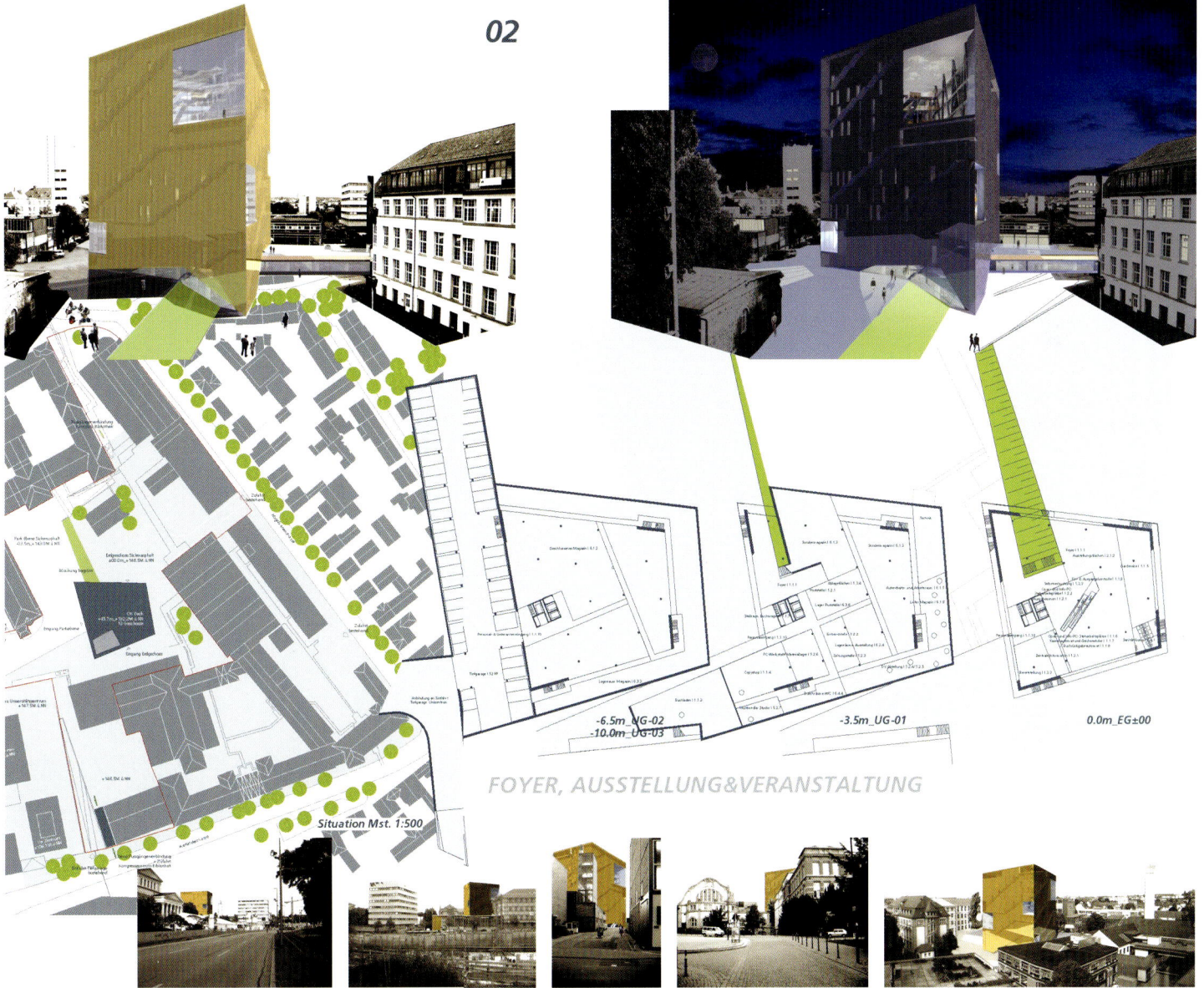

Situation Mst. 1:500

-6.5m_UG-02
-10.0m_UG-03

-3.5m_UG-01

0.0m_EG±00

FOYER, AUSSTELLUNG&VERANSTALTUNG

Realisierungswettbewerb   2. Phase   November 2005

N

147258

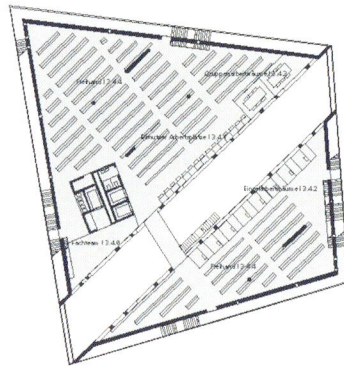

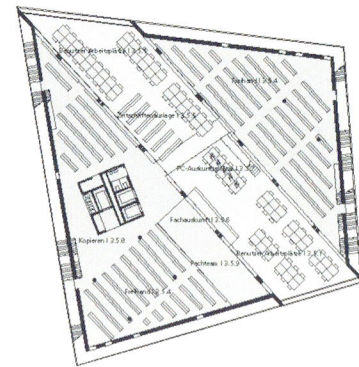

24.5m_OG+07

28.0m_OG+08

NATUR & TECHNIKWISSENSCHAF

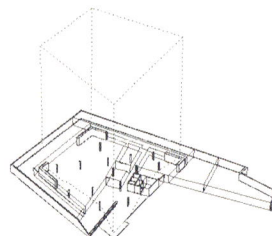

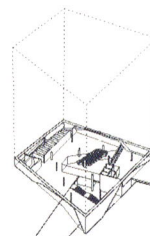

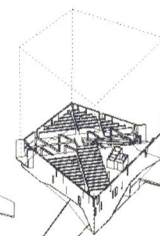

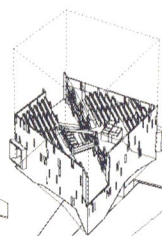

Untere Eingangshalle

Obere Eingangshalle

Allgemeine Bibliothek

Lesesaal Recht&Wirtsch.

Universitäts- und Landesbibliothek Darmstadt　　　*Realisierungswettbewerb 2. Phase*　　　*November 2005*

45.7m_OK Dach

Cafeteria

42.0m_OG+12

Lesesaal

28.0m_OG+08

Bibliothek

24.5m_OG+07

Bibliothek

21.0m_OG+06

Bibliothek

Schulungsraum I 2.1.4
Technik
Besprechungen &
Sozialraum el 2.6.1/2.5.2
Lager Cafeteria I 6.2.2
Cafeteria 1 1.1.2

10.5m_OG+03

Bibliothek

35.0m_OG+10   38.5m_OG+11   42.0m_OG+12

7.0m_OG+02

Bibliothek

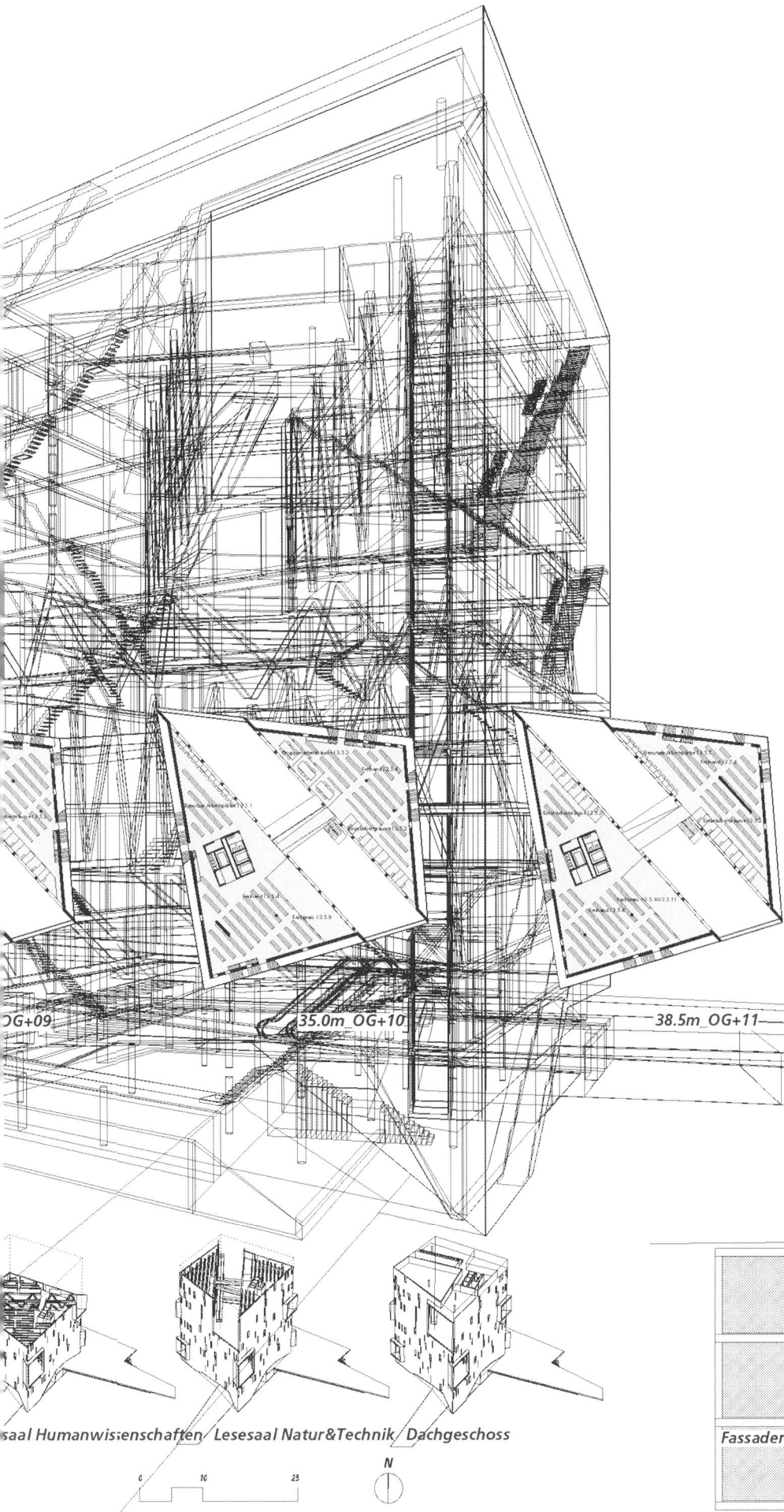

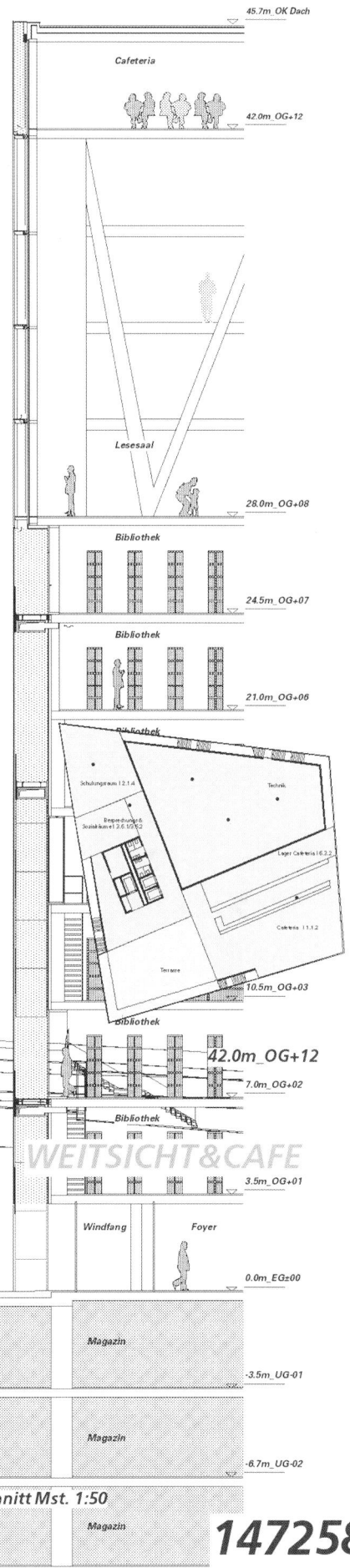

WEITSICHT&CAFE

3.5m_OG+01

Windfang   Foyer

0.0m_EG±00

Magazin

-3.5m_UG-01

Magazin

-6.7m_UG-02

Fassadenschnitt Mst. 1:50

Magazin

147258

OG+09

saal Humanwissenschaften Lesesaal Natur&Technik Dachgeschoss

0   10   25

N

SCHULTES FRANK ARCHITEKTEN Berlin
SCHULTES FRANK建筑事务所 柏林

"'Order is' – Louis Kahn is right, and even if he were wrong, here, in the chaos around the old TUD building, order is definitely the planner's foremost obligation."

"'秩序使然'——路易斯康是正确的，即使在达姆施塔特老楼里他错了，秩序无疑是规划人员最重要的义务。"

作者 Axel Schultes, Charlotte Frank   合作伙伴及建筑系在校生 Fritz Lobeck, Andreas Schuldes   专家 building services: HL-Technik GmbH, Prf. Dr. Klaus Daniels, Munich

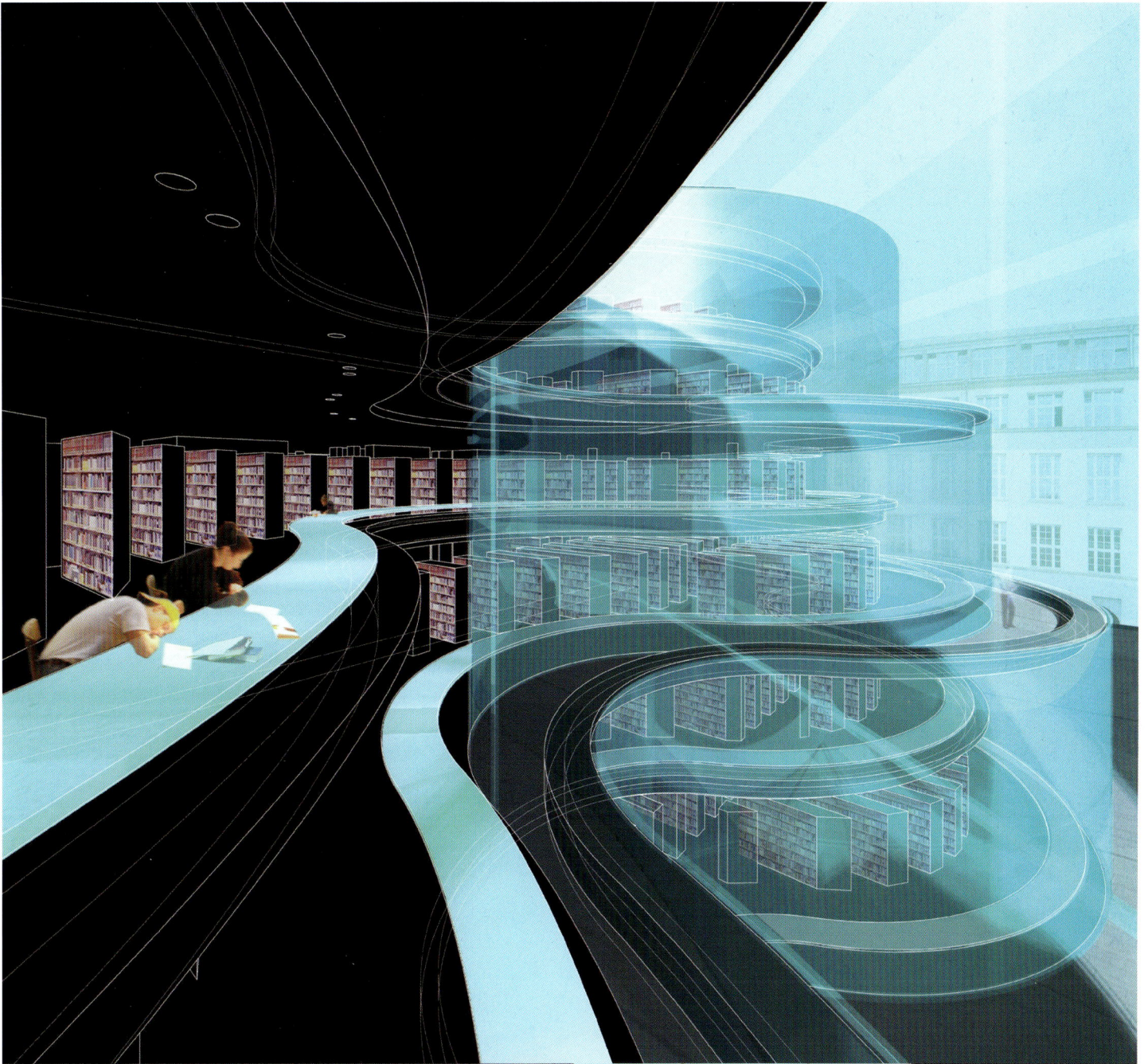

Fassade Schnitt, Ansicht 1:20

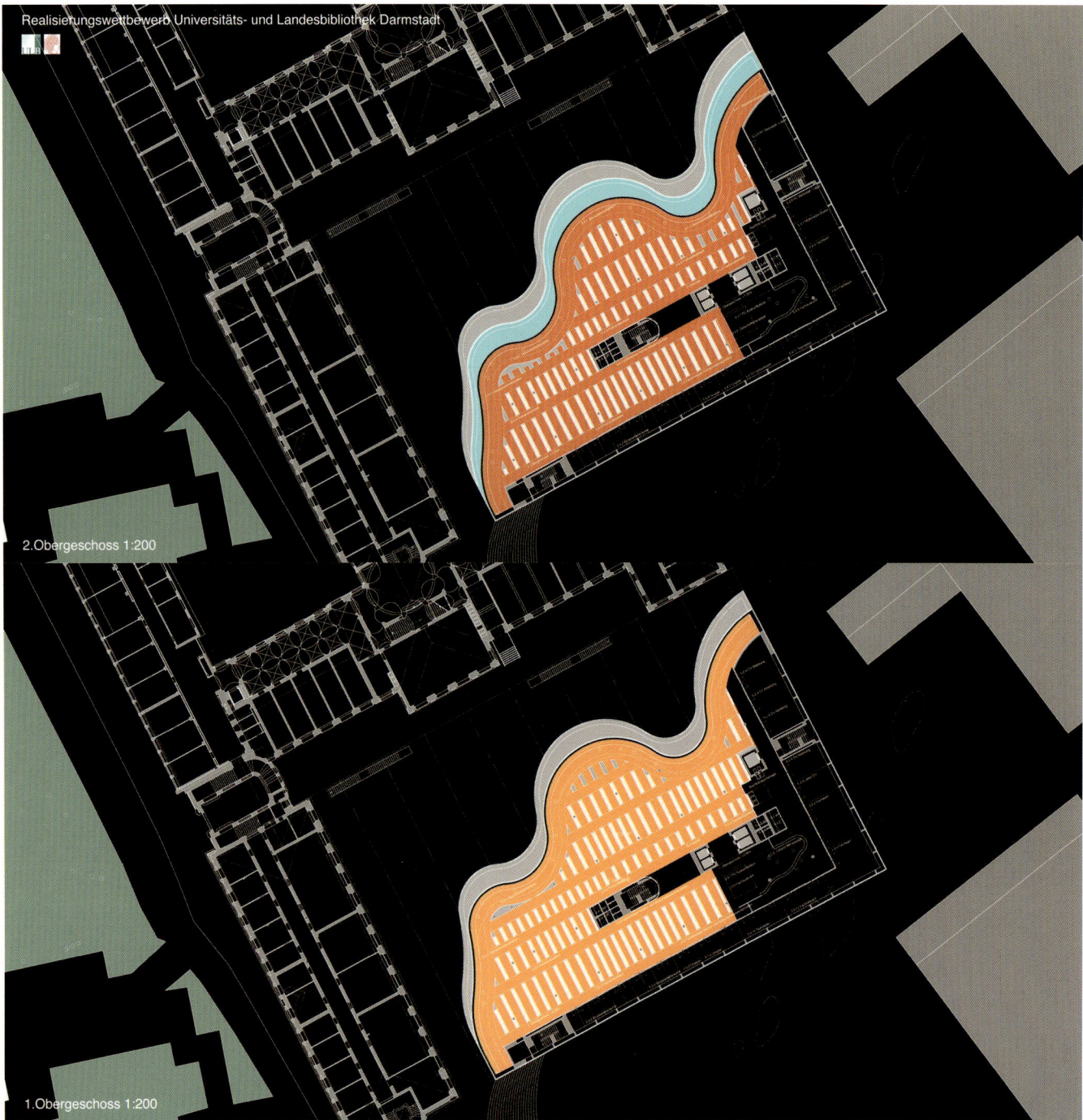

2.Obergeschoss 1:200

1.Obergeschoss 1:200

Ansicht von Süd-Westen 1:200

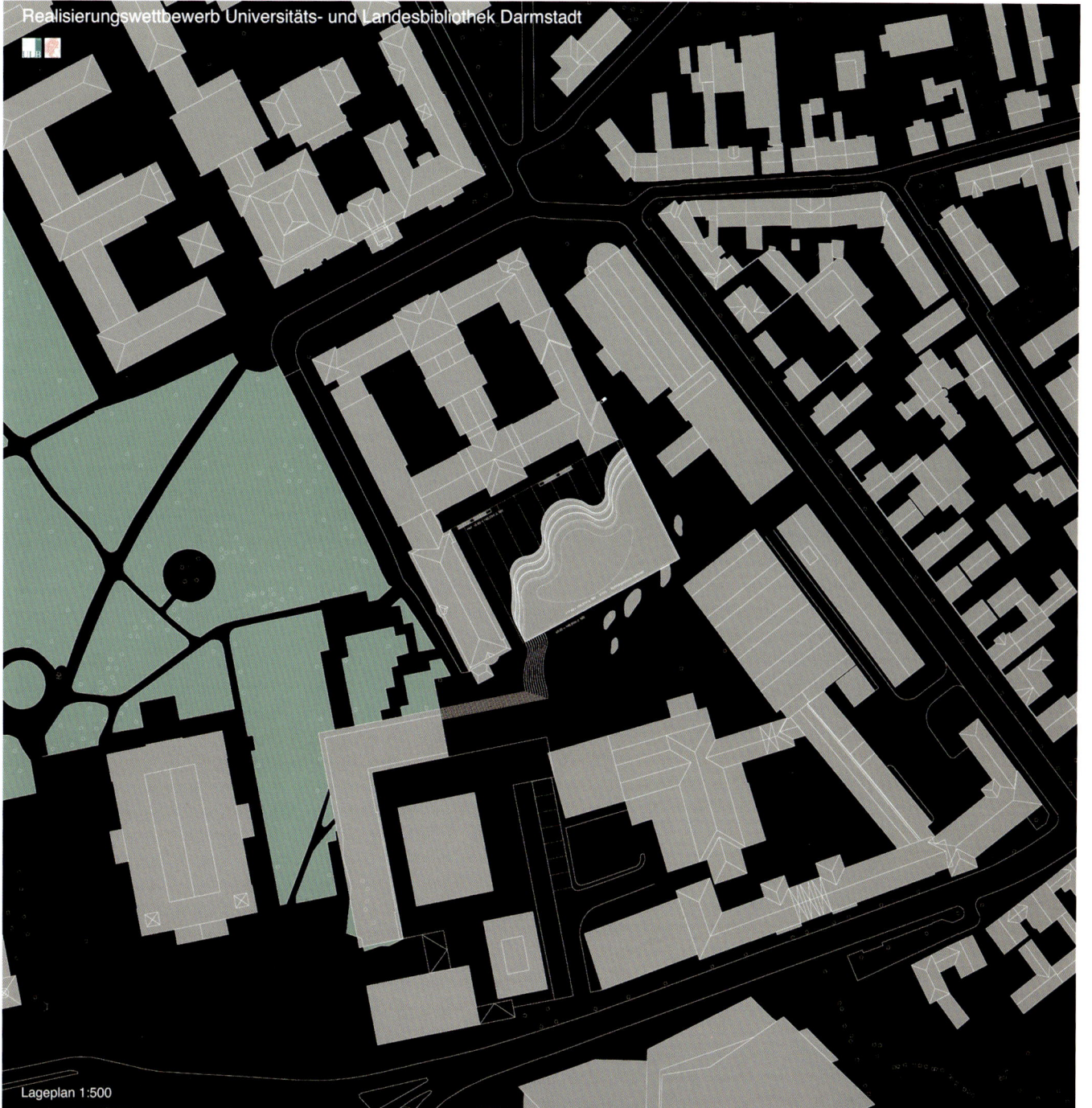

Lageplan 1:500

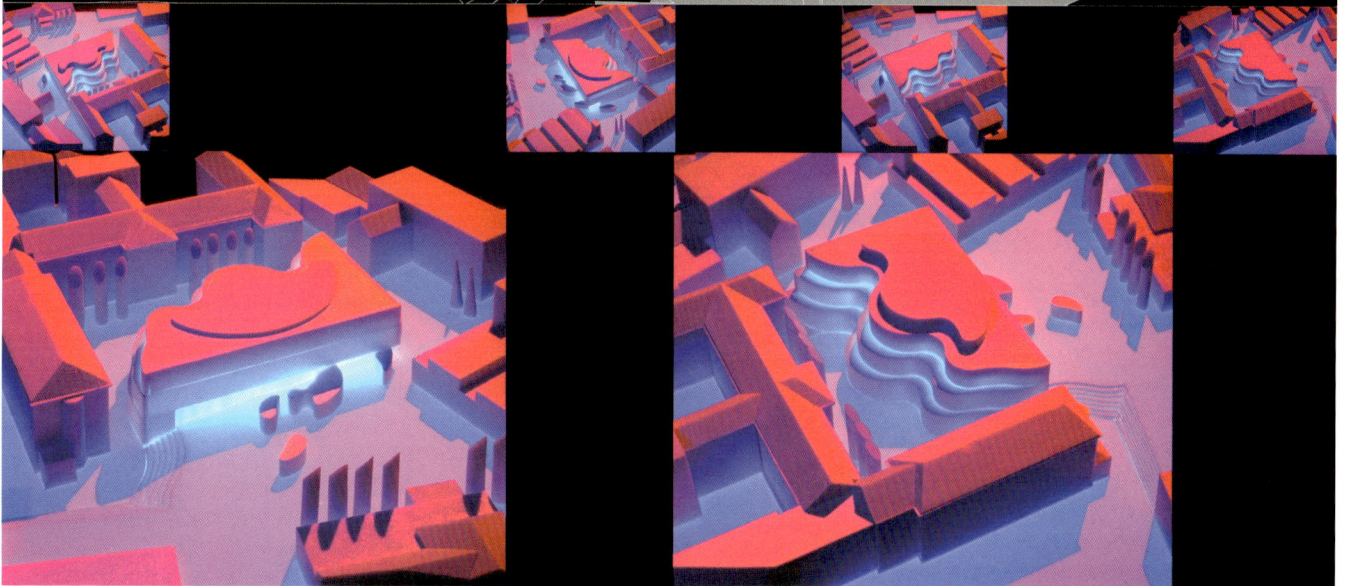

LOVE architecture an

Martin Mechs
Architekturbüro U

thread colle

Southbank Project in Stellenbosch
南部海岸项目，南非斯泰伦博斯

Southbank Project Stellenbosch, South Africa
南部海岸项目 南非斯泰伦博斯

Open, two-stage project competition
公开项目竞赛，分为两个阶段。

Competition site

地点 Stellenbosch, South Africa 时间 07/2006-09/2006 主办方 Spier Holdings 参赛者 1st stage: 96; 2nd stage: 6
面积100 ha 竞赛费用 225,000 USD 专业评奖委员会 Adrian Enthoven, Tanner Methvin, Spier Holdings, South Africa; Ikem Stanley Okoye, Nigeria/USA; Luyanda Mphalwa, South Africa; Michael Keniger, Australia; Mike Rainbow, UK; Salah Hassan, curator and art historian, Sudan/USA; Anne Lacaton, France

As yet, Africa has no single institution representing the culture of the entire continent under one roof. When the sponsors of this competition, old-established wine-makers in South Africa's western cape province, decided to build "Southbank", a new residential development, they also set themselves the ambitious goal of making the "Africa Centre" its centerpiece. The obvious plan was to run the procedure for this project as an open competition; the management was entrusted to [phase eins]. and the University of Witwatersrand, Johannesburg.

The plan called for a usable area of 150,000 square metres, mainly for various forms of residential uses, on an 80-hectare site embedded in a valley landscape. The Africa Centre proper was to cover approx. 11,000 square metres of the usable area. Furthermore there would be a guesthouse, retail, sports facilities, school, nursery school, etc. The project committed itself to a high standard: A vibrant blend of residential, artistic and cultural functions, a meeting place for traditional and contemporary art and culture, an amalgam of museum and culture centre set in a community where artists can reside temporarily or permanently. To view the result, please visit www.southbank.co.za.

至今非洲还没有一个建筑物能够代表整个非洲大陆的文化的。当竞赛主办方，南非西部开普省的一家历史悠久的葡萄酒制造商决定建起一个新的居住区"南部海岸"的时候，他们决心把"非洲中心"做成一个中心。确定项目竞赛采取公开的形式，管理协调工作委托[phase eins].公司和约翰内斯堡金山大学进行。

该项目规划需要在藏于山谷风景中一块80公顷的地上留出15万平方米的使用面积，主要用作各种住宅使用。非洲中心拟占1.1万平方米的使用面积。此外，还会建一家宾馆、商店、运动设施、学校、幼儿园等。项目要求极高，是一个集住宅、艺术和文化功能为一体的充满活力的建筑，是一个传统和现代的艺术和文化的碰撞处，也是一个置身风景如画社区里的博物馆和文化中心的混合物。若要浏览项目结果，请访问网站www.wouthbank.co.za。

Aerial view　鸟瞰图

Competition site　竞赛地点

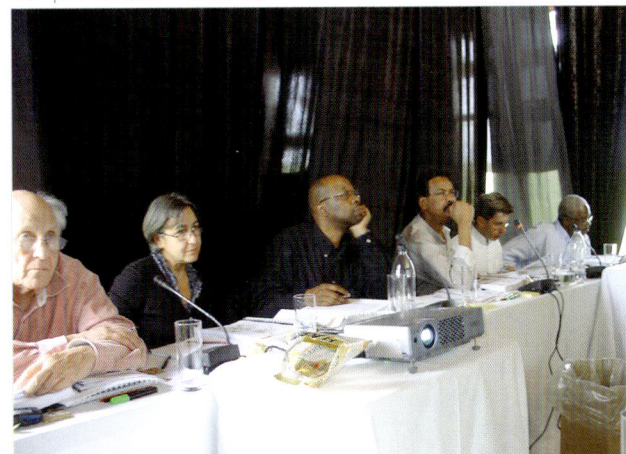

H. Prins, A. Lacaton, L. Mphalwa, S. Hassan, M. Rainbow, I.S. Okoye

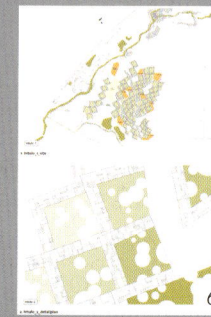

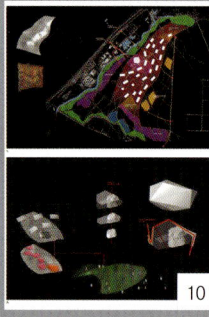

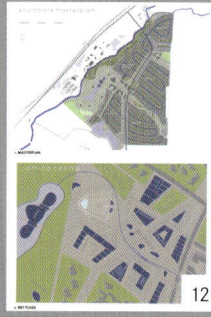

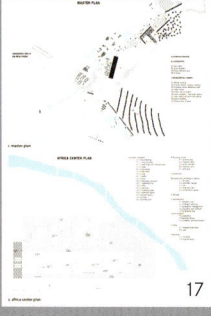

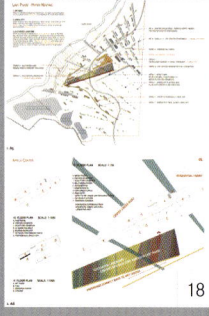

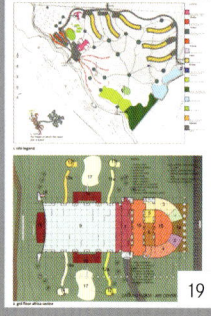

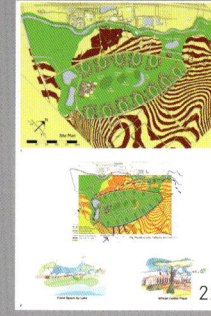

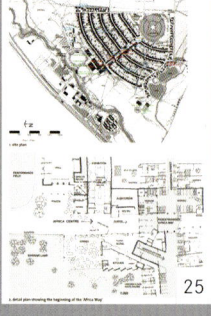

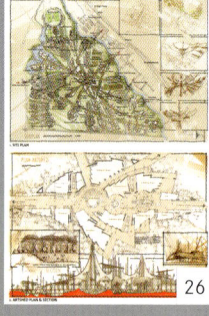

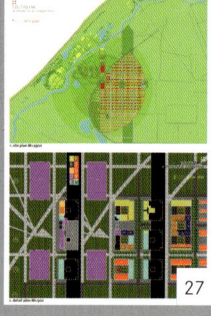

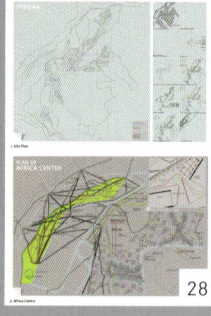

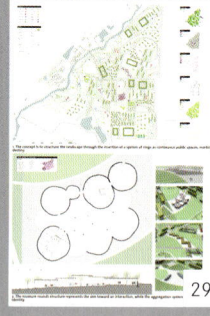

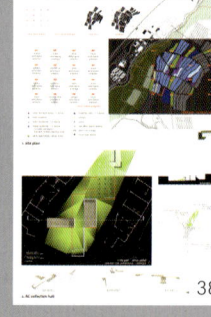

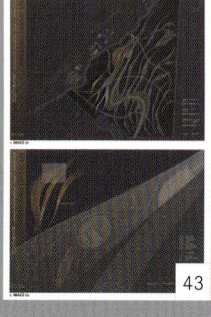

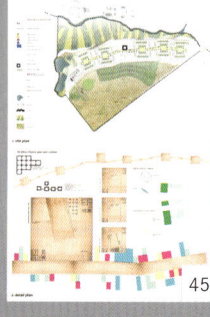

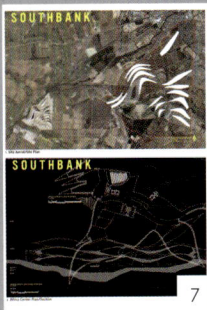

7

8

15

16

23

24

31

32

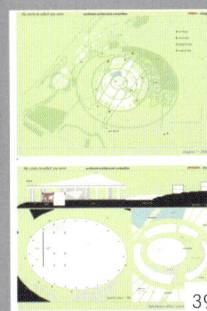

39

40

47

48

Participants 1st stage 第一阶段参赛者

1 constructconcept, Berlin, Germany 2 LOVE architecture and urbanism, Graz, Austria 3 Martin Mechs with Architekturbüro Uli Tischler, Graz, Austria 4 Gerber Architekten, Dortmund, Germany 5 MATSUOKASATOSHITA MURAYUKI, Tokyo, Japan 6 1j2b architects, Fribourg, Switzerland 7 Qua'Virarch, Chicago, Illinois 8 MODIStudio_Associati, Campobasso, Italy 9 GreenhilLi Design PTE LTD, Singapore 10 SHEPPARD ROBSON LTD, London, UK 11 gaudlarchitekten, Dessau, Germany 12 Studio 3 Architects/Planners, Hout Bay, South Africa 13 James Atkinson, Edinburgh, UK 14 Technum NV, Hasselt, Belgium 15 Eugen Ulirsch, Zurich, Switzerland 16 Red Landscape Architects (Pty) Ltd, Pretoria, South Africa 17 Christophe Hutin, Bordeaux, France 18 Hsin-Hung Tsao, Astoria, New York 19 r. van wezel and associates, Joburg, South Africa 20 Bollati architects, Montevideo, Uruguay 21 Mashabane Rose Associates cc; Johannesburg, South Africa 22 Dalhousie University, West Pennant, Canada 23 Luca De Gol, Hugo Castaneda, Alvaro Corredor, Helsinki, Finland 24 The Workplace Development Firm CC, Port Elizabeth, South Africa 25 Harber & Associates, Durban, South Africa 26 AJ Architects, Kapstadt, South Africa 27 Louis Krüger, Adelfia, Italy 28 Trace and associates, Johannesburg, South Africa 29 A+P architettura, Rom, Italy 30 Nightingale Associates, Kapstadt, South Africa 31 Edina Osmanovic, Mannheim, Germany 32 Total Design + Associates, Castries, Saint Lucia 33 STUDIO EGRET WEST, London, UK 34 Andrade Morettin, Sao Paulo, Brazil 35 Verzone Woods Architectes, Rougemont, Switzerland 36 Santos Prescott & Associates, San Francico, USA 37 Junya Toda Architect & Associates, Osaka, Japan 38 normaldesign/thread collective, Brooklyn, New York 39 rabaschus und rosenthal, Dresden, Germany 40 Dodi Moss Srl, Milan, Italy 41 sitengineering srl, Vigevano, Italy 42 SLAG, Firenze, Italy 43 Frédéric Haesevoets, Brüssel, Belgium 44 the idom group, London, UK 45 Magdalena Szczypka, Tychy, Poland 46 Worldlab, Århus, Denmark 47 Masauso Branda, Harare, Zimbabwe 48 Architecton Design Studio, Harare, Zimbabwe

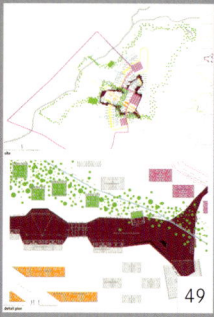
49

50

51

52

53

54

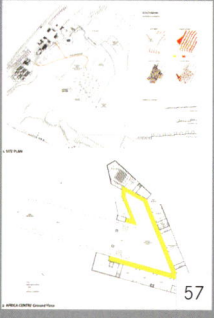
57

58

59

60

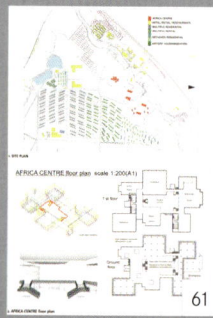
61

62

65

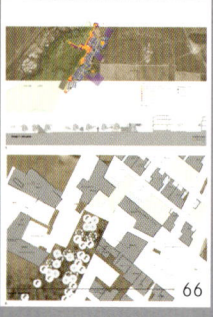
66

67

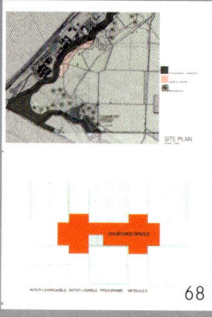
68

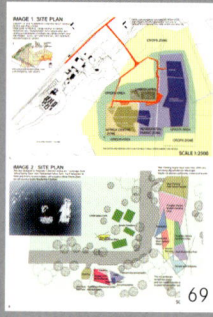
69

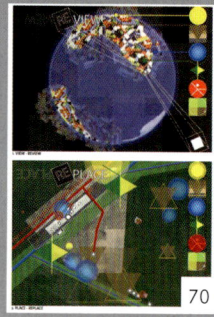
70

73

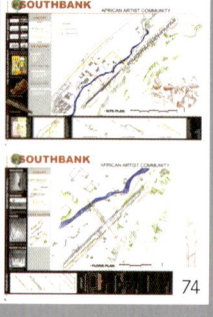
74

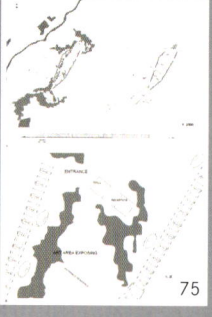
75

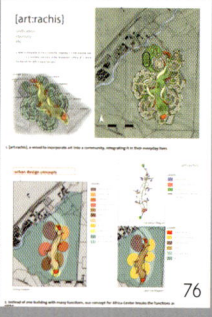
76

77

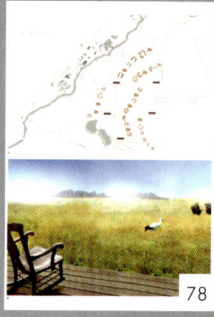
78

80

81

82

83

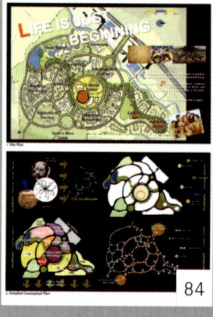
84

85

87

88

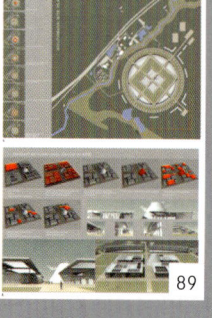
89

90

91

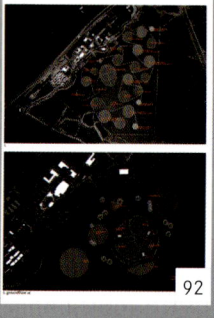
92

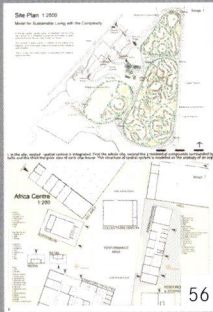

55   56

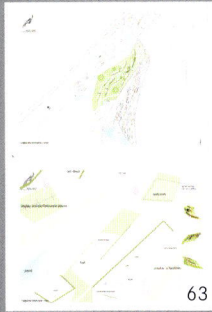

63   64

71   72

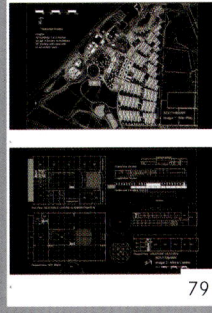

79

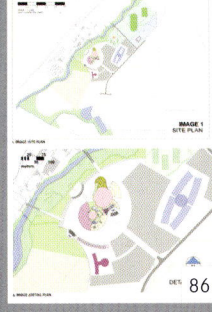

86

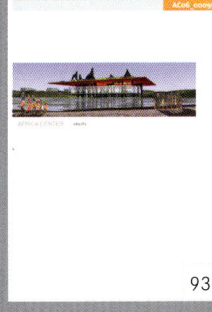

93

49 buero blickpunkt, Berlin, Germany **50** konyk architecture pc, Brookly, New York **51** GROUP A, Rotterdamm, Netherlands **52** Takayuki Kamei, Fukuoka, Japan **53** motsepe architects, Johannesburg, South Africa **54** Aliso Odinyo, UNIVERSITY OF NAIROBI – architecture department, Nairobi, Kenya **55** Antarctica Group, Melbourne, Australia **56** Daisuke Matsushita, Kyoto University, Japan **57** François Machado, Poilhes, France **58** Atelier Sanna Lahti, Helsinki, Finland **59** Albonico Sack Mzumara Architects & Urban Designers, Johannesburg, South Africa **60** GLAM+; Johannesburg, South Africa **61** arcmatic, Johannesburg, South Africa **62** studio83, Jakarta, Indonesia **63** MGF ARCHITEKTEN GMBH, Stuttgart, Germany **64** BH Architects, Pretoria, South Africa **65** Pieter Jooste, Joannesburg, South Africa **66** Wesley Hindmarch, Hobart, Australia **67** Bentley Faulmann Architects, Yellowknife, Canada **68** Habib Chanzi, Melbourne, Australia **69** Patchara Jumpangen, Nonthaburi, Singapore **70** SCAD, Savannah, South Africa **71** Savannah college of Art and Design, Savannah, South Africa **72** Smit & Serfontein Architects Pty (Ltd), Santa Carina, Brazil **73** umar.pareja architecture, Alexandria, South Africa **74** Gary White and Associates, Pretoria, South Africa **75** Juan Carlos Moreno Coriat, Santiago de Cali, South Africa **76** studioMAS, Johannesburg, South Africa **77** Tim Wippich, Hannover, Germany **78** Manuela Perz, Munich **79** D Hirschman, Architects, Kapstadt, South Africa **80** Wiswedel Architects, Düsseldorf, Germany **81** Wolf & Wolf Architects, Kapstadt, South Africa **82** Patrick Kathuli, Mwingi, Kenya **83** Zhou Zhenxiong; Shenzhen, China **84** Realrich Sjarief, London, UK, UK **85** Philip Kilian, Pretoria, South Africa **86** Thomas C Falck Architects, Worcester, UK **87** CMSP Arquitetura + Design, Florianópolis, Brazil **88** ecoedge architecture, Richmond, Australia **89** Valentin Oleynik, St. Petersburg, Russia **90** Quagga Group, Kapstadt, South Africa **91** Sean Mahoney, Kapstadt, South Africa **92** Kohlmayer Oberst Architekten, Stuttgart, Germany **93** Mohsen Sane'i, Tehran, Iran

thread collective & normaldesign Brooklyn, USA
Thread collective & normaldesign设计工作室 美国布鲁克林

"The collage integrate existing objects, materials and fragments, creating new forms, facilitating a dialog between parts, and revealing unexpected meanings; it is this concept that spatially organises and links the three elements of the Southbank design: residential fabric, landscape and the Africa Centre."

"拼贴把现有物体、材料和片段合为一体，创造出新形式，利于部分之间的对话，也表示意想不到的意义；正是这种概念把南部海岸项目设计的三个要素，即居住建筑物、景观和非洲中心从空间上组织起来。"

作者 thread collective: Elliott Maltby, Mark Mancuso und Gita Nandan; normaldesign: Matthias Neumann 建筑系在校生 Nazia Aftab 专家 modelmaking: Michael Caton + Jerome Barbu; renderings: John Watson; green consultant: Lauren Gropper, solar + wind calculations: solar e systems

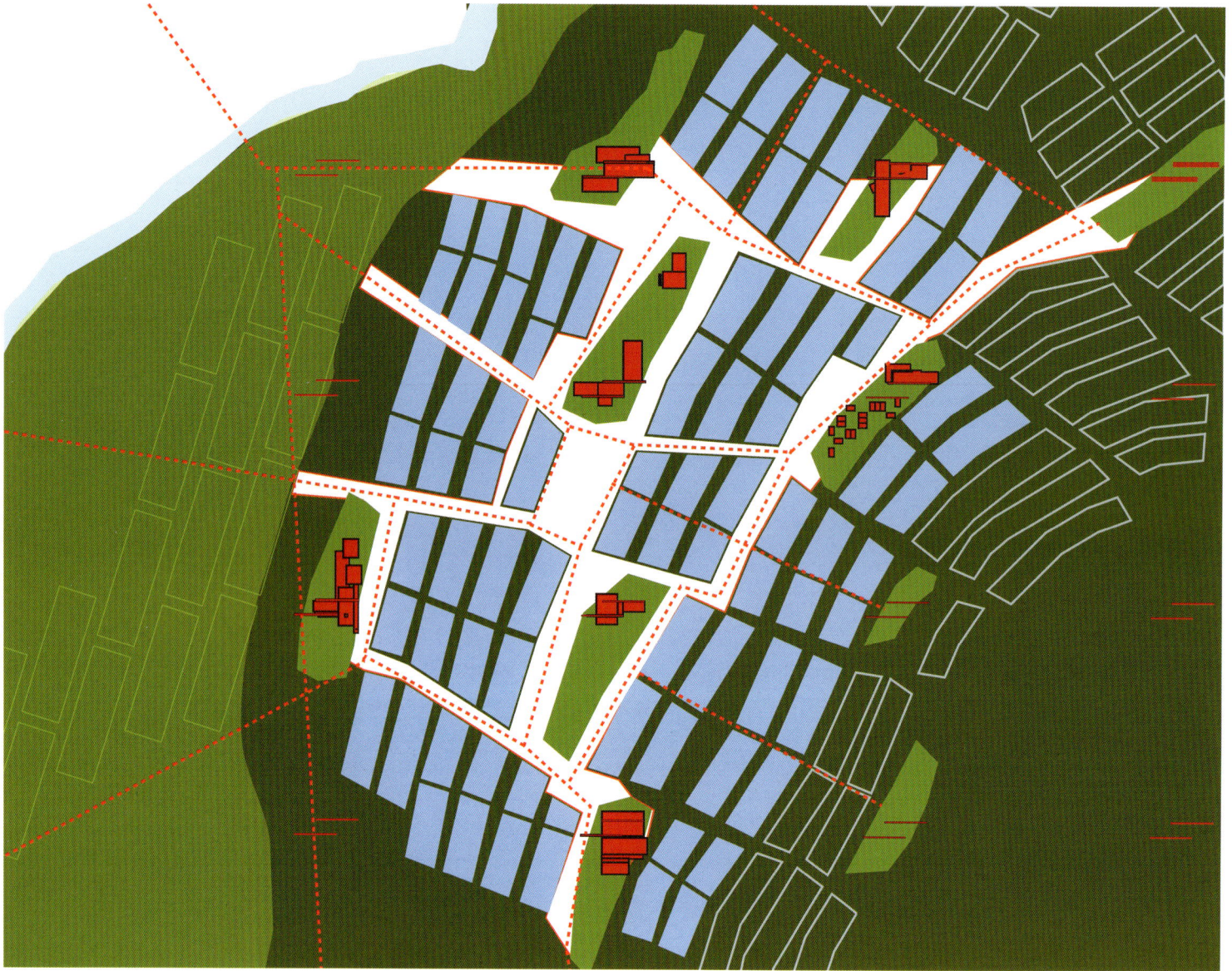

sustainability diagram

in contrast the africa center, the residential fabric is subtle in its display of sustainability, while maintaining the same degree of energy efficiency and water conservation

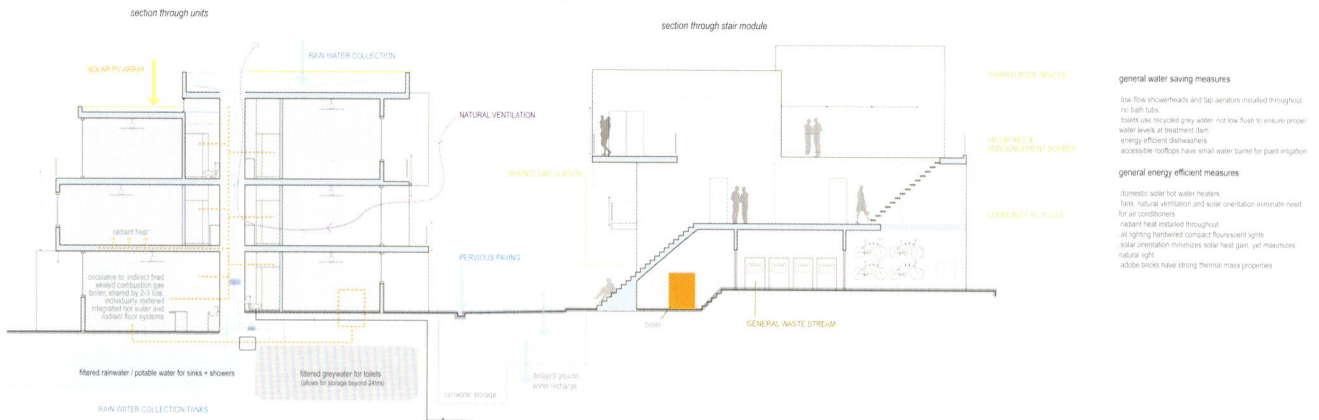

section through units

section through stair module

SOLAR PV ARRAY

RAIN WATER COLLECTION

NATURAL VENTILATION

SHARED GREEN SPACE

SHARED CIRCULATION

LIGHTWELL SKYLIGHT

PERVIOUS PAVING

radiant heat

circulates to indirect fired sealed combustion gas boiler, shared by 2-3 flats, individually metered integrated hot water and radiant floor systems

GENERAL WASTE STREAM

filtered rainwater / potable water for sinks + showers

filtered greywater for toilets (allows for storage beyond 24hrs)

collector storage

delayed ground water recharge

RAIN WATER COLLECTION TANKS

general water saving measures

low-flow showerheads and tap aerators installed throughout
no bath tubs
toilets use recycled grey water, not low flush to ensure proper water levels at treatment dam
energy efficient dishwashers
accessible rooftops have small water barrel for plant irrigation

general energy efficient measures

domestic solar hot water heaters
fans, natural ventilation and solar orientation eliminate need for air conditioners
radiant heat installed throughout
all lighting hardwired compact flourescent lights
solar orientation minimizes solar heat gain, yet maximizes natural light
adobe bricks have strong thermal mass properties

LOVE architecture and urbanism Graz
LOVE建筑及城市规划事务所 格拉茨

"Each of the built volumes is wing-shaped, thus creating a feeling of innumerous wings sweeping the vast natural landscape. Together, they form an urban pattern reflecting the expanse and vigor of the landscape."

"每个建筑物外形都很像翅膀，因此创造出无数翅膀飞过广表自然景观的感觉。他们一起构成一种城市模式，反映景观的广阔和活力。"

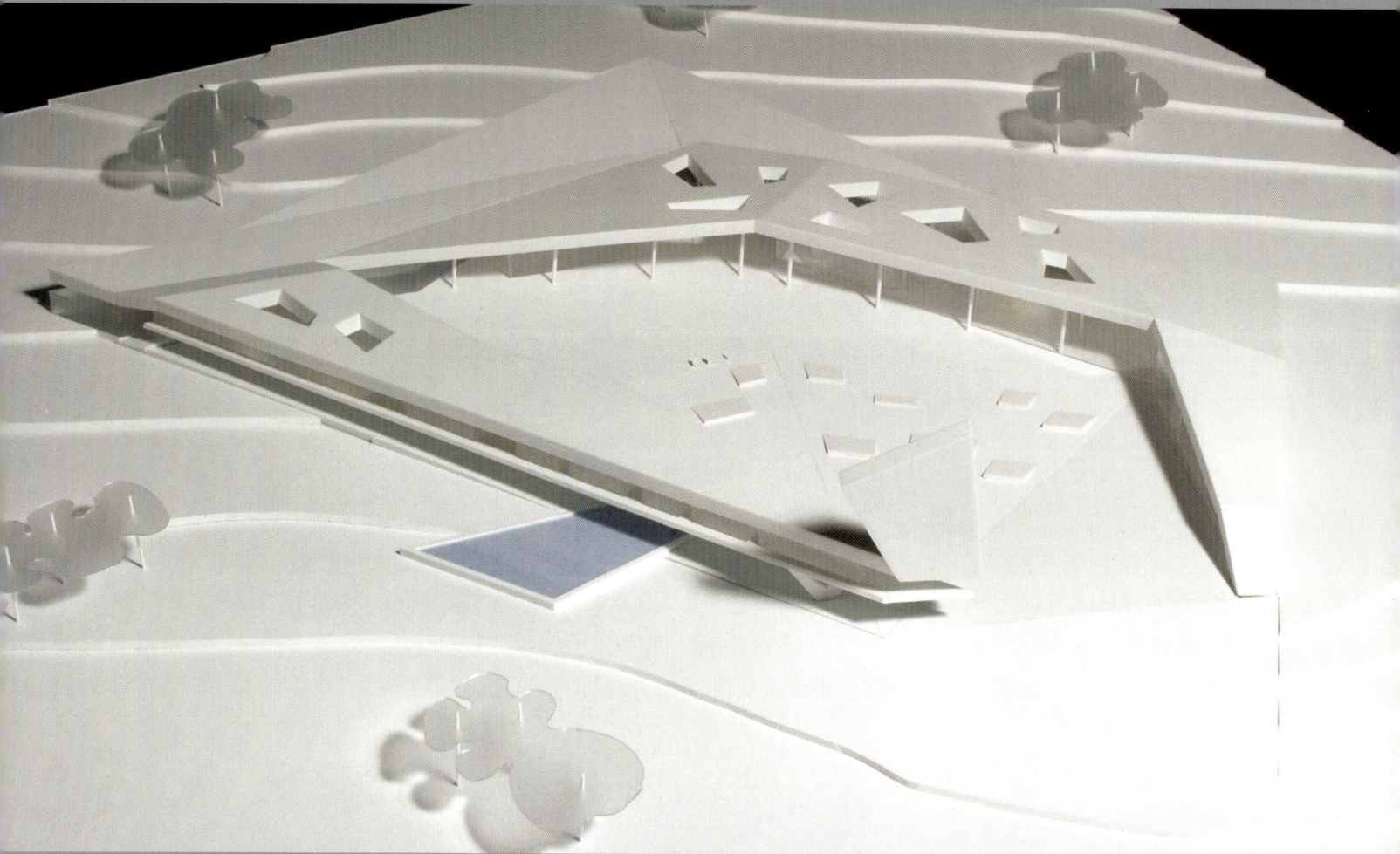

作者 LOVE architecture and urbanism 合作伙伴 DI Gerald Brencic, Iulius Popa 专家 HKLS planer: La TEC GmbH Nfg Keg; landscape architecture: Koala; building physics: Müller; structural analysis: Ingenieurbüro Petschnigg

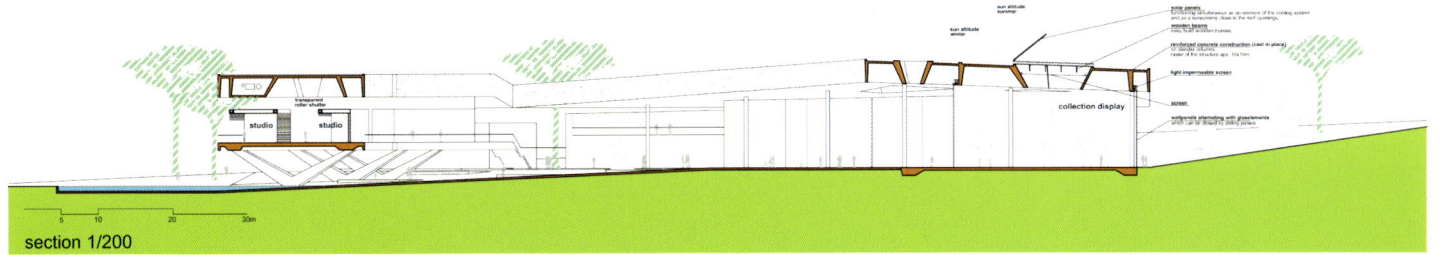

section 1/200

5   10   20   30m

Africa Centre/Art Shet
1 africa centre/art shet

Wings: Residental/Buildings
2 shopping, bars and restaurants   4 school versions   5 site office
3 possible hotel locations: a) art hotel, b) sport hotel, c) hotel with a view

Squares
6 squares: a) africa square, b) southbank square
sculptures/spaces for art interventions

Landscape-Wings
8 sports facilities
9 recreation

Parking
10 main parking   11 optional parking

Water   Main Pedestrian Routes

site access

main entrance pedestrians

entrance pedestrians

park

recreation

trees

site access

landscape wing

community

private gardens

agriculture

Martin Mechs with Architekturbüro Uli Tischler Graz
Martin Mechs with Architekturburo Uli Tischler建筑事务所 格拉茨

"The 'colored strips', while signaling the acceptance of existing differences, are moved toghether closely to allow reciprocal perception 'across the alley' – and the potential for communication."

" '彩色纸条' 在标志对存在差异的认可的同时被移近，允许 '穿过小巷' 这一互惠和沟通潜力的看法。"

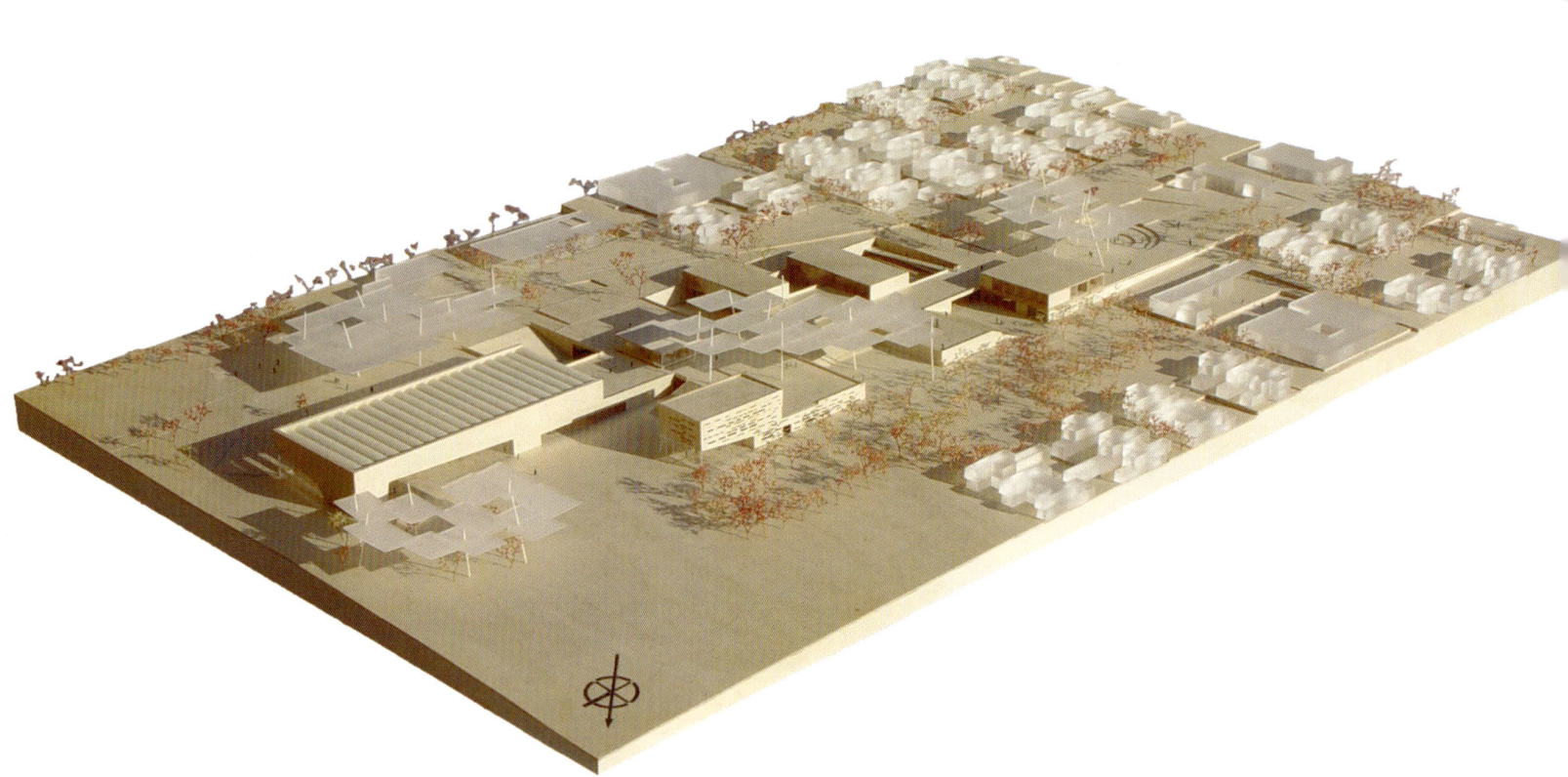

作者 Martin Mechs with Architekturbüro Uli Tischler   合作伙伴 consultant: Felicitas Konecny; phase 1: Wolfgang Isopp, Christina Kimmerle; phase 2: Markus Hopferwieser, Christoph Wiesmayr, Herwig Marx, Jakob Winkler (renderings)   专家 landscape architecture: Thomas Proksch "Land in Sicht" Vienna; energy- and water supply; sustainability: Arge Energie AEE INTEC (Charlotta Isaksson, Christian Platzer); modelmaking: Patrick Klammer

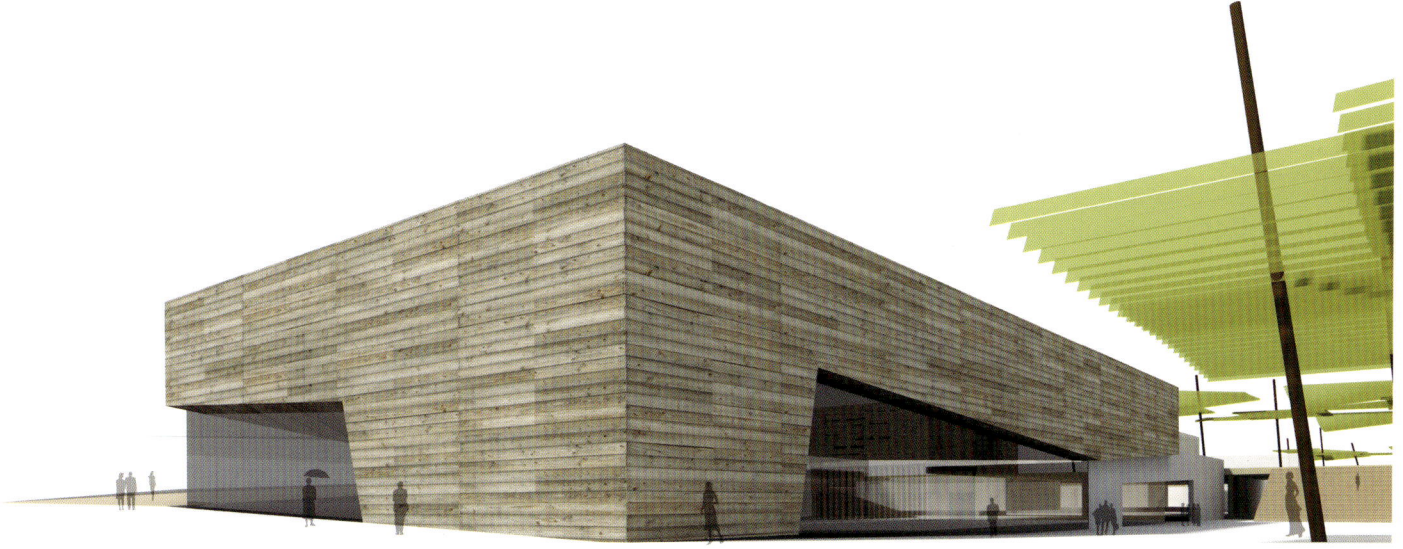

Martin Mechs with Architekturbüro Uli Tischler, Graz

Southbank Project, Stellenbosch

Ateliers Lion archite

Jafar Tukan Architects

Kisho Kurokawa

L

SCHULTES FRANK ARC

von Gerkan, Marg un

Z

tes urbanistes

Administration Complex in Tripoli

行政办公建筑区，利比亚的黎波里

rchitect & associates

eon Wohlhage Wernik

HITEKTEN

d Partner

aha Hadid Architects

Administration Complex Tripoli
行政办公建筑区，利比亚的黎波里

Project competition for architects and city planners preceded by an application procedure
建筑师和城市规划人员的项目竞赛，开始前有申请程序。

Agricultural Land

Proposed Third Ring Road

Airport Highway

Forest

Proposed Train Station

Islamic World Call Society

Railway Track

Proposed Metro Line

Forest

G.M.M.R. Water Pipeline

Competition Site

Building
Forest
Agricultural Land

地点 Tripoli   时间 12/2006-6/2007   主办方 ODAC   参赛者 8   面积 615,000 sq m   竞赛费用 300,000 USD
专业评奖委员会 Craig Dykers, Oslo/New York; Guido Hager, Zurich; Prof. Rodolfo Machado, Boston;
Prof. Matthias Sauerbruch, Berlin/London; Prof. Peter Zlonicky, chairman, Munich   其他专业评奖委员会 Prof. Bruno Sauer, Valencia
专家评奖委员会 Ali I. Dabaiba, managing director, organisation for development of administrative centers, Tripoli;
Dr Mostafa Al Mezughi, chairman of general corporation for housing and infrastructure, Tripoli; Anwar A. Sassi, chairman of
Brega & Ras Lanuf higher committee, Tripoli; Dr Ali Shebani, chairman of national consulting bureau, Tripoli   其他专家评奖委员会 Mohsen M.
Ben Halim, national project coordinator, national consulting bureau, Tripoli; Dr. Ahmed M. Shembesh, director general,
Libyan national center for standardisation and metrology, Tripoli

The governmental complex which will be the hub of a new business and administrative district in the years to come, will be situated approx. 7 kilometres south of the city center. Next to the site is where the highway to the international airport crosses Tripoli's third main ring road. The first task was to draw up a master plan (including landscaping and transportation grids) for a spacious open quarter that Libyans can identify with, complete with high-performing internal and external access routes, well connected to adjoining green spaces and other districts the city plans to develop in the future. Next, the project's most important buildings were to be designed in a style that would reconcile the regional vernacular (e.g. manual construction) with the visual language of contemporary architecture and at the same time address climatic challenges. The overall program called for a gross floor area of some 70,000 square metres including as key elements the parliament building, a conference palace with room for 1,500 delegates, and the buildings for the secretariat of the general people's committee, and the coordinating council of the people's leaderships. These are Libya's highest-ranking political functions after that of the top leader-ship. Another element of the program was a fivestar, 540-bedroom hotel where guests of the state and, during parliamentary sessions, the people's representatives will be lodged. Lastly, it called for designing buildings to house 20 government departments. In keeping with the comprehensive and ambitious nature of the program, a procedure to select candidates internationally preceded the actual competition which offered generous prizes to the winners.

该行政办公建筑区位于市中心南部7公里的地方。项目地点旁边是国际机场高速公司与的黎波里三环路的交叉处。竞赛任务之一是起草一份总平面图。其次，项目的重要建筑物的设计风格应该与该区的本土特点和谐一致，运用现代建筑的视觉艺术语言，同时也要适合当地气候。
整个项目规划需要约7万平方米的建筑总面积，包括议会大楼、能容纳1500名代表的会议厅、人民委员会秘书处和人民领导权协调委员会的办公大楼。规划的另一内容是一家有540间客房的五星级酒店。最后，项目还要求设计20家政府部门的办公大楼。
该项目综合性强、气势恢宏，因此在实际竞赛前有一个在全球范围内选择参赛候选人的程序，竞赛向获胜者提供丰厚的奖品。

Aerial view　鸟瞰图

Competition site　竞赛地点

Competition site　竞赛地点

View of the city　城市景色

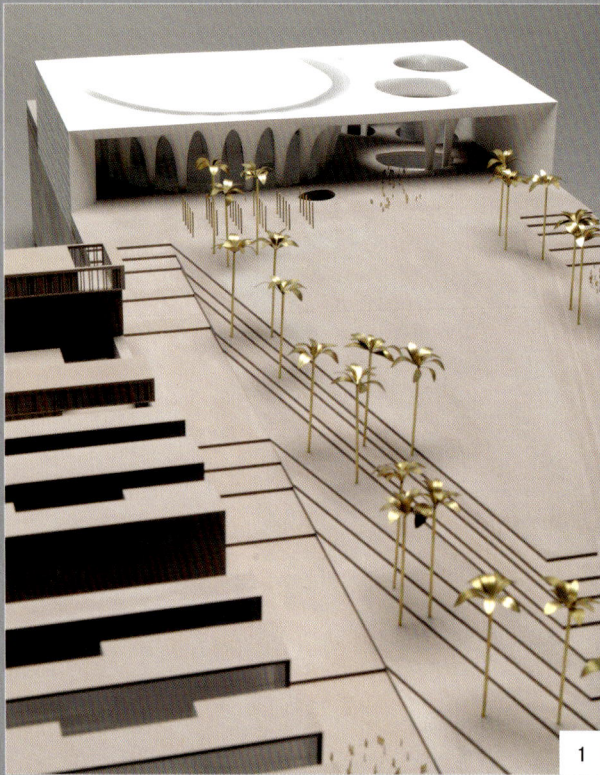

1

2

5

6

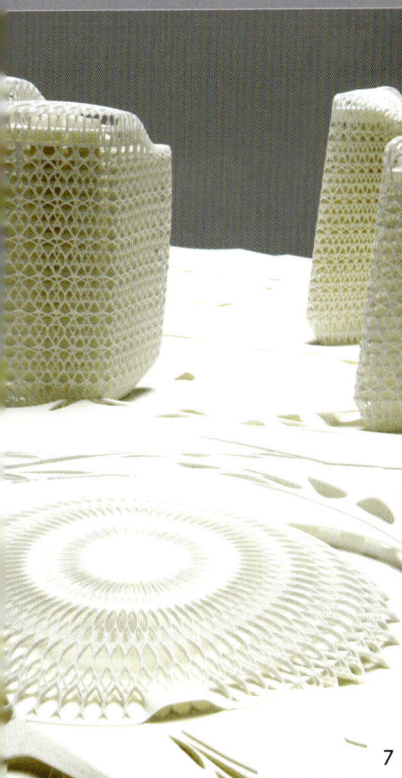

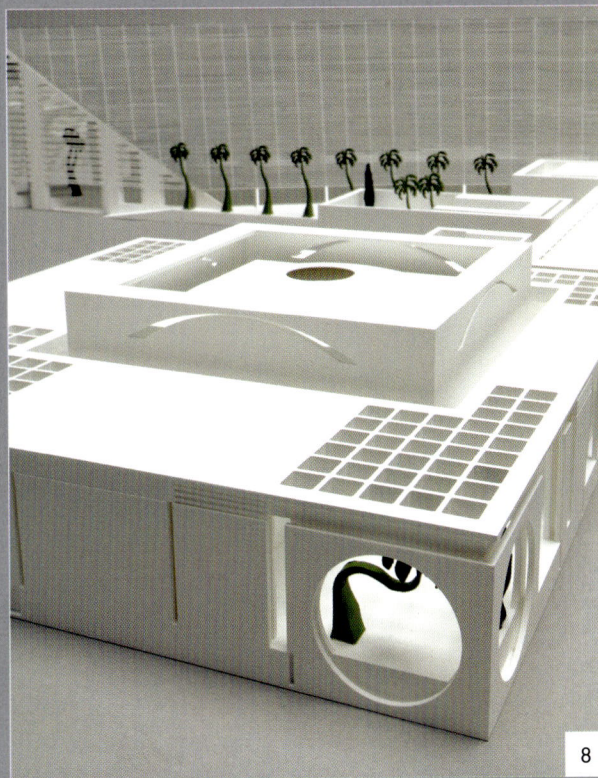

Qualified participants
合格的参赛者

1
1st prize　一等奖
Léon Wohlhage Wernik Architekten,
Berlin

2
2nd prize　二等奖
Burckhardt + Partner, Zurich

3
3rd prize　三等奖
von Gerkan, Marg und Partner
Architekten, Hamburg

4
4th prize　四等奖
Ateliers Lion architectes urbanistes,
Paris

5
Further participant　其他参赛者
Kisho Kurokawa architect &
associates, Tokyo

6
Further participant　其他参赛者
Consolidated CE – Jafar Tukan
Architects, Amman

7
Further participant　其他参赛者
Zaha Hadid Architects, London

8
Further participant　其他参赛者
SCHULTES FRANK ARCHITEKTEN,
Berlin

Léon Wohlhage Wernik Architekten Berlin
Leon Wohlhage Wernik建筑事务所 柏林

"Tripoli Greens – fulfills the Organisation for Development of Administration Centers' desire for a bold and visionary symbol of modern Libya. Our concept is iconic, unique and it will create an identity for the site that will be recognisable from afar."

"的黎波里绿色实现了行政办公中心开发建设现代利比亚一个大胆的、有远见的标志的愿望。我们的概念是标志性的、独特的，从远处就能认出的竞赛地点的身份标志。"

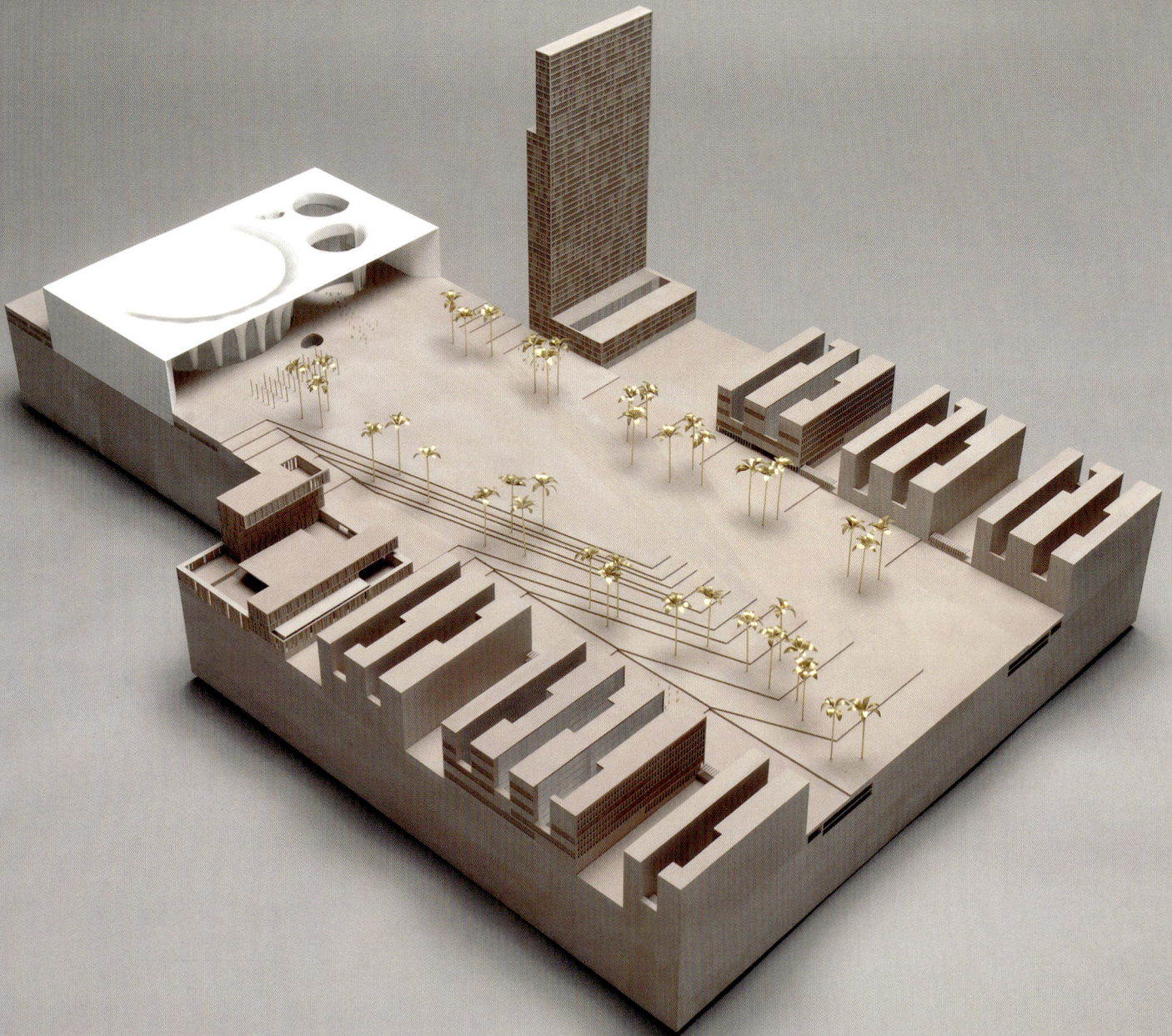

作者 Hilde Léon, Konrad Wohlhage (†), Siegfried Wernik   合作伙伴 Klaus-Tilman Fritzsche, Sebastian Lippok, Julius von Holst, Marius Mensing, Florian Dreher, Tim Lindner, Adrian König, Hans-Josef Lankes, Gerrit Neumann, Jutta Kliesch, Thiele Nickau, Miriam Göllner
专家 structural analysis: Werner Sobek Ingenieure GmbH, Stuttgart; sustainability: Happold Ingenieurbüro GmbH, Berlin; landscape planning: ST raum a. Landscape architecture; real estate identity: MetaDesign AG, Berlin

Léon Wohlhage Wernik Architekten, Berlin

Administration Complex, Tripoli

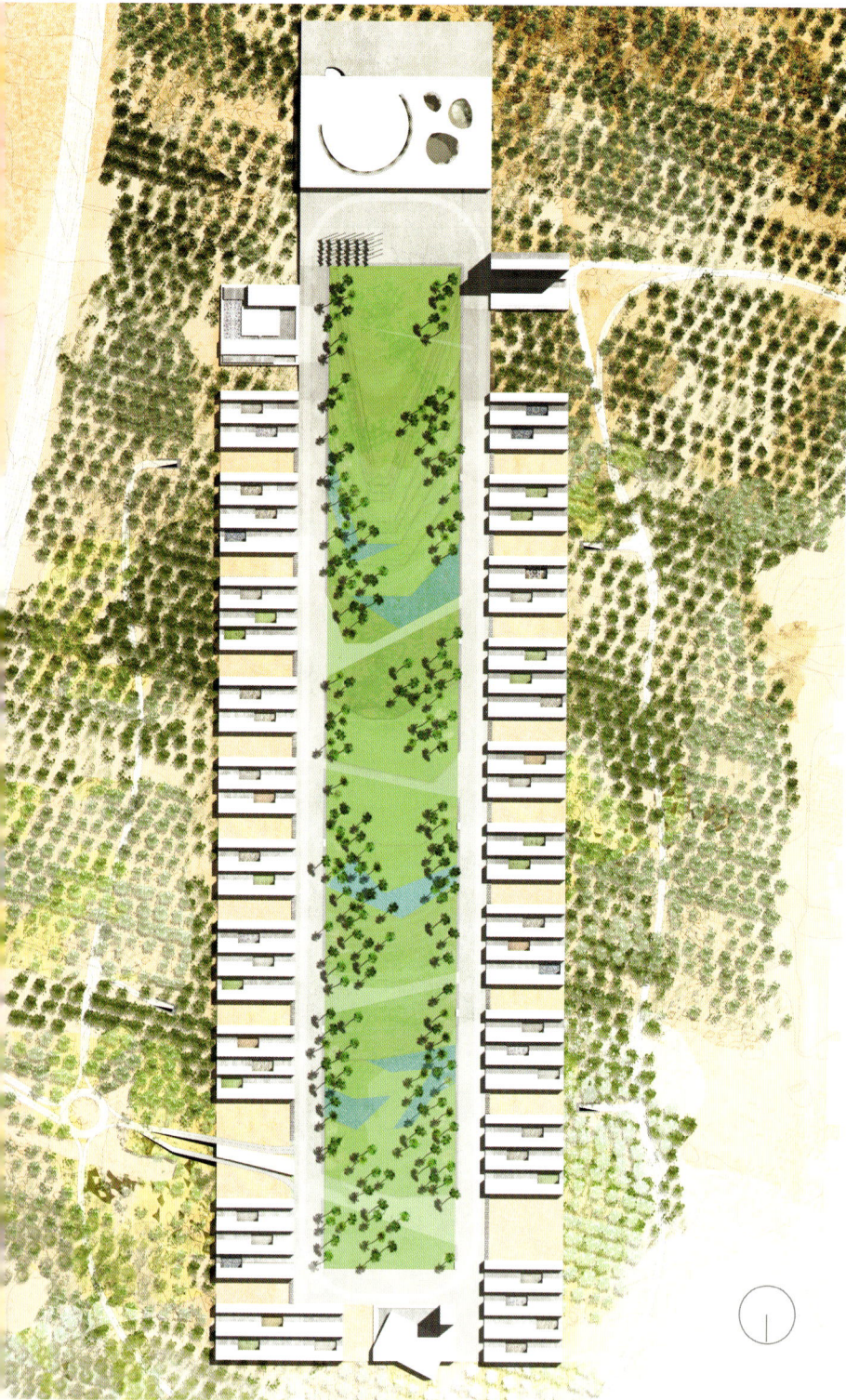

Léon Wohlhage Wernik Architekten, Berlin

Administration Complex, Tripolis

The
Administration
District

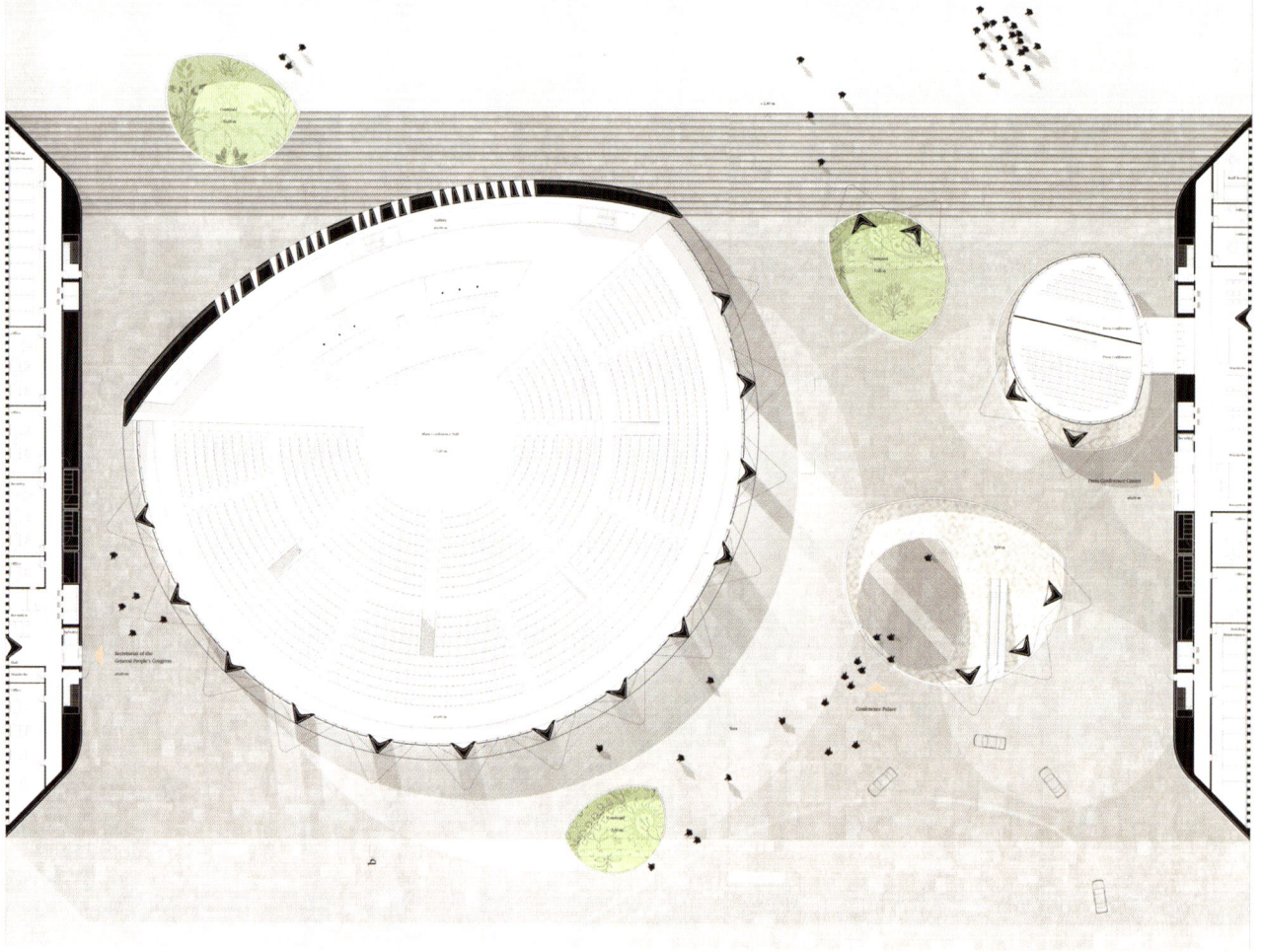

**CONFERENCE PALACE
AND SECRETARIAT OF THE GENERAL PEOPLE'S CONGRESS**

Floor plans, scaled 1:200 / 1:500

The followed vision is that the building illustrates the political culture of Libya in a sculptural form. Representatives of the people meet under its big roof to discuss and shape the future of the country. The roof provides protection from the sun while allowing cool breezes to waft through. The atmosphere is that of a great open foyer allowing one to feel rather than see the expansive volume of the generous Main Conference Hall of the Conference Palace through its translucent membranes. The special design of the load-bearing structures through their sculptural moulding creates an uplifting effect. The mix of varying volumes, courtyards and roof openings dramatises the interplay of space, light and shadow under the roof. Several expansive staircases, escalators and elevators lead one from the foyer in to the lower level of the plinth.

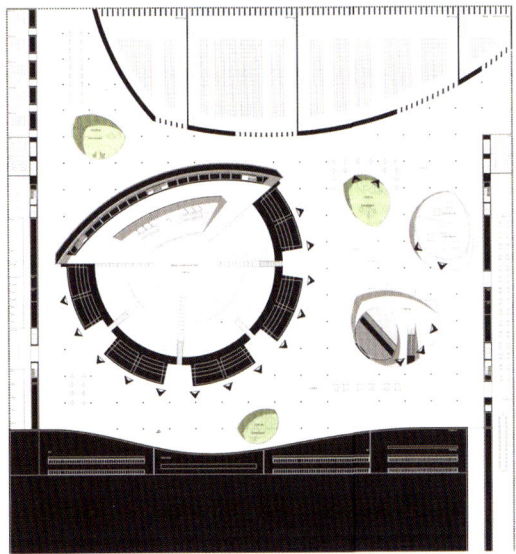

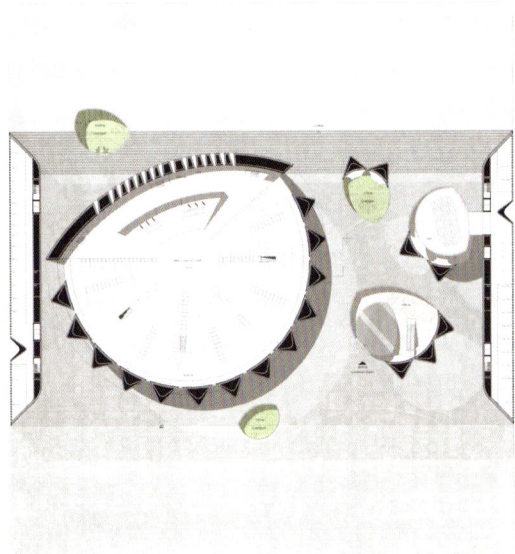

**CONFERENCE PALACE
AND SECRETARIAT OF THE GENERAL PEOPLE'S CONGRESS**

Floor plans, conceptual plan views and sections

All required functions of the administration, including the press centre, are housed in the sides of the building, the giant pillars supporting the roof. At the moment, the roof is reserved for technical and construction purposes, with an option for accommodating other special functions. The goal was to propose and develop a concept based on a very strong and central core.

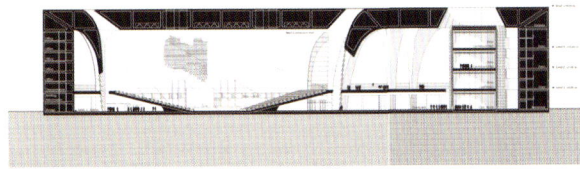

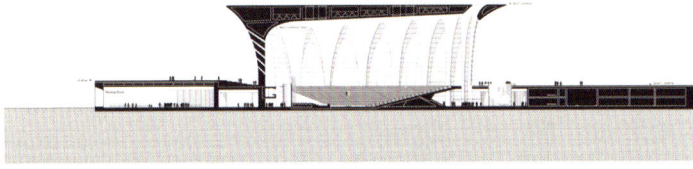

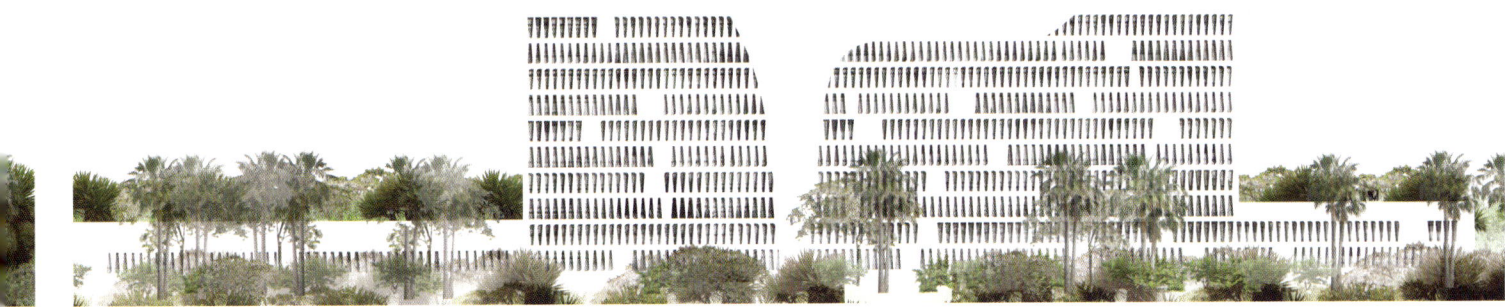

Léon Wohlhage Wernik Architekten, Berlint

Administration Complex, Tripoli

The
Administration
District

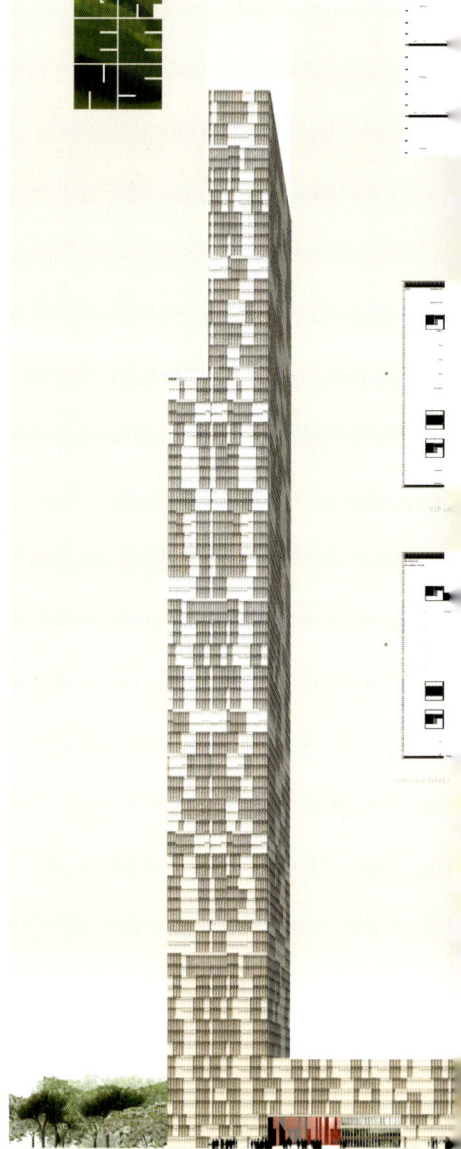

GENERAL PEOPLE'S COMMITTEE'S BUILDINGS AND VIP HOTEL
floor plans, conceptual plan views and sections

The programme calls for three varying sizes of ministries - small, medium and large. One basic type was developed allowing for the rational and economic use of space for offices on the upper floors, with special functions being accommodated on the ground floor or even the first floor. The requirements for natural daylight, ventilation, and shade as well as a strong architectural expression determined the design. The east-west oriented long narrow stretches within the individual building blocks provide shade to neighbouring entities while allowing a cool breeze to flow. These are interspersed by courtyards in differing positions. They cut into the building sometimes going down to the first floor thus adding character to the ground floor. The façade is perforated with recesses to reduce the heat from the southern light.

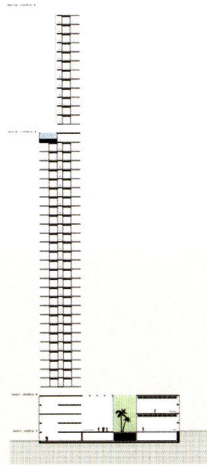

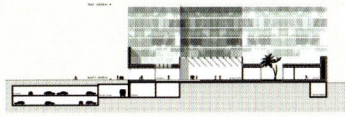
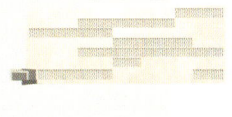

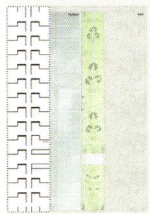
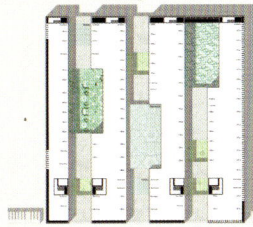
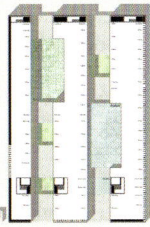
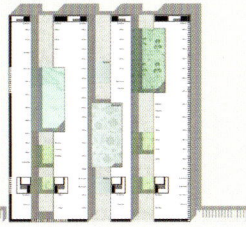

The VIP Hotel takes on a sculptural form like the other buildings. The recessed part of the tower starts above the wellness floor, which runs along its entire length while providing outside access and higher views. Rooms are generously proportioned with luxury private baths. The plinth level rooms have been designed and arranged to cut out any disturbances. There is a direct and secure connection to the Conference Palace via the plinth level. Two separate entry ways cater to the public spaces of the hotel. The deep three-dimensional relief cladding of the hotel's façade diffuses the light evoking an image of a sculptured wall.

Léon Wohlhage Wernik Architekten, Berlin

Administration Complex, Tripoli

Burckhardt + Partner Zurich

Burckhardt + Partner建筑事务所 慕尼黑

"The new Government district for the state of Libya is placed in a garden. It distinguishes itself in its form and arrangement from the surrounding urban pattern."

"利比亚的新政府区位于一个花园里，在形式和安排上与周围城市模式不同。"

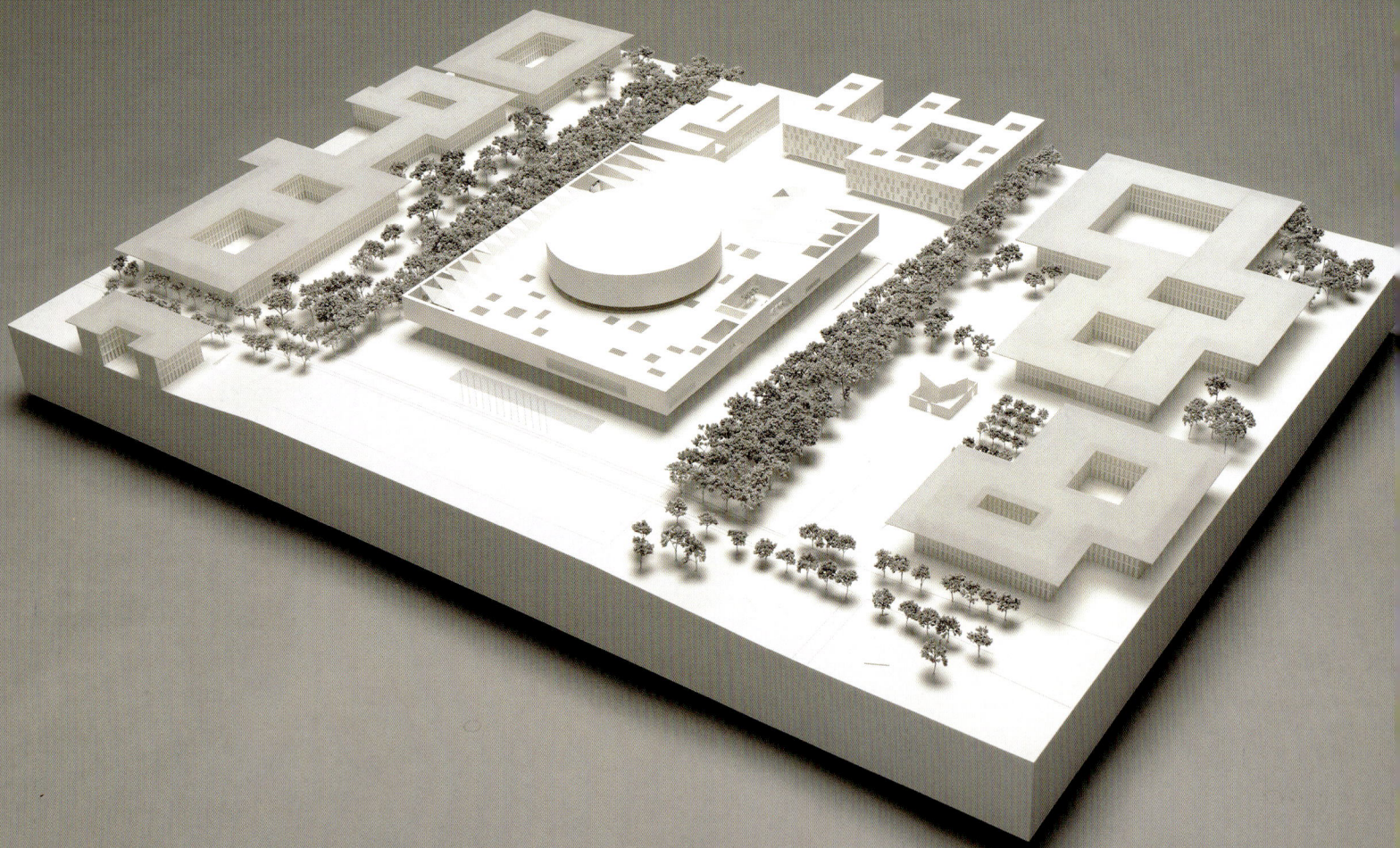

作者 Mathis Simon Tinner 专家 landscape architecture: Vogt Landschaftsarchitekten, Zurich; mechanical & MVAC: HL-Technik AG, Zurich; traffic: Ernst Basler + Partner, Zurich; renderings: Raumgleiter GmbH, Zurich

Urban Strategies

Garden Landscape

The new Government district for the state of Libya is placed in a garden. It distinguishes itself in its form and arrangement from the surrounding urban pattern. On one hand it creates a green island in the outskirts of Tripoli. On the other hand the new neighbourhood connects with its surrounding by means of pedestrian and bicycle paths that pick up existing street patterns and form a continues pattern on the site and beyond. The national forest on the south end of the site will merge into the garden landscape and will become an integral part of the new administration complex.

Urban Pattern

The urban pattern proposed for the site has the potential to extend beyond the actual competition site. In a first phase the different ministries embrace the government palace, the VIP hotel and the office of the general secretary in a protective ring. A second phase allows for an extension along the ring road. If desired even across the freeway. The existing apartment block on the North West end of the site becomes an integral part of the garden. Current uses on the site such as the ostrich farm and the camel farm can be integrated and will remain.

Height Development

The new government headquarter consists of mainly low buildings with a maximum height of four stories. Except the government palace distinguishes itself in height from the surrounding structures. Its shape is visible and recognisable from far away. Be it during the day or in the night. Great importance is given to the visibility from the airport highway.

The Alleys

Both sides of the Airport Highway will have landscaped flat hills that create a different spatial experience along the airport highway. They also serve as acoustical barriers and are made out of the excavation from the buildings on the site. The ring road and the streets to the existing neighbourhood are expressed by the use of clearly delineated alleys of trees.

The Park

The Urban Plazas

The Ancillary Uses

Spread throughout the park and placed along the pedestrian paths are numerous public uses. Sports facilities invite everyone to use the park. Restaurants, café and playgrounds animate and vitalize the park. The ostrich farm and the camel herd can remain in its current place. The concept proposes to split the mosques in different entities that can be erected near the program parts they are needed for.

The Courts

Social Sustainability

Urban Scale

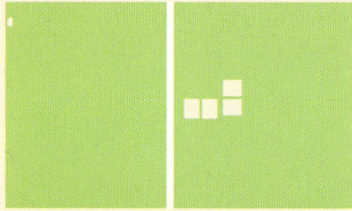

The garden landscape ➡ The green island in the urban fabric

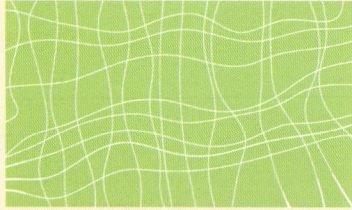

Pedestrian and bicycle paths ➡ The web to the existing surrounding

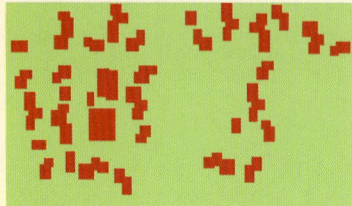

The low rise buildings ➡ Generous open spaces at a human scale

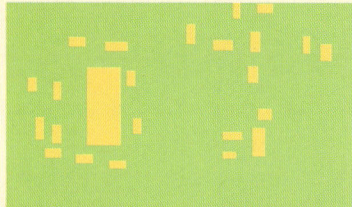

Urban public plazas ➡ The link and the identity

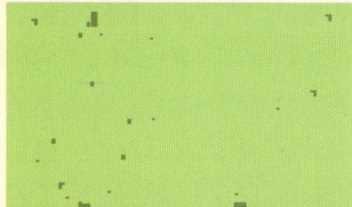

The outdoor facilities ➡ An invitation to the garden for everyone

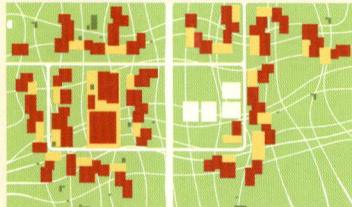

The garden landscape ➡ The green island in the urban fabric

Building Scale

The committee buildings, the Government Palace and the VIP Hotel are based on courtyard schemes. In contrast to the generous public garden, the inner landscaped courts offer more private places to enjoy. A series of common uses are organised around the courtyards on ground level of the committee buildings. Lounge areas and communication spots are spread throughout the regular office floors.

- 🟥 Entrance hall
- 🟦 Common uses
- 🟩 Landscaped courtyards

Environmental Sustainability

Garden Landscape

The urban pattern of the buildings allows for a large area to be landscaped. Only one third of the entire site is occupied with built structures, urban plazas or streets. The entire project is developed to avoid heavy excavation. The excavated material will be re-used on site to form the acoustical "land art" barriers on both sides of the airport highway. Trees and plants will be local.

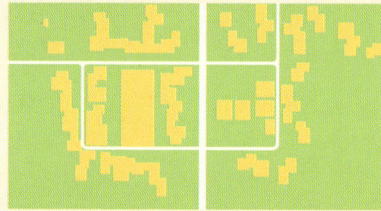

- 🟨 Built area = 27%
- 🟩 Green area = 73%

Facades

All facades are developed to avoid the extensive use of air conditioning. Natural fixed, shading elements in the horizontal direction and in the vertical direction (in the façade) will provide sufficient shading. Users will have the option to either employ natural ventilation or air conditioning. All buildings are in light colour to minimize heat gain through the façade.
The façade design and the building mass is optimized to minimize "heat loss" in the winter months and to reduce cooling loads in the summer months. All buildings will have

Committee buildings and the Secretariat of the General People's Congress

The façade is based on a prefabricated, insulated concrete panel that alternates with operable windows. The form and the arrangement of the panels provide sufficient shading and allows for optimum natural lighting for the interior.

VIP Hotel

The façade of the VIP Hotel is based on a prefabricated panelling system that will provide sufficient shading. In contrast to the committee buildings the outer skin is more open to allow for views into the garden landscape and the courtyards.

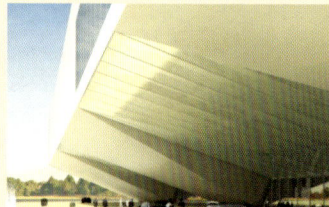

Conference Palace

The construction of the Government Palace is based on a structural steel grid system with prefabricated concrete panels. Natural light for the meeting spaces within the Palace will be mainly achieved with a system of Northern lights on the roof.

Economical Sustainability

Phasing

The urban scheme and the general layout of the buildings allows for a high flexibility for the construction and the organization of the Administration Complex. In a first phase the core of the complex, consisting of the Government Palace, the VIP Hotel and the Secretariat of the General People's Congress can be built. The positioning of the ring road allows for further phases of the complex to be added to the scheme as needed.

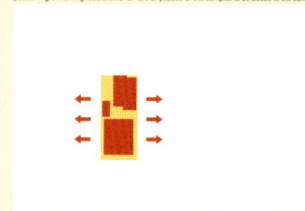

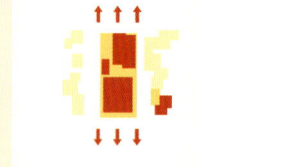

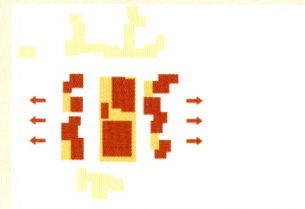

Saving Energy
The building envelope and the building mass are designed to minimize "heat loss" in the winter months and to reduce cooling loads in the summer months. All buildings will have sufficient insulation and will be equipped with fixed shading devices where necessary.

Cooling
The buildings will be cooled with air-conditioning systems (input of fresh air) and static cooling systems. The static cooling systems are calculated with totally 32.34 MW. The air-handling units need a total cooling capacity of 12.66 MW.
The cold water of all chillers will be served on the basis of 12/16 °C. Each building has its own cooling exchange system to serve cold water for the coils of the air-conditioning systems (12/16 °C) or cooling ceilings (16/20 °C). The buildings will be cooled only at outside temperatures over 24 °C. Otherwise natural ventilation provides for a pleasant climate. In order to reduce the gas or oil consumption to run boilers or EPS, a vacuum pipe collector system should be installed on the surface of the water tanks or elsewhere appropriate.

Electrical power supply
2 electrical supply nets will serve the building complex. A second electrical power supply will be served by gas or oil-driven motors with generators. To reduce the electrical power, run by the public network or EPS-system, a photovoltaic system (integrated in the shading elements on the roofs or in the landscape) can serve a peak power supply.

Boiler System
The boiler system with a total capacity of 17.85 MW will serve a maximum heating capacity of 7.55 MW in winter. The total heat loss of all buildings is calculated with 6.1 MW, hot water for all bathrooms (especially hotel and lavatories 0.65 MW) and hot water for all kitchens (0.8 MW).

Fresh Water and Waste Water
The calculation for the buildings shows that only 30 % of the complete water consumption will need to have drinking water quality. A biological cleaning system with water tanks, which will also be served by storm water or rainwater, can be used for flushing toilets. The water tanks can also serve sprinkler systems which might be needed in different parts of the Administration district.

Principles Energy - Systems   Total cool.cap. 40.500 kW$_{TH}$

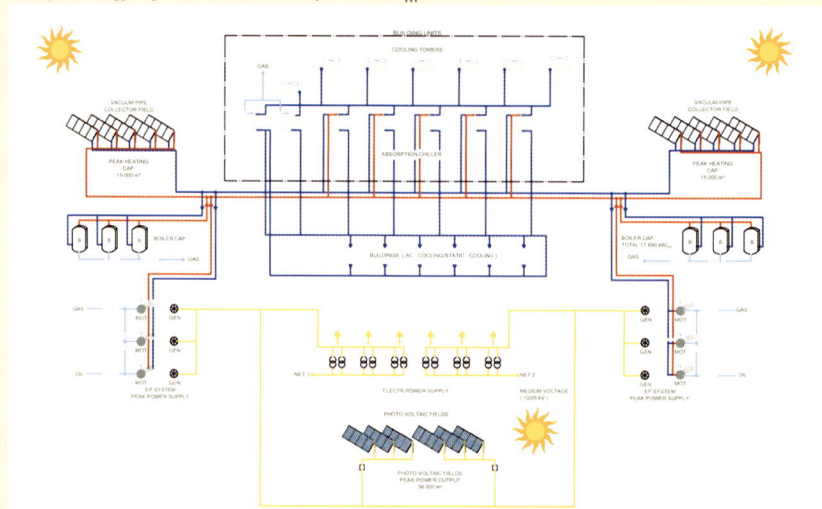

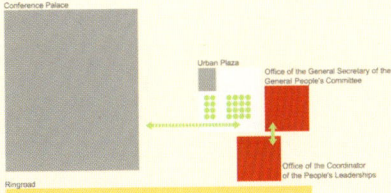

Concept

The Office of the Coordinator of the People's Leadership is based upon a standard courtyard structure. It shares an urban plaza with the Office of the General Secretary of the General People's Committee. The program is organized on two floors around a landscaped courtyard. Two additional floors allow for an extension within the same building.

The Office of the Coordinator of the People's Leaderships is placed in immediate proximity to the Office of the Secretary of the General People's Committee and the Conference Palace.

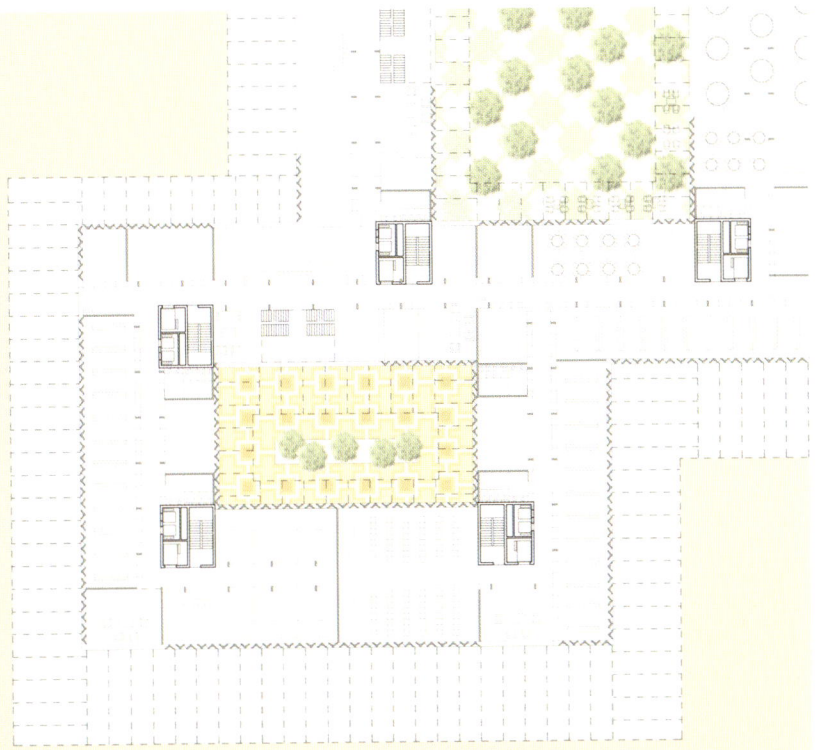

Conference Palace

Urban Plaza

Office of the General Secretary of the General People's Committee

Office of the Coordinator of the People's Leaderships

Ringroad

Social space

- Entrance hall
- Common uses
- Landscaped courtyards

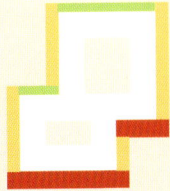

- South 10m
- East / West 7.5m
- North 5m

Roof Shading Concept

The fixed roof shading devices are designed with a permanent, perforated membrane that duplicates the "shadow of natural leaves" by using a geometric, repetitive pattern. The facades are developed to avoid the extensive use of air conditioning. Natural fixed, shading elements in the horizontal direction and in the vertical direction (in the façade) will provide sufficient shading. Users will have the option to either employ natural ventilation or use air conditioning.

Courtyard

The interior court of the Offices of the Coordinator of the People's Leaderships is part of the overall landscape scheme for the courtyards. A pattern of geometrically arranged tiles of different colours will contrast with a canopy of trees that will offer additional natural shading.

von Gerkan, Marg und Partner Architekten Hamburg
von Gerkan, Marg und Partner建筑事务所 汉堡

"A beautiful landscape with a rich natural scenery will be created as a park for the new government district."
"为新政府区创造出一个像花园一样的美丽的自然风景的景观建筑。"

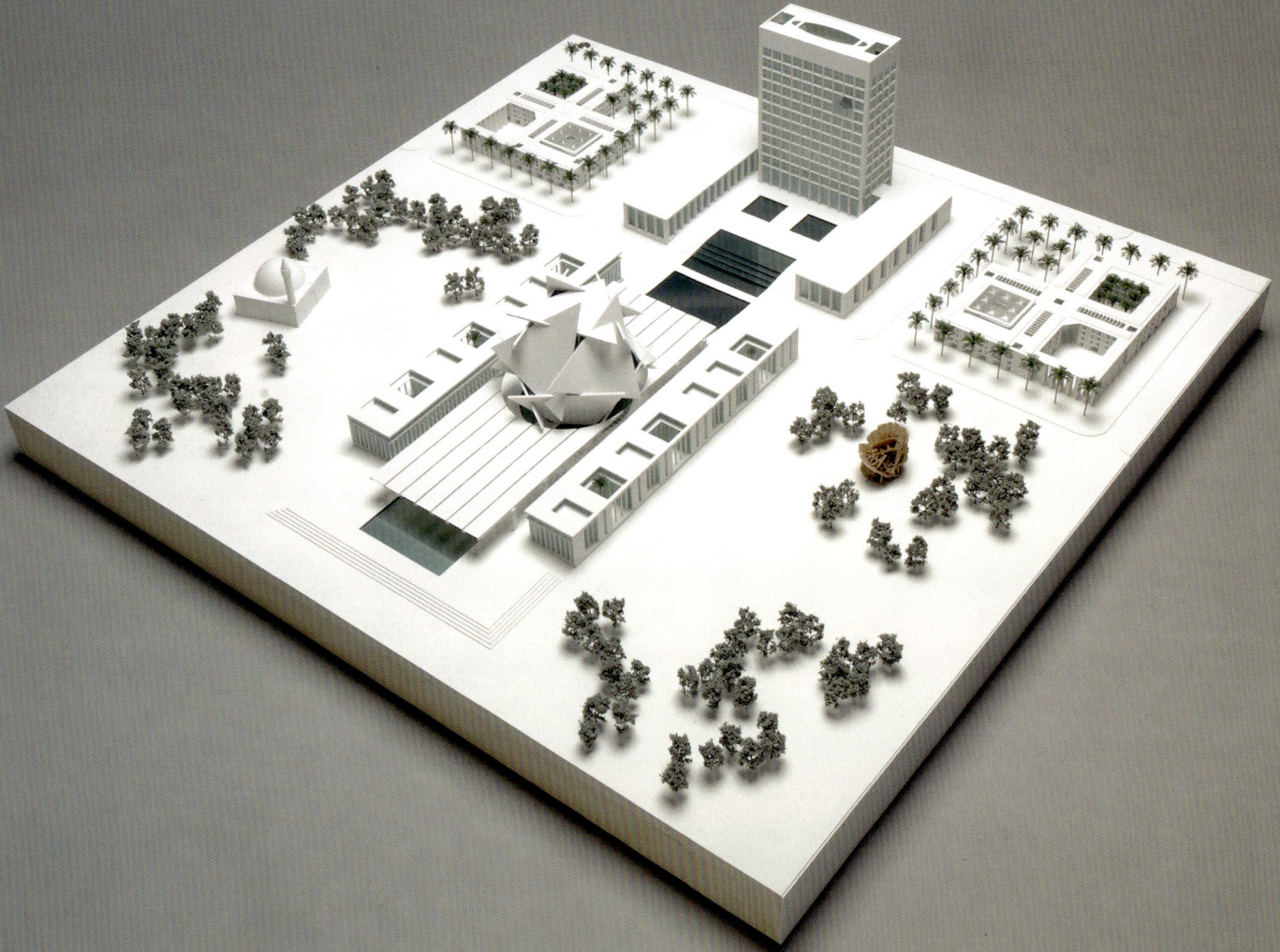

作者 Meinhard von Gerkan, Jürgen Hillmer　专家 landscape architecture: Breimann Bruun, Hamburg;
energy planning: Transsolar, Stuttgart

Description Administration Complex in Tripoli

The area chosen for the development of the Administration Complex is located at the Airport Highway south of the proposed Third Ring Road. A beautiful landscape with a rich natural scenery will be created as a central park for the new Government District. The iconic building is located in the center part of the park in order to create a powerfull visual ascer-tainability.

The concept is based on 7 positions

Structural Order
As well as every living existence a proposed health park also depends on principles of structural order. Therefore each structural element has to be defined in relation to the whole. The principle has to structure the building volumes and the uses, the plan and elevation have to be derived out of the same order. Analoque to the grammar of language the structural order has to take over a rational steering-function.

The overruling concerns of structural order are:
1. Serenity of architectural form
2. Hierarchy
3. Visual ascertainability
4. Orientation in space

Individuality
In traveling over the world, we experience locations, which stick to our memory and others, which mingle with other pictures very fast. The more specific and explicit the appearance of a certain location is, the more our memory will be touched by it. Most of the contemporary city planning is characterized by ran-domness and commutability. There are only a few plannings, which are equipped with a unique and specific identity.
The objective for the new health industrial park should be to develop the specific identity out of the building-task and regional situation.

Grid
The building- and infrastructure are due to the ap-proach of antique Greek cities based on a grid. The traffic-grid serves with several categories of facilities.
1. Walkways
2. Bicycle lanes
3. Streets for cars and buses
4. Mixed used areas for pedestrians, cyclists, delivery and the fire brigade

The grid offers alternations of narrow alleys and wide squares which are also intersections in the network. Different themes of landscape design should create a distinct character for each area.

Public space
One main issue is the defined quality of public space. The design of street-objects and sections, as well as the early realisation of parks and plazas is as important as the layout and design of the buildings for the success of the development.

Balance of determination and freedom
The premise for a lively Health Park is a balance of determination and freedom of architectonic articulation. The determination is given through the definition of an urban structure with public space, plots and building highs based on a measure. Within this system of structural order the freedom of finding an individual form for each building has to be kept.

Simplicity
The command of simplicity does not mean primitive, banal or unimaginatively. Meant is simplicity in the sense of plausible, self-evident and clear. Architec-tural solutions which take on arbitrary forms and need complex structures to be interesting are less suitable to meet the requirements of the new Software Park, neither in form nor in content.

Balance of variety and unity
The uncomfortableness of our environment is very often based on an excess of sameness, which is perceived as monotony or an excess of variety, which is registered as chaos. How important the balance of variety and unity is can be emphasized by analysing architectural history. The recipe for the highly appreciated fabric of ancient cities is exactly rooted in this balance. It is necessary to analyse these examples to transform the principles into the future, but under no circumstances, to copy them.

The above mentioned principles are used to create a set of different geometrically defined areas which contrast the natural landscape. Each area responds to the specific function and condition of its location.

Landscape

The setting of the architecture within the city, bordering a greened area, allows this green to surround and float through the new building complex. The entire site is embedded in a forested area of palm trees and eucalyptus, continuing the existing plant-ings to its south.

When driving towards this new governmental city you have to pass through this thick green belt. Once you reached the first ring road, the green opens up for the outer ring of governmental buildings. The spaces between the buildings are planted with palm trees. They grow up between the parking lots, from one floor below. They spread out in an even grid, according the architectural grid, all the way to the curb, so they function as plaza trees as well as street trees. The inner ring road surrounds the central park of the site.

The par-iament house is the central part of this park. It sits in the middle axis on a slightly raised stone plaza. A large water basin marks its middle axis. The surrounding park works with the structures and patterns of the surrounding landscape. Paths divide the area into amorphous shapes. Wherever the cross each other a small plaza opens up and creates spaces to rest, for pavilions, water plays, etc. On a second layer trees stand loosely on the entire park. The surrounding landscape interfuses this newly created open space.

The inner courtyards of the buildings are places to take a break and enjoy the fresh air under the translucent fabric shade roofs. Additional to these roofs, some palm trees provide a second roof layer. They "dance" playful around a central round pool that sits in the centre of each courtyard, directly under the circular opening in the roof structure. The entire surface is paved in a light colored stone, with paths that run towards the pool in the centre. The theme of the central park is interpreted in these much smaller open spaces.

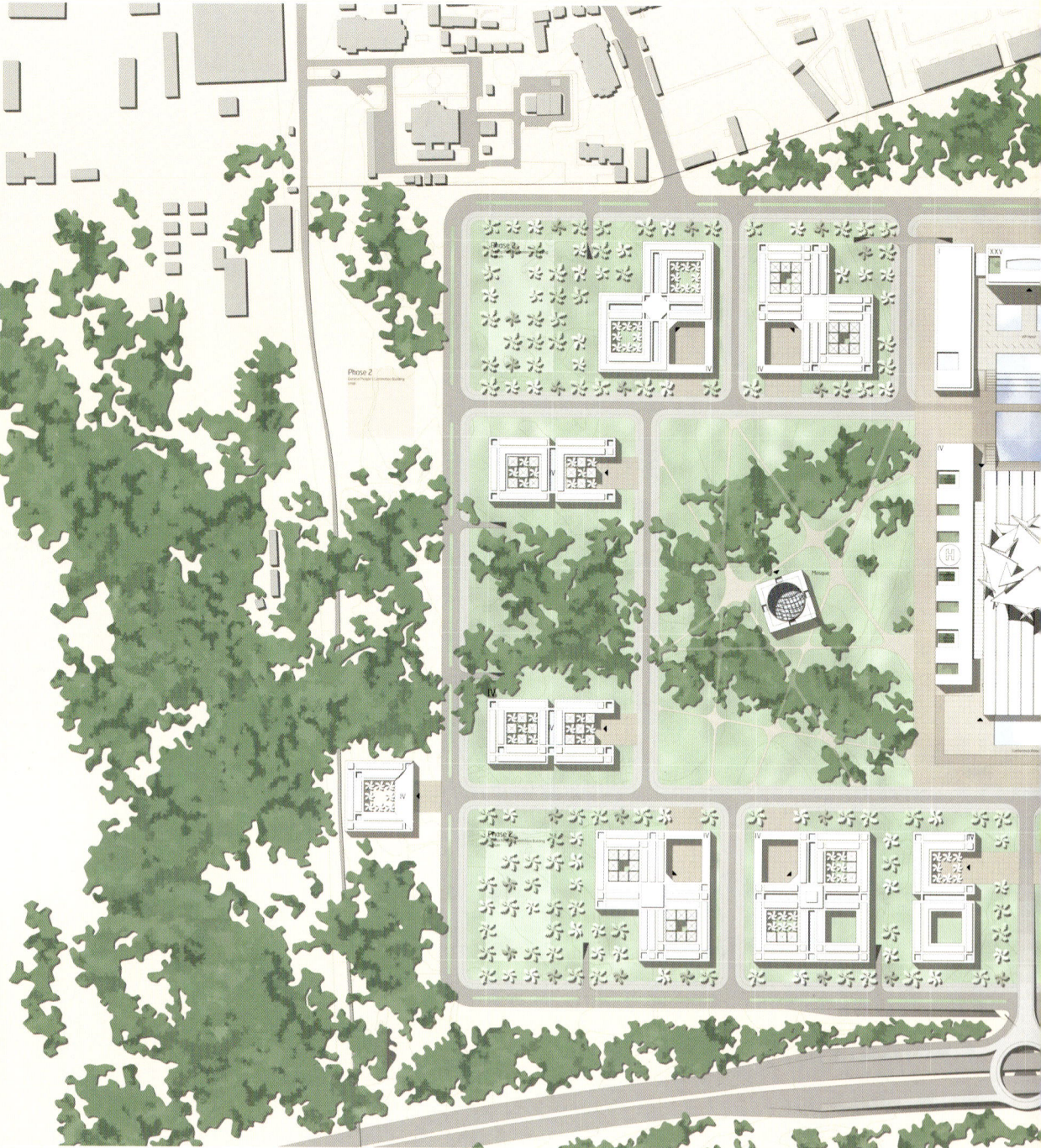

administration complex in tripoli

administration complex in tripoli

distribution

traffic

parking

pipe plan

site plan . scale 1:1000   0   40   100   200

administration complex in tripoli

Administration Complex, Tripoli

conference palace and secretariat of the general people`s congress and other facilities

administration complex in tripoli

Ateliers Lion architectes urbanistes Paris
Ateliers Lion建筑及城市规划事务所 巴黎

"... what can be more important in the symbolic issue than confronting the political institution with the Libyan land and its climatic contrasts?"

"在象征问题中，什么可能比面对在利比亚土地上气候对比强烈的政府办事机构更重要的事情呢？"

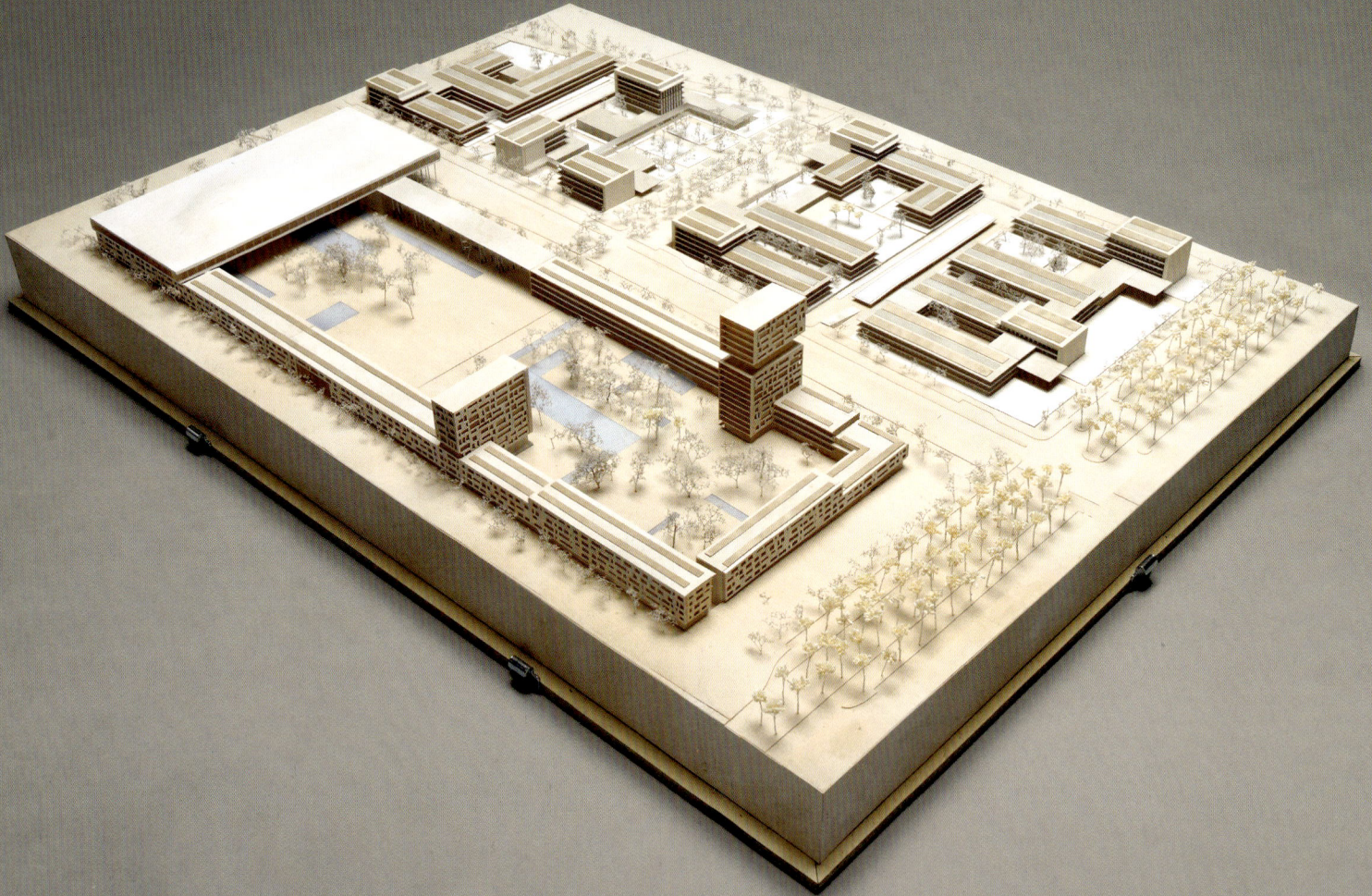

作者 Yves Lion, Francois Leclercq, Claire Piguet　合作伙伴 Delonne Léonard, Kim Hyon Seok, Le Minh Triet, Mahajan Reena,
Ramone Laurianne, Ré Christelle　自由建筑师 Nicolas Laisné, architect, Christophe Roussel, architect　专家 energy planning:
Transsolar, Thomas Auer, Stuttgart; traffic: Citec, Philippe Gasser, Geneva

South facade

Conference palace & Secretariat of the General people's Congress West facade

Main conference hall

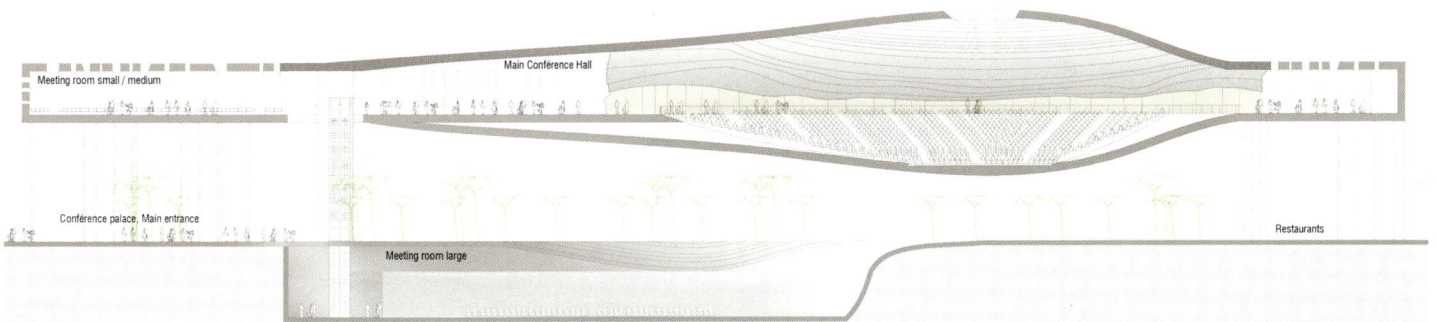

Meeting room small / medium

Main Conférence Hall

Conférence palace, Main entrance

Meeting room large

Restaurants

Longitudinal section   1:200

SITE PLAN

N

0 10 50 100 200

1 : 1000

Kisho Kurokawa architect & associates Tokyo
Kisho Kurokawa建筑事务所 东京

"Iconic central government dome, introduction of a cultural hub, linking these two main elements by the 'urban axis'"
"通过'城市轴'把标志性的中央政府圆顶和一个文化区这两个主要元素联结起来。"

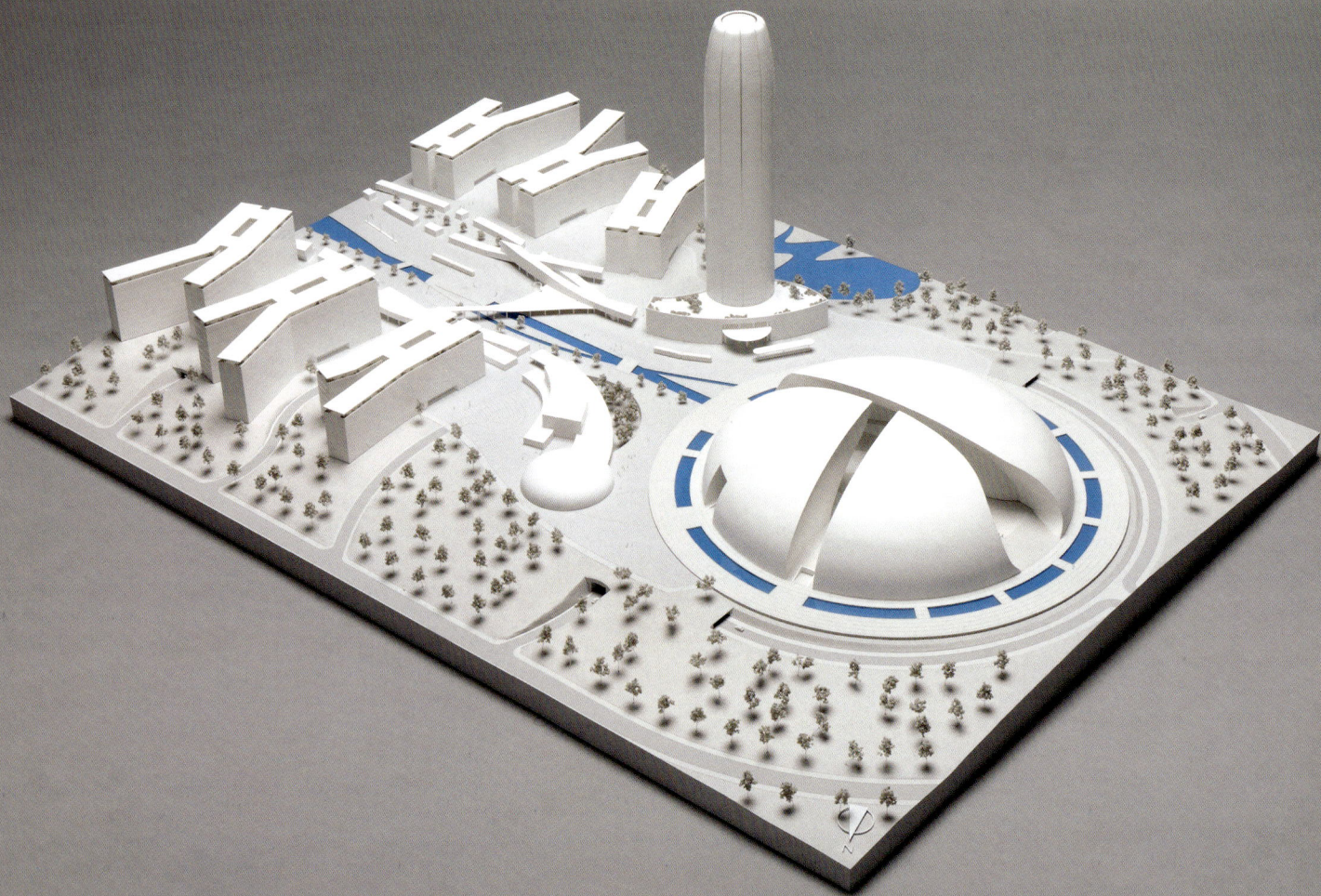

作者 Kisho Kurokawa (†)

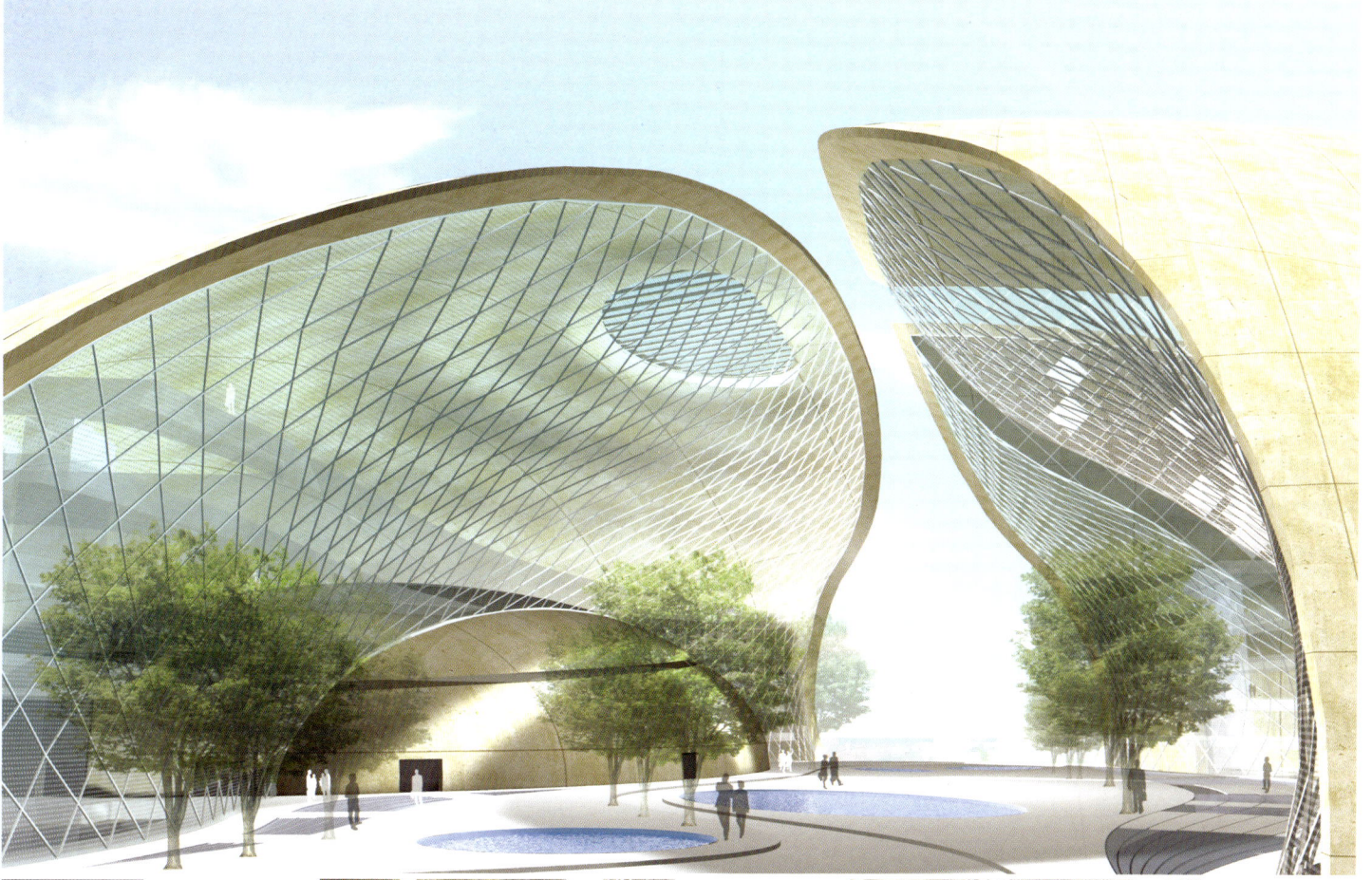

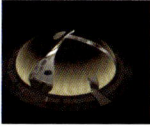

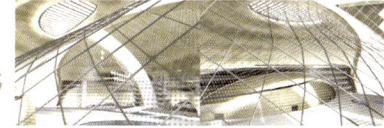

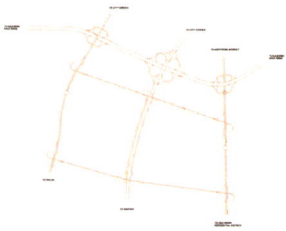

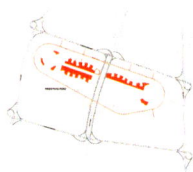

ROADS

GREEN BUFFER

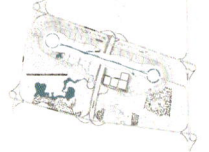

WATER SYSTEM

PROGRAM

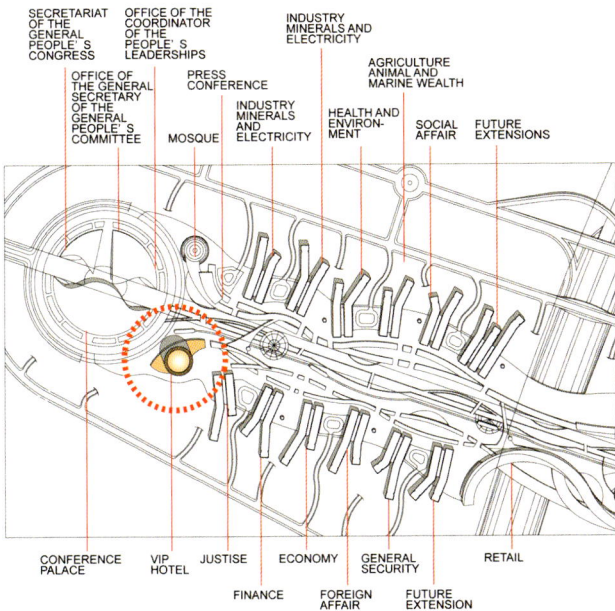

Connection to Adjacent Buildings
Scale1:6000

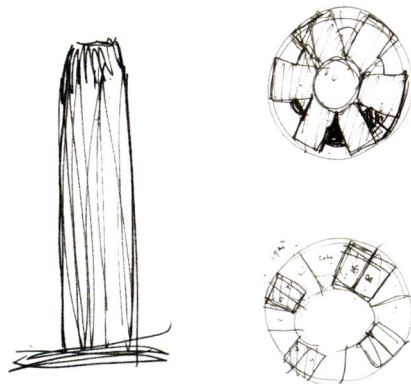

SECRETARIAT OF THE GENERAL PEOPLE'S CONGRESS
OFFICE OF THE GENERAL SECRETARY OF THE GENERAL PEOPLE'S COMMITTEE
OFFICE OF THE COORDINATOR OF THE PEOPLE'S LEADERSHIPS
PRESS CONFERENCE
MOSQUE
INDUSTRY MINERALS AND ELECTRICITY
INDUSTRY MINERALS AND ELECTRICITY
HEALTH AND ENVIRONMENT
AGRICULTURE ANIMAL AND MARINE WEALTH
SOCIAL AFFAIR
FUTURE EXTENSIONS

CONFERENCE PALACE
VIP HOTEL
JUSTISE
FINANCE
ECONOMY
FOREIGN AFFAIR
GENERAL SECURITY
FUTURE EXTENSION
RETAIL

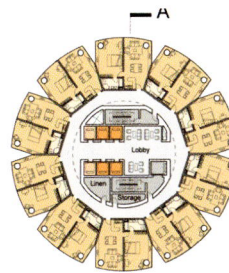

Typical Plan
Boarding Suites   Scale1:600

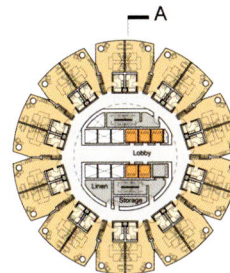

Typical Plan
Double Bed Rooms   Scale1:600

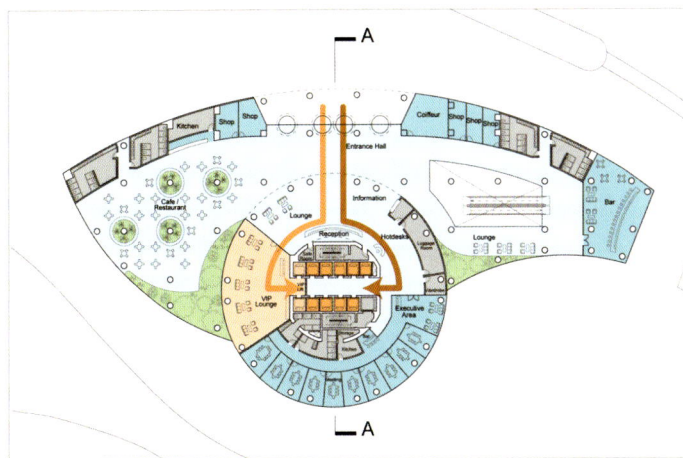

Ground Floor Plan
Hotel Entrance Lobby   Scale1:600

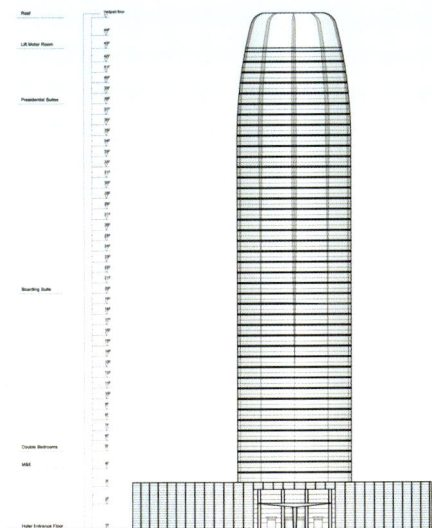

Sketches

Front Elevation
Scale 1:1200

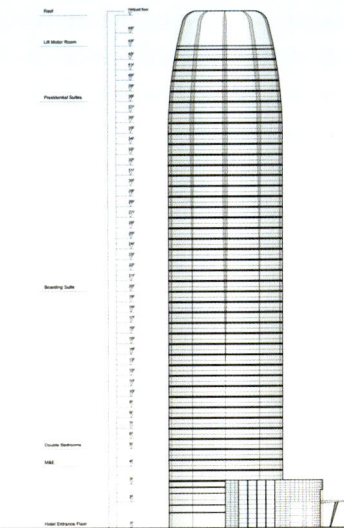

Side Elevation
Scale 1:1200

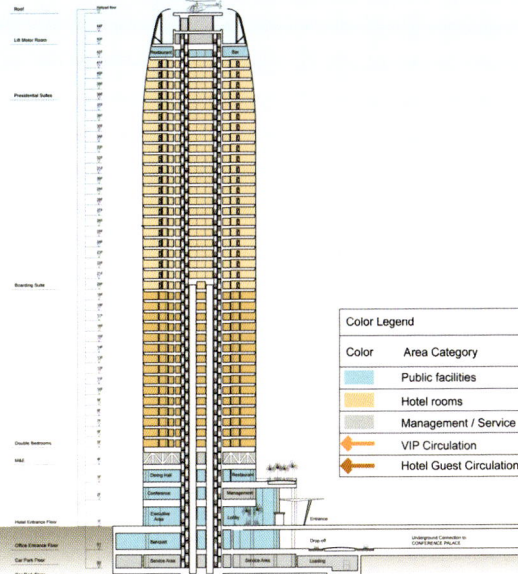

Section A-A
Scale 1:1200

Color Legend

Color	Area Category
	Public facilities
	Hotel rooms
	Management / Service
	VIP Circulation
	Hotel Guest Circulation

Consolidated Consultants Engineering & Environment Jafar Tukan Architects Amman

工程环境综合咨询公司和Jafar Tukan建筑事务所 安曼

"Rhythm of the city flows into the site and collides with nature."

"城市节奏流入竞赛地点，与自然产生碰撞。"

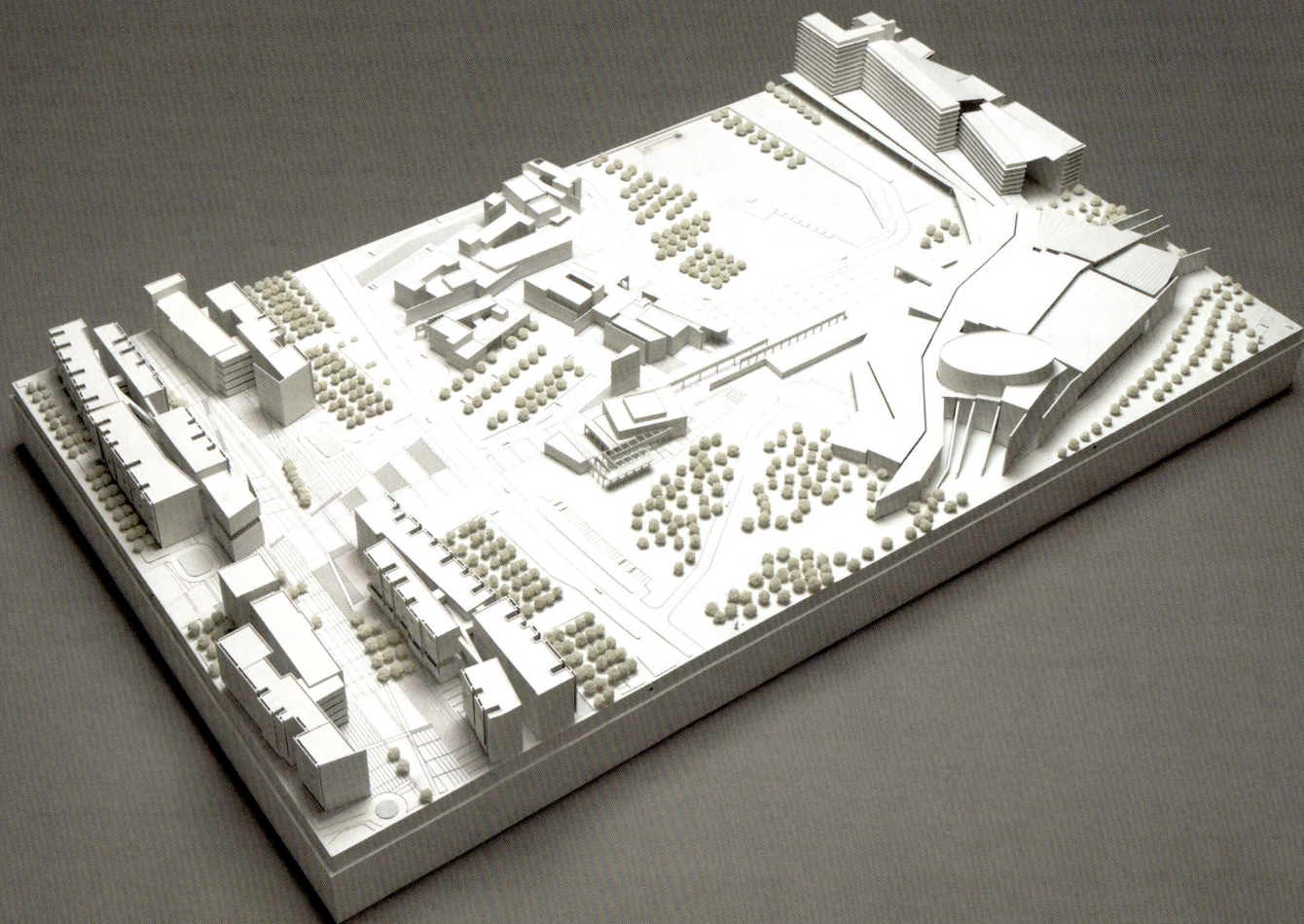

作者 Jafar Tukan, Yasser Darwish, Mohammad Abbas   合作伙伴 Tareq Ghanam, Saba Innab, Andaleeb Bizreh, Hadeel Hamad, Jumana Hamadani, Ahmad Seyam, Nemeh Mansour   专家 traffic: Consolidated Consultants Engineering & Environment –Jafar Tukan – Architects, Issam Bilbesi, Amman

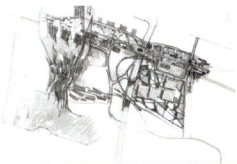

Libya is embarking on new era where massive construction will be one of its major traits. Large projects will most likely take place outside the city along the highways. Sprawling is undesirable pattern of growth and will strip Tripoli of its unique spatial qualities

Such large projects are vital to the city and can contribute a lot to its character if planned properly. We believe that the location of the Governmental HQ on this site is one step in the right direction but should be inscribed as part of a global vision of the city. The site is located on a pivot area right between the city and the countryside. The latter is very present due to the relatively well conserved forest. Our strategy consists of locating future large projects along the third ring road. These projects can extend all the way east to the old airport thus creating a well defined zone with a transitional typology. This zone is beneficial in two ways; first, it has the merit of preserving and encouraging the densification of the existing city. Second, it acts as a solid base for future developments that are to take place beyond it. This zone is hybrid between city and nature; its genesis is the forest; its heart is the hill. Rhythm of the city flows into the site and collide with nature.

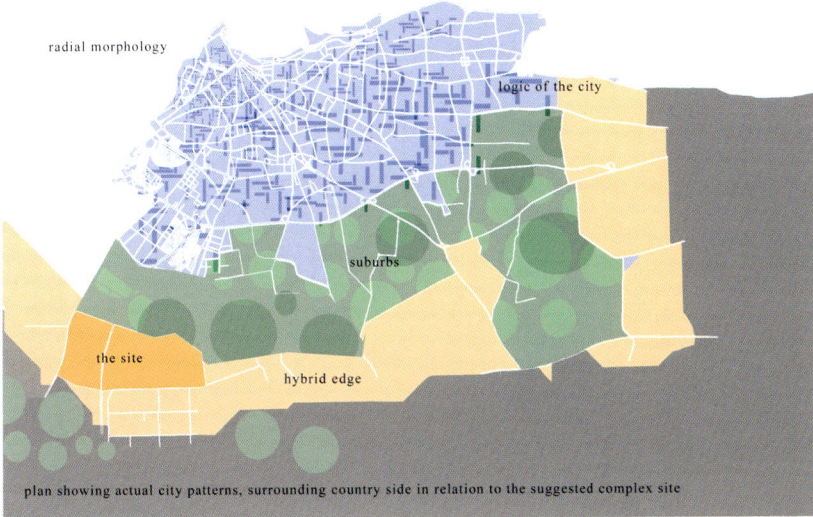

radial morphology

logic of the city

suburbs

the site

hybrid edge

plan showing actual city patterns, surrounding country side in relation to the suggested complex site

existing

sprawl

sprawl vs. planned growth

densification

channeling interaction

defining the edge

filtering the expansion

vip hotel and conference center

the edge, hybrid pivot

collision of green and city

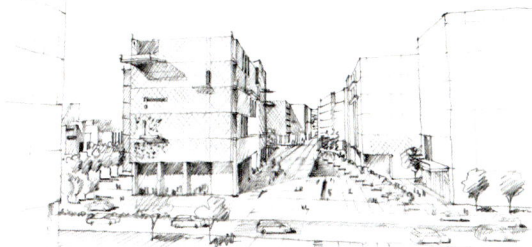

main spine

interrupted by car, linked by green

New Hybrid Typology, responding to ground situation

city

Linear structures, spine

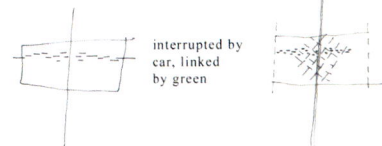

congress from main spine

organic structures

view from airport highway

forest

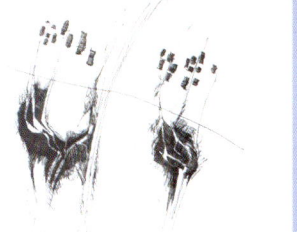

The General people's committees are configured in a linear arrangement producing a main spine that cuts through the site longitudinally. The spine is interrupted by the existing highway but connected through the emerging green platforms climbing up to hover over just above the highway. Besides "linking" the two sides of the spine, the green hill also defines the hierarchy created in the planning of the committees by opening up from the political to the more public as we move from extreme west to extreme east. As we move along the axis we see a clear defined spine intersecting with another flow coming from The Grand Plaza. This defined spine gradually climbs the green hill up opening up to a more fluid configuration where the logic of the organic structure intersects with the city logic. The committees in this part hold the more open to the public role, for example; culture, agriculture, etc...in a more interactive way and unfold into supporting s e r v i c e s .

hill/ bridge

spine

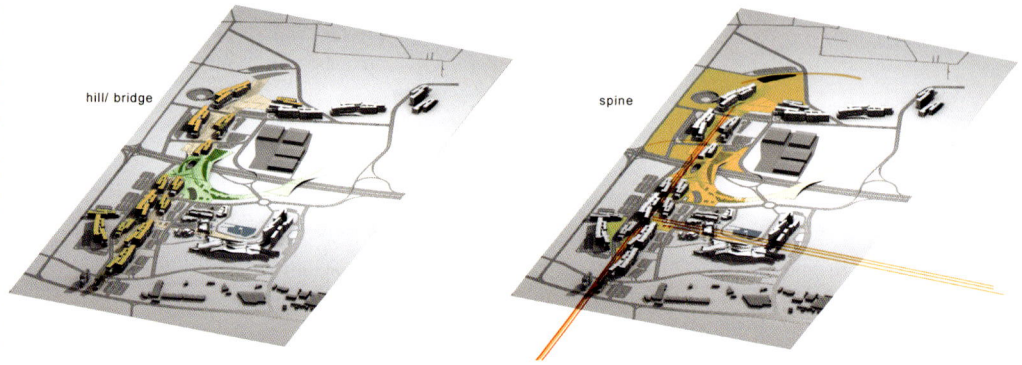

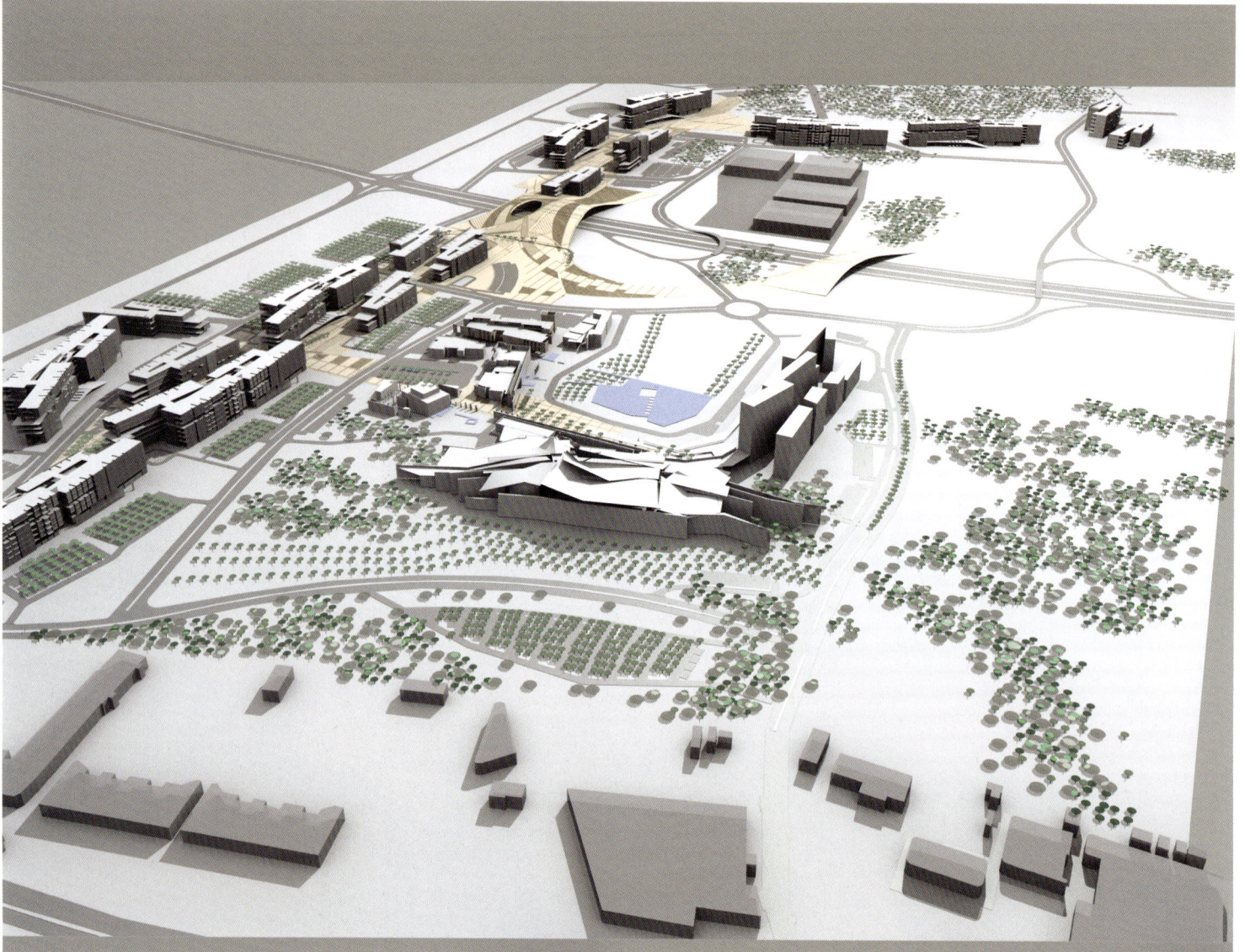

Administration Complex, Tripoli

Consolidated CE – Jafar Tukan Architects, Amman

Zaha Hadid Architects London
Zaha Hadid建筑事务所 伦敦

"The new seminal government buildings will be situated in a national botanic garden representing the geographic of Libya."
"有创意的新政府大楼将坐落在一个表现利比亚地理特点的国家植物花园里。"

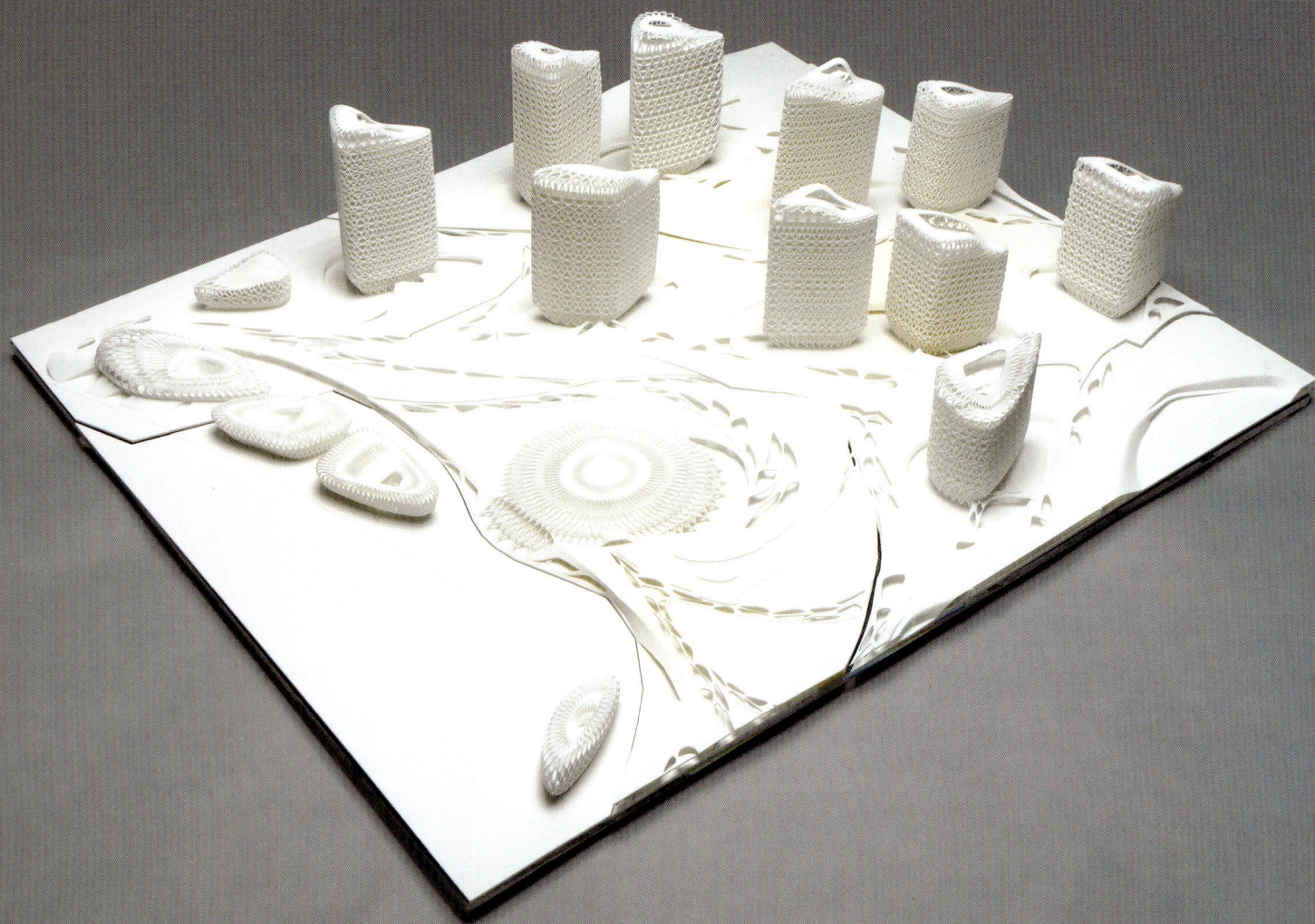

作者 Zaha Hadid, Patrick Schumacher   合作伙伴及建筑系在校生 Helmut Kinzler, Ebru Simsek, Lars Teichmann, Enrico Kleinke, Brian Dale, Lauren Barclay, Deniz Manisali, Dadatsi Dominiki, Emily Chang, Dimitris Akritopoulos, Hala Sheikh, Kelly Lee, Lillie Liu, Oznur Erboga, Pavlidou Eleni, Shih-Chin Wu, Shiqi Li, Nantapon Junngurn, Sylvia Georgiadou, Theodora Ntatsopoulou, Eirini Fountoulaki, Claus Voigtmann   专家 structural analysis: ARUP Engineering, Keith Jones, London; landscape architecture: Gross.Max, Eelco Hooftman, London; traffic: ARUP Engineering, Tim Gabbitas, London

MASTER PERSPECTIVE VIEW TO THE NORTH

Landscape Concept

The new seminal governmental buildings will be situated in a unique 200 hectare National Botanic Garden representing the distinct geographic regions in Libya. The National Botanic Garden will become a striking symbol and representation of the country as a whole. The three main geographical regions in Libya which will be represented are Tripolitania, Cyrenaica and Fezzan. These regions consist of a variety of landscape typologies ranging from coast, desert and mountain. The gardens will have a public and recreational dimension and will become a public resource through conservation, education and scientific research representing the face of modern Libya for the 21st century. The dynamic configuration of the governmental quarter will be extended into the site and across the motorway by means of an undulating, free flowing, botanic spine whose shape and forms are inspired by various geomorphologic landscape formations. A circuit of curvilinear walkways provides connectivity, linkages and circulation loops. Emphasis is placed on the dynamics of ecosystems rather than to horticultural classification using native species and representing plant communities and associative habitats. The two patches of existing woodland are extended and incorporated into a botanical forest. Importantly the whole site forms part of a wider green structure and ecological corridor. The woodland can provide a landscape framework for eventual future expansion of built fabric within a coherent whole.

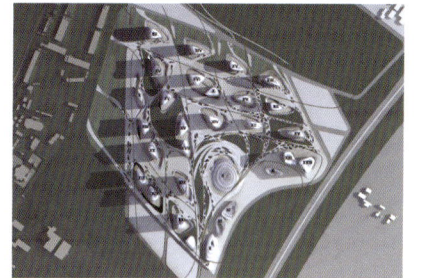

MASTER PLAN WITH NUMBER OF FLOORS

LANDSCAPE CROSS SECTION

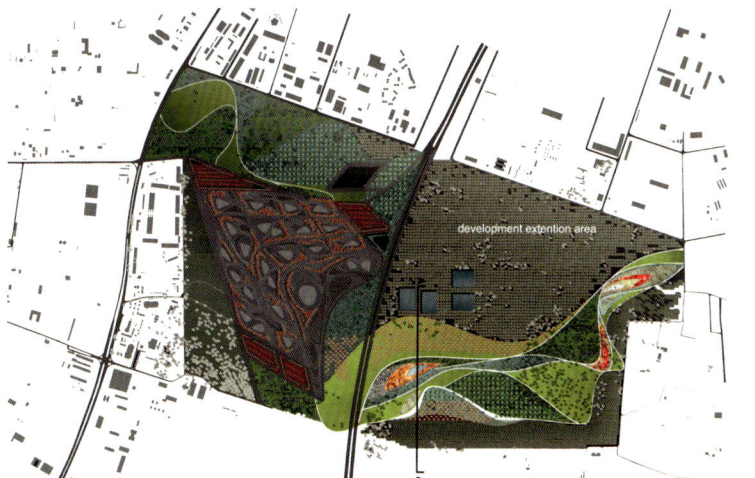

MASTER PERSPECTIVE VIEW TO THE SOUTH

BIRD'S EYE VIEW OF OVERALL MASTER PERSPECTIVE

OVERALL LANDSCAPE DEVELOPMENT

MASTER PERSPECTIVE

ADMINISTRATION COMPLEX IN TRIPOLI

SHEET NO.1

CONGRESS PALACE VIEW

CONGRESS HALL INTERIOR VIEW

CONFERENCE PALACE COURTYARD VIEW

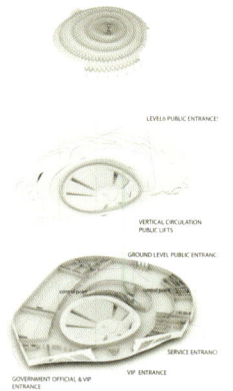

LEVEL 6 PUBLIC ENTRANCE

VERTICAL CIRCULATION
PUBLIC LIFTS

GROUND LEVEL PUBLIC ENTRANCE

SERVICE ENTRANCE

GOVERNMENT OFFICIAL & VIP
ENTRANCE

VIP ENTRANCE

CONGRESS HALL ACCESS DIAGRAM

ENTRANCE_PUBLIC AREAS
DINING AND KITCHEN
CORE
OFFICES
MEETING ROOMS
BOARD ROOMS
MAIN CONFERENCE HALL
GENERAL BUILDING SERVICES

PROGRAM DIAGRAM

GROUND FLOOR

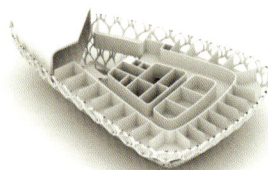

FIRST FLOOR

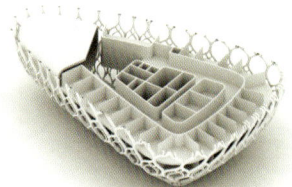

SECOND FLOOR
PROGRAM USAGE DIAGRAM

SITE PLAN   LEVEL +6.00 m

ADMINISTRATION COMPLEX IN TRIPOLI
SHEET NO.3b

SCHULTES FRANK ARCHITEKTEN Berlin
SCHULTES FRANK建筑事务所 柏林

"A 'megaform as urban landscape', the village at the foot of the mountain, a symbiosis of opposites, may well serve the need for an icon of the Libyan constitution, may well serve as a symbol of Libya's turn to openness, inside and to the world abroad."

"一个'作为城市景观的巨型建筑',山脚下的村庄,对立面的互生现象很好地为利比亚政府标志建筑的需要服务,也成为利比来内部和对外开放的标志。"

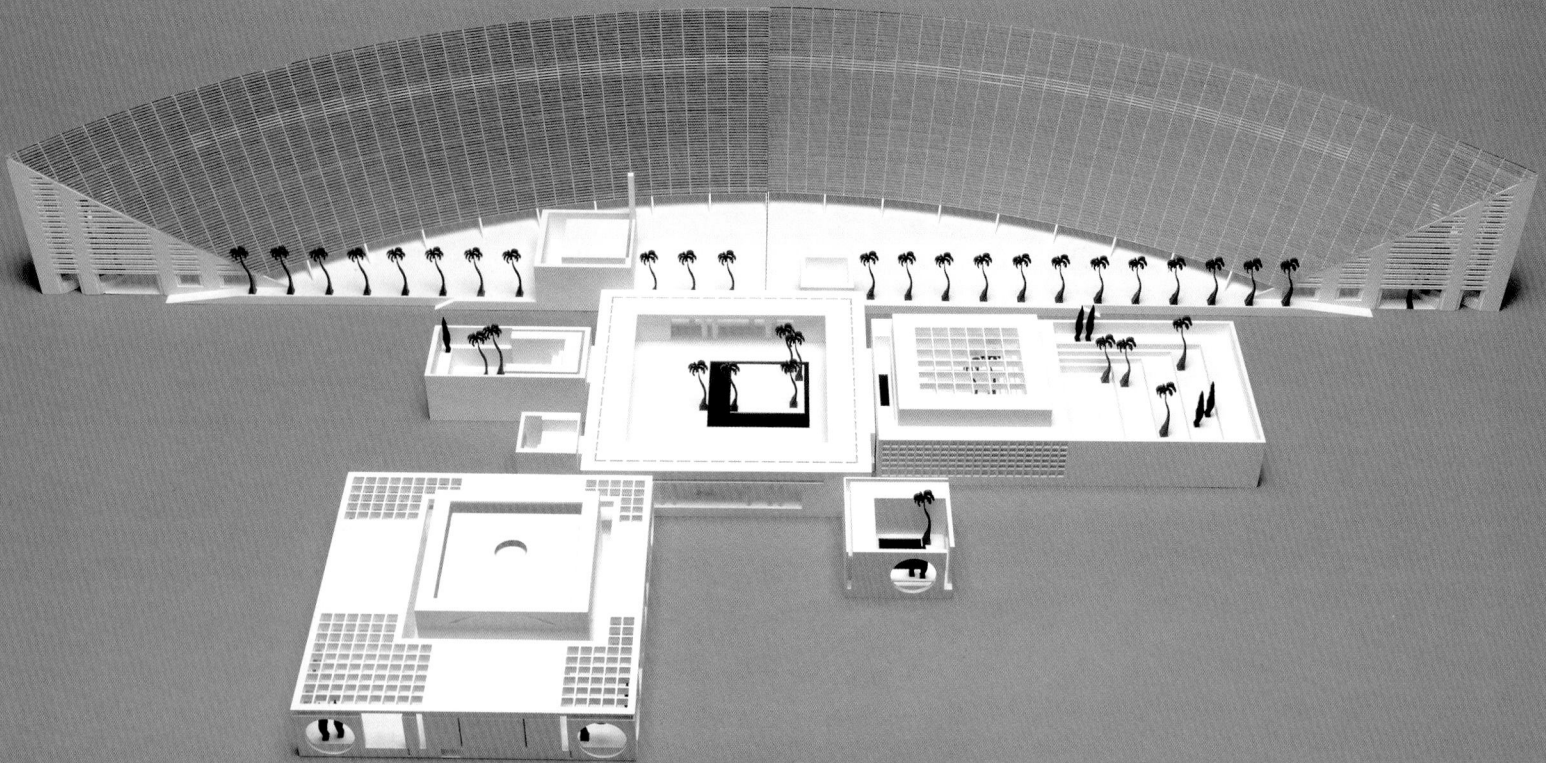

作者 Axel Schultes, Charlotte Frank   合作伙伴及建筑系在校生 Monika Bauer, Fritz Lobeck, Sören Timm, Christian Laabs, Frithjof Kahl   专家 landscape architecture: Büro Thomas, Kirsten Thomas, Berlin; building services engineering: HL Consulting and Shareholder Comp. Ltd., Klaus Daniels, Munich; traffic planning/infrastructure: GRI, The Company for Traffic Infrastructure Regionalisation and Infrastructure, Bodo Fuhrmann, Berlin

Administration Complex Tripoli

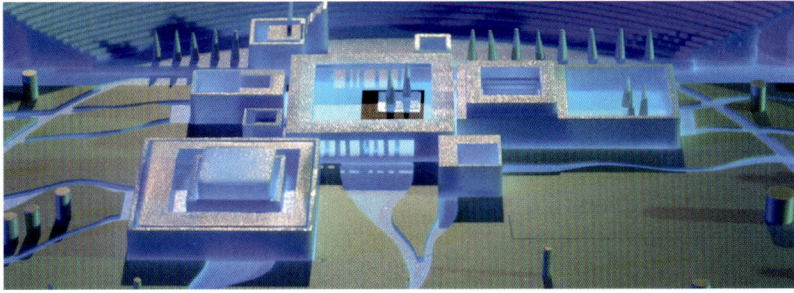

Administration Complex Tripoli

445

Administration Complex, Tripoli

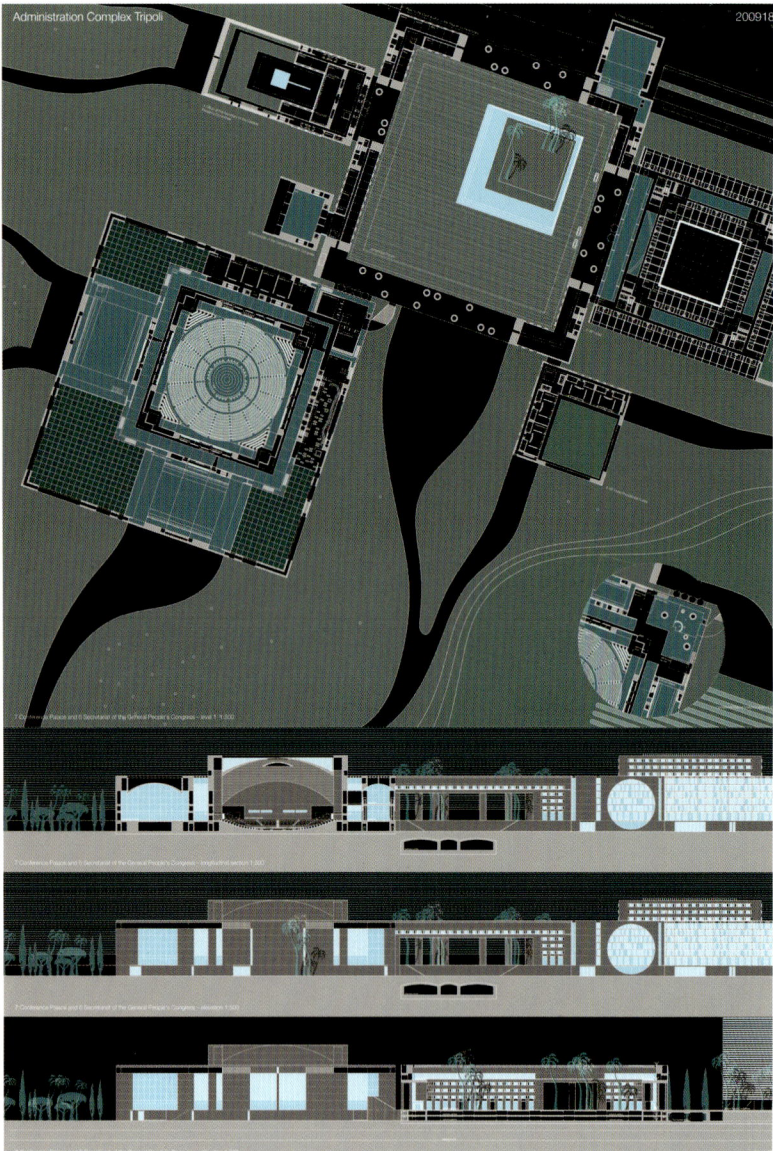

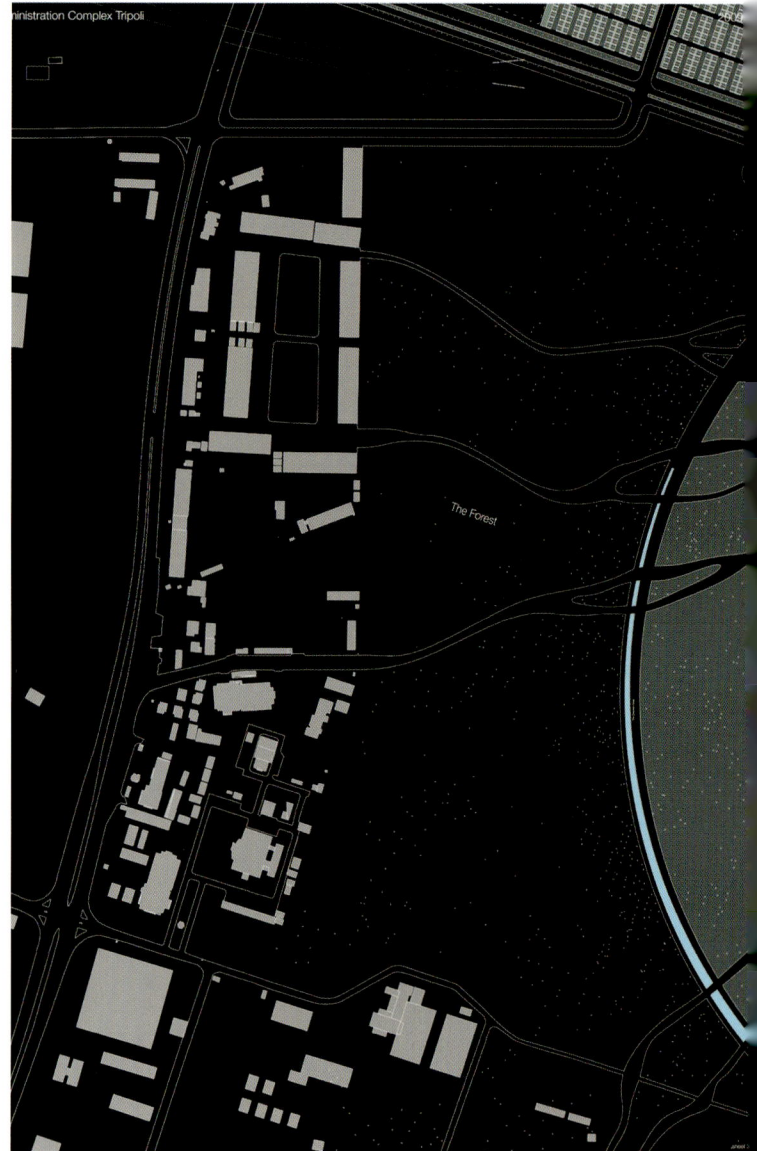

The Forest

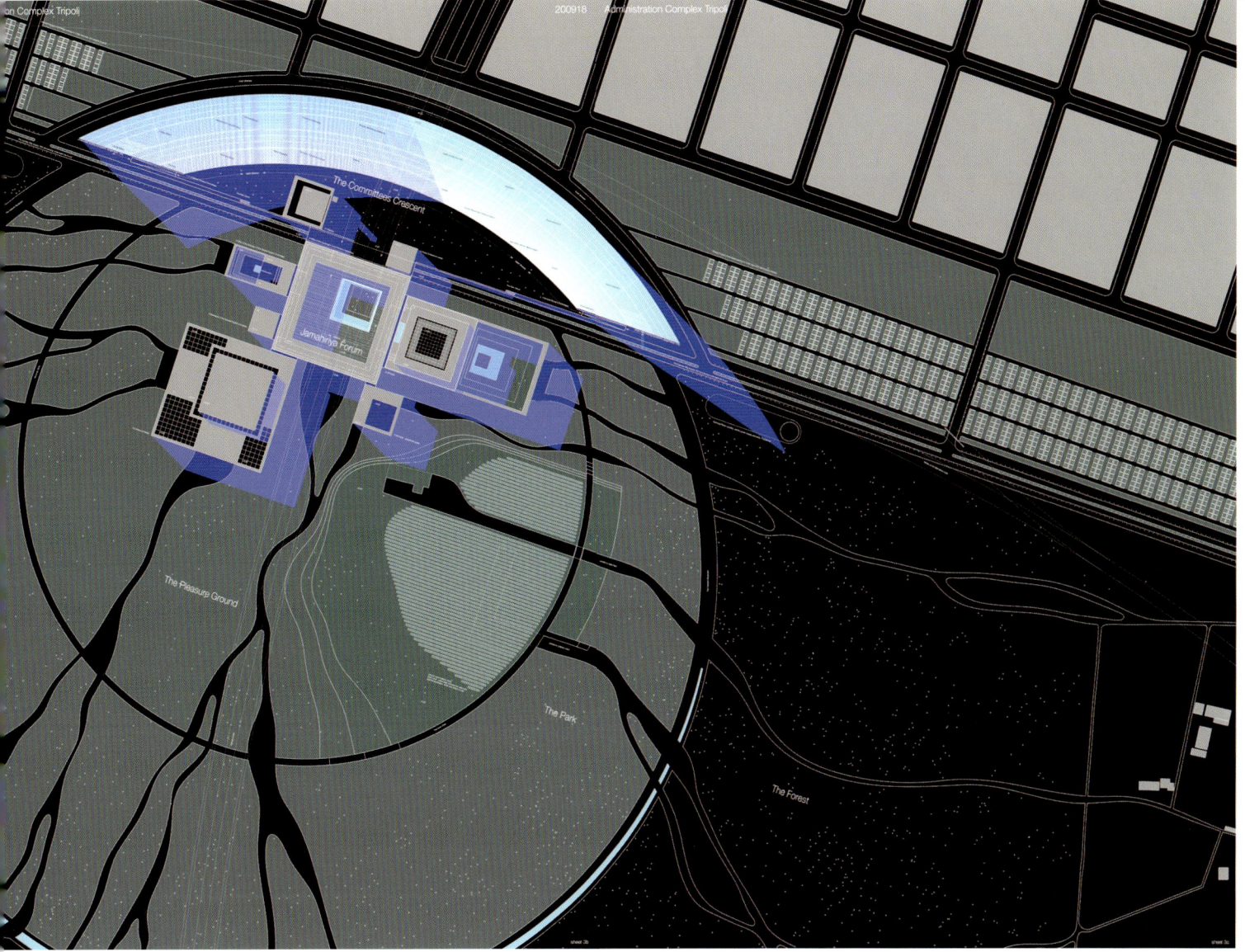

The Committees Crescent

Jamahiriyah Forum

The Pleasure Ground

The Park

The Forest

Aedas

K

COOP HIMMELB(L)AU

Harry Seidler an

Peia As

Massimiliano Fuksas

Tower at Suk Al Thalath Al Gadeem in Tripoli

Suk Al Thalath Al Gadeem高楼
的黎波里

eihues + Kleihues

Associates

sociati

Tower at Suk Al Thalath Al Gadeem Tripoli
Suk Al Thalath Al Gadeem高楼 的黎波里

Restricted project competition preceded by an application procedure
限制严格的项目竞赛，开始前有申请程序。

Mediterranean Sea

Corinthia Bab Africa Hotel

Dath Al Imad Towers

Abou Rgeaba Square

Mosque

Al Rasheed

Ebn Majid

Ebn Anaffice

Ebn Aroumi

Almajed

Al Ahouass

Mosque

Ebn Uniss

Al Fateh Tower

Al Mari

Competition Site

Tariq Ben Ziad

Ebn Jazia

Tejani

Ebn Albaytar

Fair Grounds

Oumar Al Muktar

Planned Subway Line

地点 Tripoli
时间 09/2007–01/2008
主办方 Libya Africa Investment Portfolio (LAP)
参赛者 6
面积 8,200 sq m
竞赛费用 100,000 USD
专业评奖委员会
Faisal Khalil Al-Bannani, Tripoli;
Donald Bates, Melbourne/London;
Prof. Hilde Léon, Berlin;
Jafar Tukan, Amman
专家评奖委员会
Manfredi Anello, Dublin
Technical jurors
Bashir Saleh,
chairman of LAP, Tripoli;
Dr. Ali Shebani,
chairman of NUH, Tripoli;
Dr. Mustafa Mezughi,
chairman of NCB, Tripoli;
Ahmed Bashir Saad,
director "projects and investment", LAP;
Mohsen M. Ben Halim
"technical affairs"-coordinator,
NUH, Tripoli

Tripoli has a history of ancient building traditions and sweeping political events. It has always been a gateway to the Sahara. But years of political and economic isolation have distorted the city's image overseas: Tripoli became one of "those not-so-nice places" in travel guides. Now Tripoli reopens to the world One of the few inner-city development areas is situated around the Suk Al Thalath Al Gadeem. Today it rather looks like a small industry zone. This area, between the old town, the Italian quarter, and the coast, will become a densely built-up center marked by high-rises and housing offices, residential and retail uses. The task set for the competition was to design a high-rise ensemble (105,000 square metres of gross floor area) on a site of 8,000 square metres, for a deluxe hotel, apartments and offices. As the flagship of the project, the hotel is to be a distinctive landmark, and, public space included, the ensemble is to become the center of the neighborhood. An important element of the task was to interpret local building traditions and create a sustainable response to climatic conditions, so as to give the building an unmistakably Libyan appearance.

的黎波里建筑历史悠久。它一直是撒哈拉沙漠的门户。但是多年政治和经济上的与世隔绝歪曲了的黎波里的国际形象，使它成为旅游指南中不受欢迎的地方之一。现在，的黎波里重新向世界开放。内城开发地区之一位于Suk Al Thalath Al Gadeem。这个区域现在像是一个小型工业区，将会成为一个建筑密集的中心，有许多高层标志性的建筑和办公楼、住宅区和商店。竞赛任务是在一个8000平方米的竞赛地点上设计一个高层综合性大楼，内设豪华酒店、公寓和写字间。作为项目的旗舰，酒店将成为一个显著的地标。包括公共空间，综合大楼将成为该区的中心。任务的一个重要方面是诠释当地建筑传统，对当地气候条件作出一种可持续性的反应，赋予建筑明显的利比亚外观。

Aerial view　鸟瞰图

Competition site　竞赛地点

Competition site　竞赛地点

Roman triumphal arch in Tripoli　的黎波里的罗马凯旋门

Tower at Suk Al Thalath Al Gadeem, Tripoli

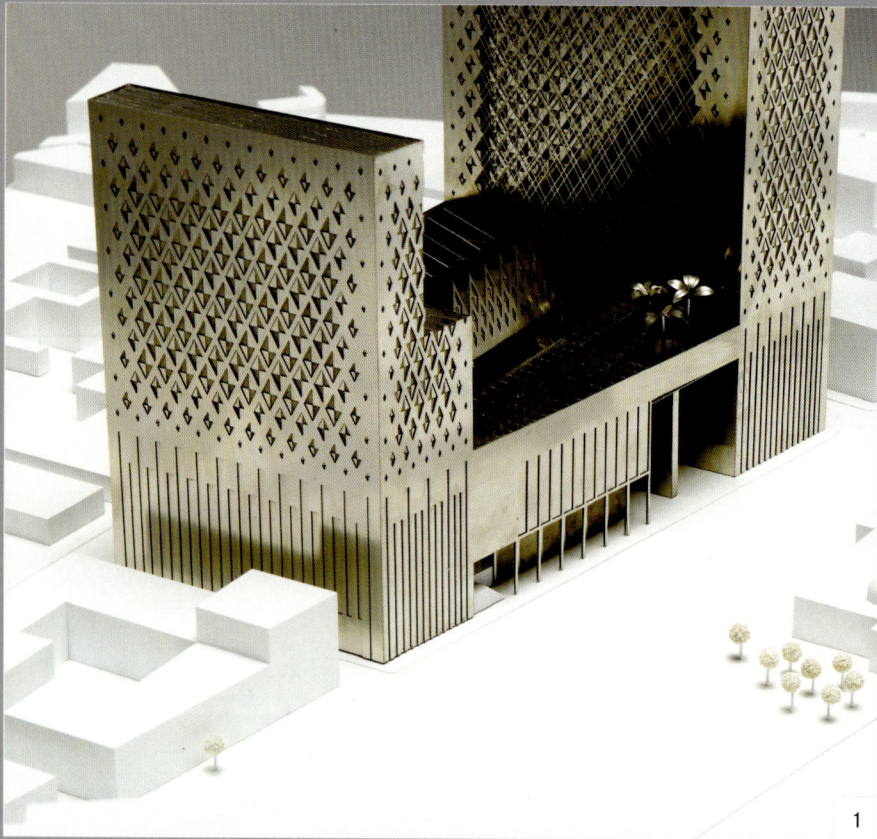

1

4

2

3

Qualified competitors
合格的参赛者

1
1st prize　一等奖
Kleihues + Kleihues Gesellschaft von
Architekten mbH, Berlin

2
3rd prize　三等奖
Peia Associati,
Milan

3
3rd prize　三等奖
Massimiliano Fuksas architetto,
Rome

4
Further participant　其他参赛者
Harry Seidler and Associates,
Sydney

5
Further participant　其他参赛者
COOP HIMMELB(L)AU,
Prix/Dreibholz & Partner, Vienna

6
Further participant　其他参赛者
Aedas, Dubai

5

6

Kleihues + Kleihues Gesellschaft von Architekten mbH Berlin

Kleihues + Kleihues Gesellschaft von Architekten mbH建筑事务所 柏林

"The 'tower of falling water' is a modern interpretation of the Arabic and Libyan court house."

"'瀑布'高楼是对阿拉伯和利比亚法院的一种现代诠释。"

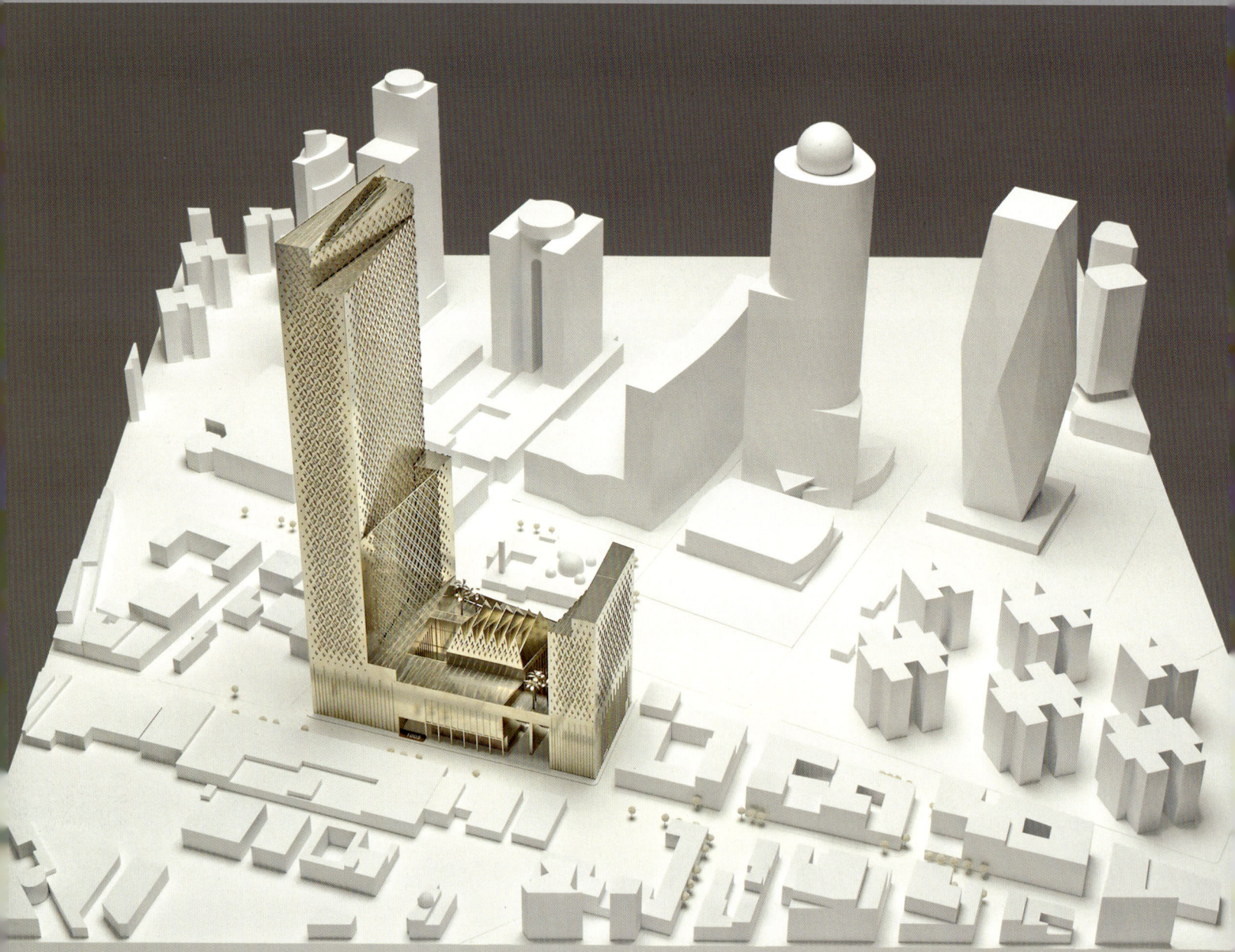

作者 Jan Kleihues   合作伙伴及建筑系在校生 Götz Kern, Anna Liesicke, Alice Berresheim, Roland Block, Daniel Horn, Philipp Zora, Sonja Grötzebach, Philipp Buschmeyer, Veronika Weber, Susanne, Thesen, Marc Helbach   自由建筑师 Marc Hensel
专家 structural analysis: HMI Hartwich Mertens Ingenieure, Berlin; fire protection: Büro für Brandschutz (Prein), Wuppertal; landscape architecture: ST Raum A Gesellschaft von Landschaftsarchitekten mbH, Berlin; façade engineering: Erich Mosbacher Beratungs- und Planungsgesellschaft für Fassadentechnik mbH, Friedrichshafen

Kleihues + Kleihues Gesellschaft von Architekten mbH, Berlin

Tower at Suk Al Thalath Al Gadeem, Tripoli

Mediterranean Sea
Abou Rgeaba Square
Al Rasheed
Ebn Anaffice
Tariq Ben Ziad
Al Mari
Hotel
Office Tower
Al Fateh Tower
Oumar Al Muktar

+255,70°

+250,25

+217,25

+100,25

+83,64°

+52,51

+70,41

±0,00

Kleihues + Kleihues Gesellschaft von Architekten mbH, Berlin

457

Tower at Suk Al Thalath Al Gadeem, Tripoli

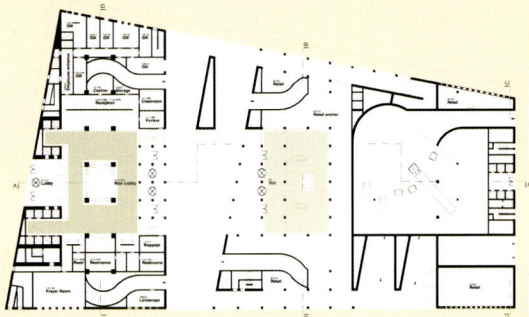

± 0,00 m / Groundfloor and -2,00 m / Delivery

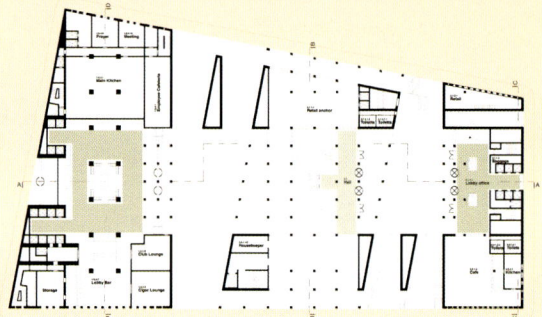

+ 3,50 m / 1st floor

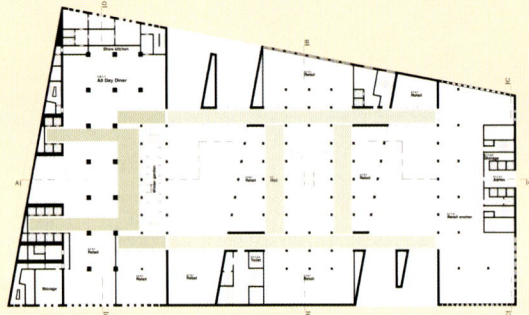

+ 8,50 m / 2nd floor

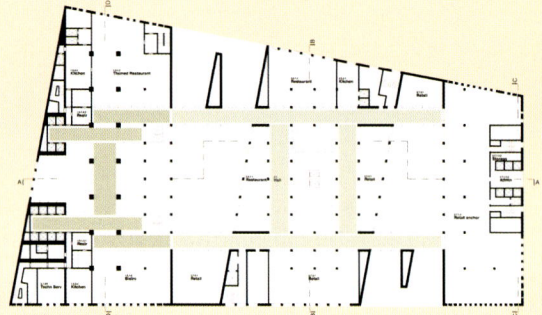

+ 13,50 m / 3rd floor

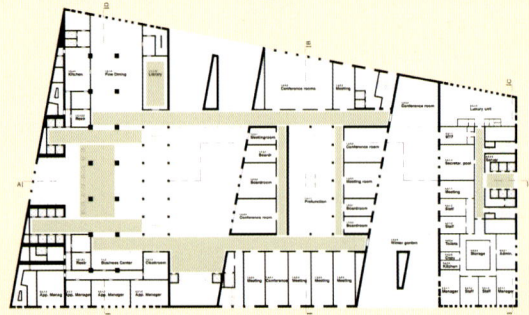

+ 18,50 m / 4th floor

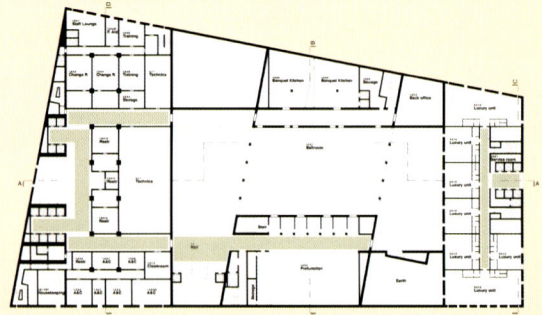

+ 23,50 m / 5th floor

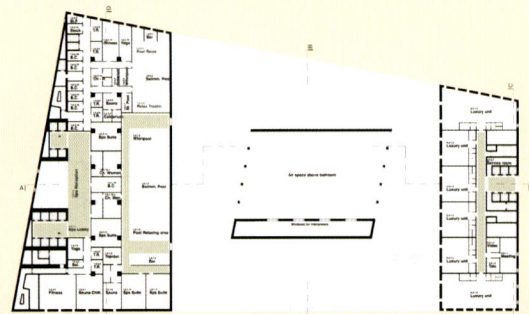

+ 28,50 m / 6th floor

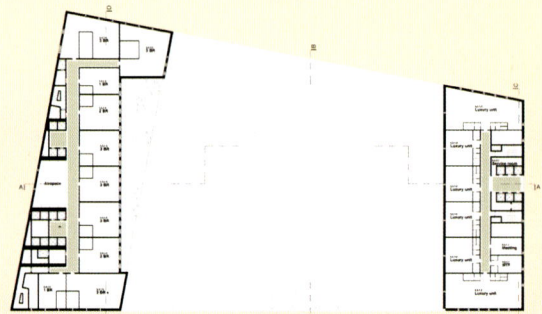

+ 35,00 m / 7th floor

Tower at Suk Al Gadeem, Tripoli, Libya

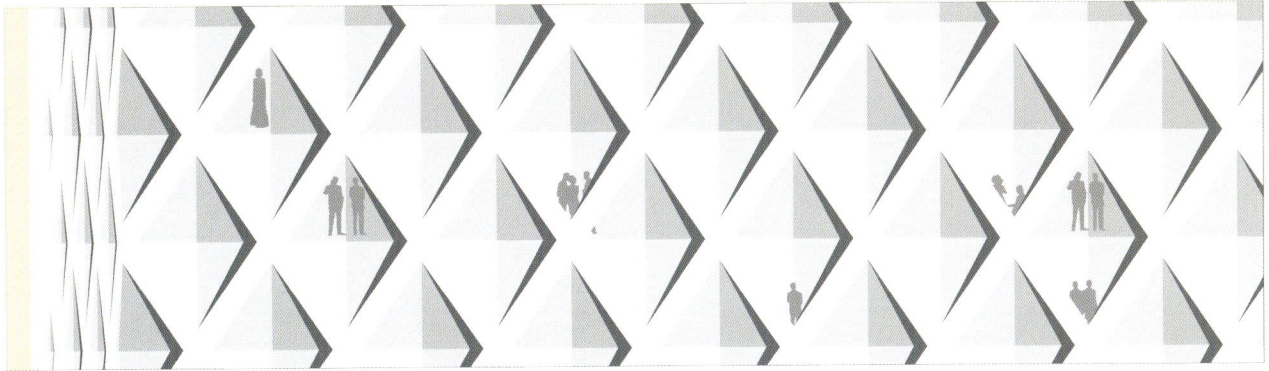

Elevation of the facade, scale 1 : 50

Detail section A, scale 1 : 50

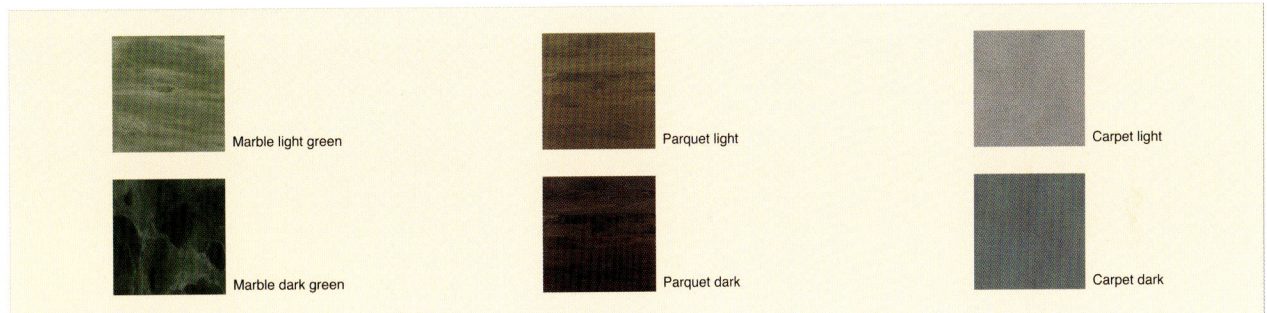

Marble light green

Parquet light

Carpet light

Marble dark green

Parquet dark

Carpet dark

Material and furniture proposal

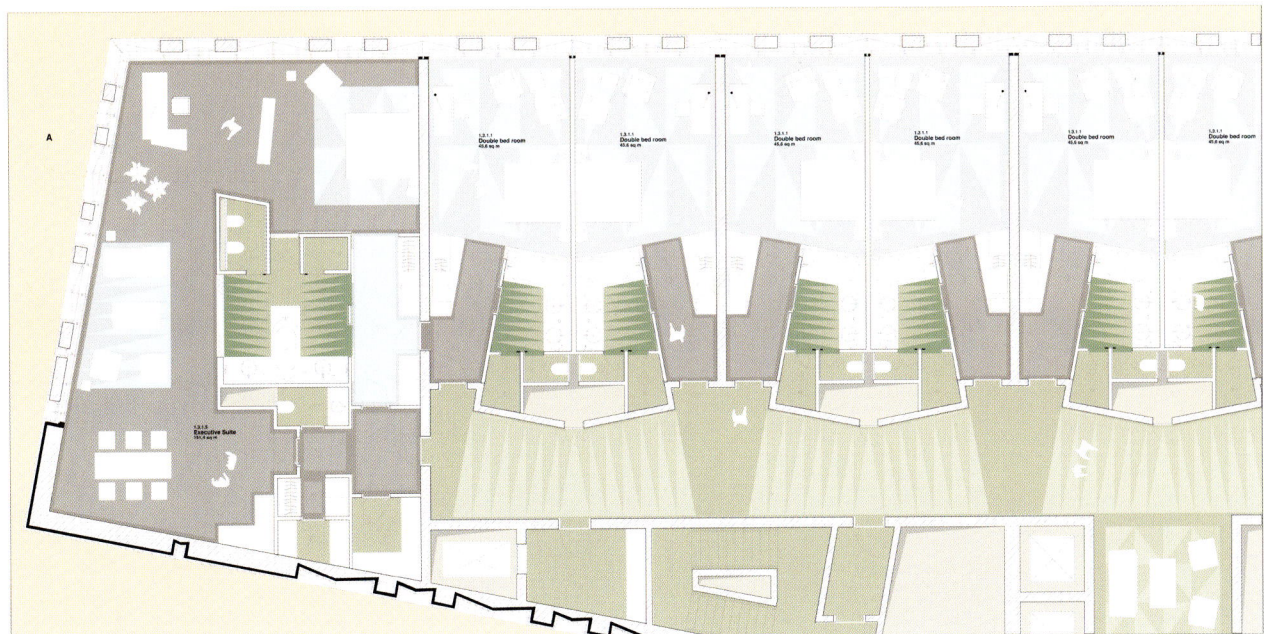

Hotel rooms, scale 1 : 50

Sheet 12

Section B - B

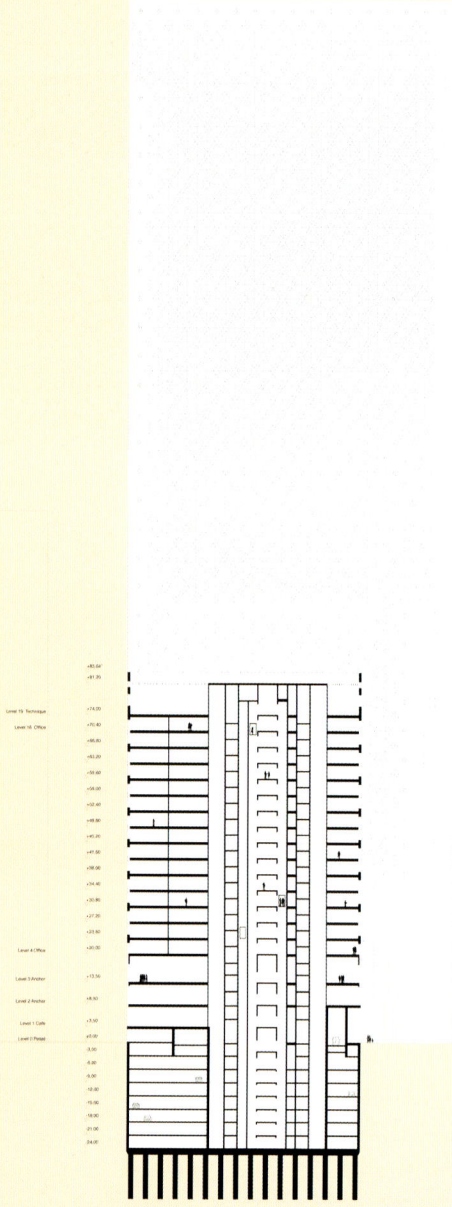

Section C - C

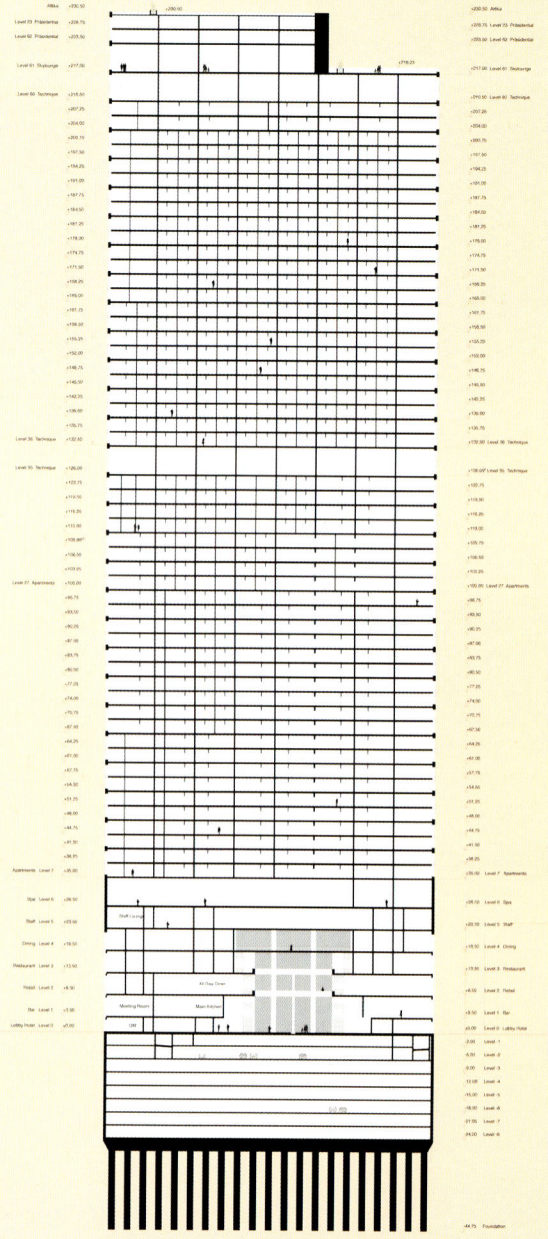

Section D - D

Standard Hotel Guest Room

Lobby Hotel

Lobby Hotel with Reception Desk and Falling Water

Elevation on Al Mari Street

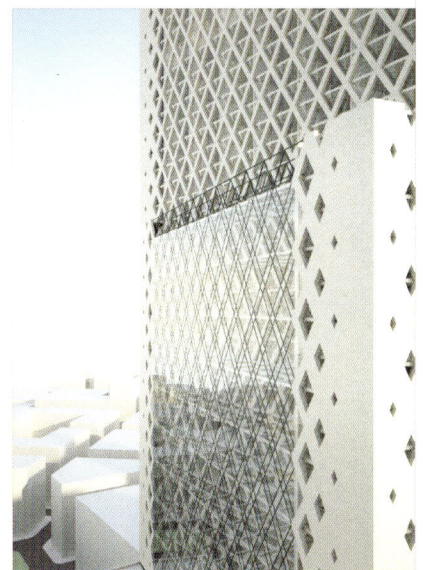
Elevation with Wintergarden   Sheet 13

Kleihues + Kleihues Gesellschaft von Architekten mbH, Berlin

461

Peia Associati Milan
Peia Associati建筑事务所 米兰

"The main concept is based on the formal abstraction of the palm tree trunk shape as the most recognisable metaphorical meaning of the African identity and unity. A perfect balance between strength and sinuosity."

"主要概念建立在棕榈树干是非洲身份和团结最易辨识的比喻意义的形式抽象上。力量和弯曲的完美平衡。"

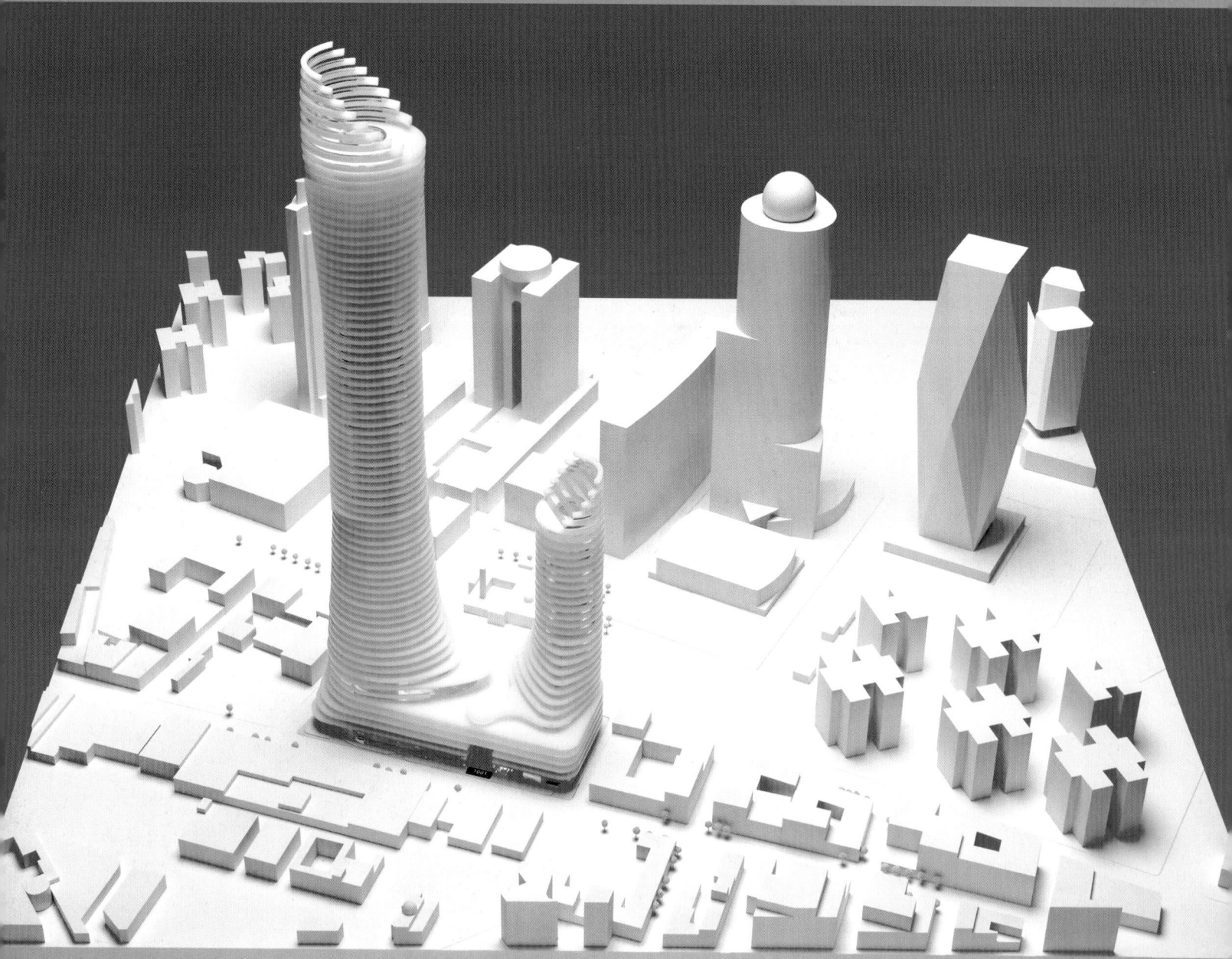

作者 Giampero Peia   合作伙伴及建筑系在校生 Luca Bonazzoli, Massimo de Mauro, Alessandro, Garzaro, Johnny Hugnot, Mayumi Kishimoto, Andrea Martelu, Lorenzo Merloni, Marta Nasazzi, Matteo, Nicotra, Francesca Patti, Anna Pavoni, Michaela, Ricciotti, Luigi Vaciago   专家 building services: Hani Awni Hawamdeh; AEB Engineering, Doha, Qatar; structural analysis: Amr el Khady, AEB Engineering Doha, Qatar; architekt: Cenon D. Ditan Jr., AEB Engineering, Doha, Qatar

Tower at Suk Al Thalath Al Gadeem, Tripoli

Ebn Jazla elevation
(South/south-east elevation)

Tariq Ben Ziad elevation
(East/south-east elevation)

Ebn Uniss elevation
(North/north-east elevation)

North
Makkah

0 5 10 15m

Mosque

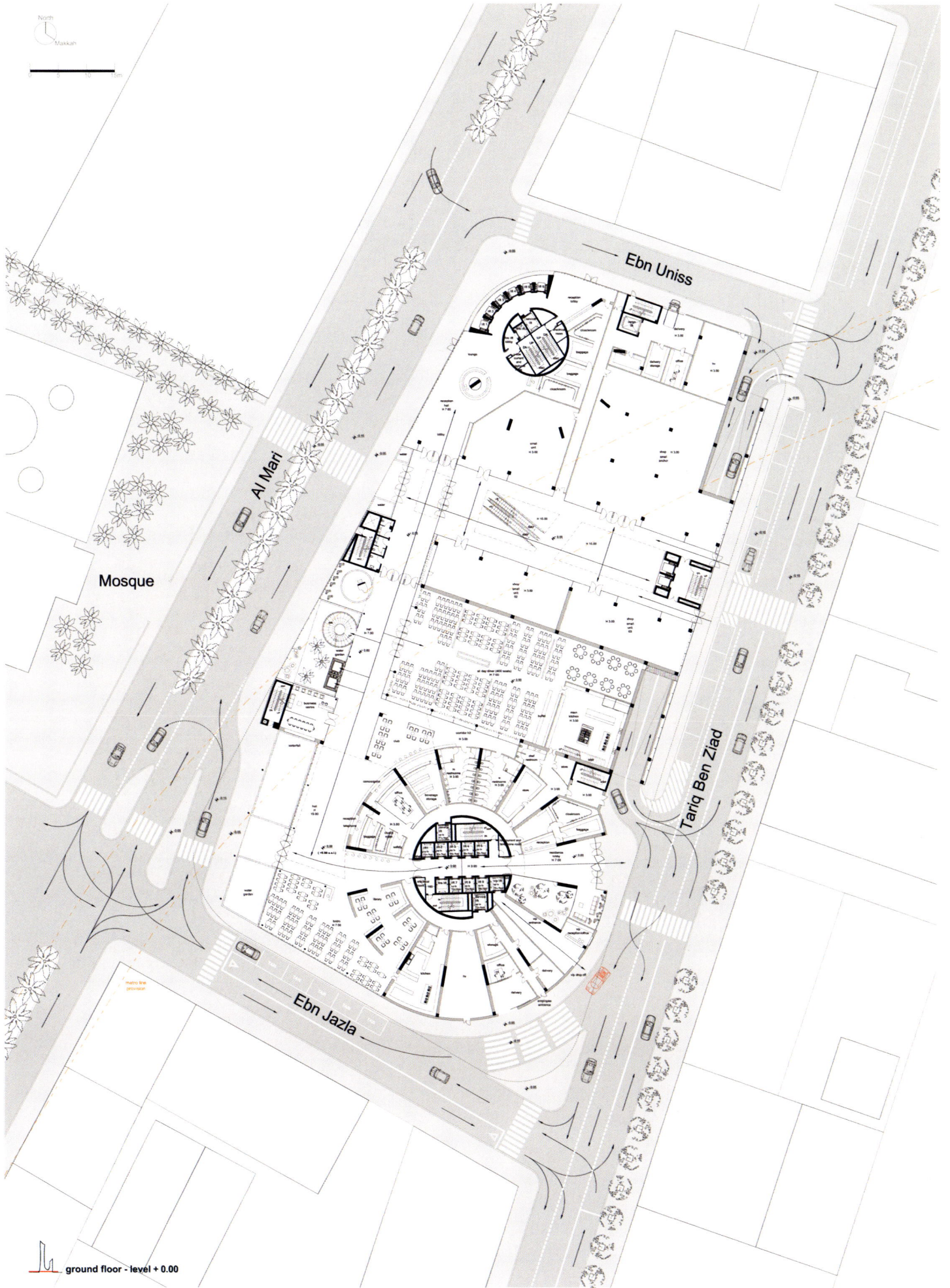

Ebn Uniss

Al Mari

Mosque

Tariq Ben Ziad

Ebn Jazla

metro line
provision

ground floor - level + 0.00

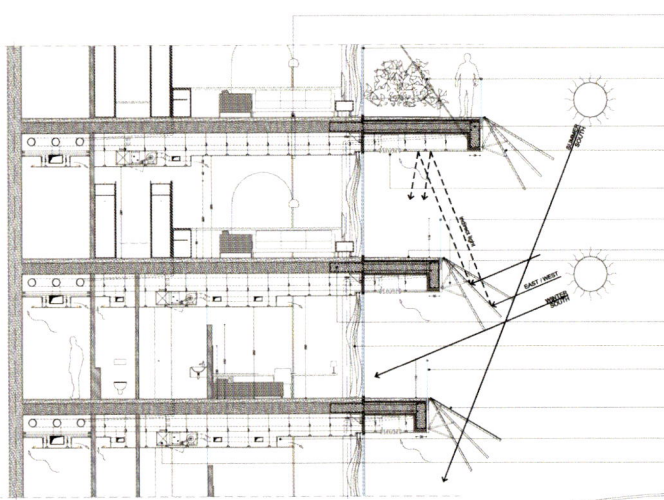

detailed longitudinal section

detailed cross section

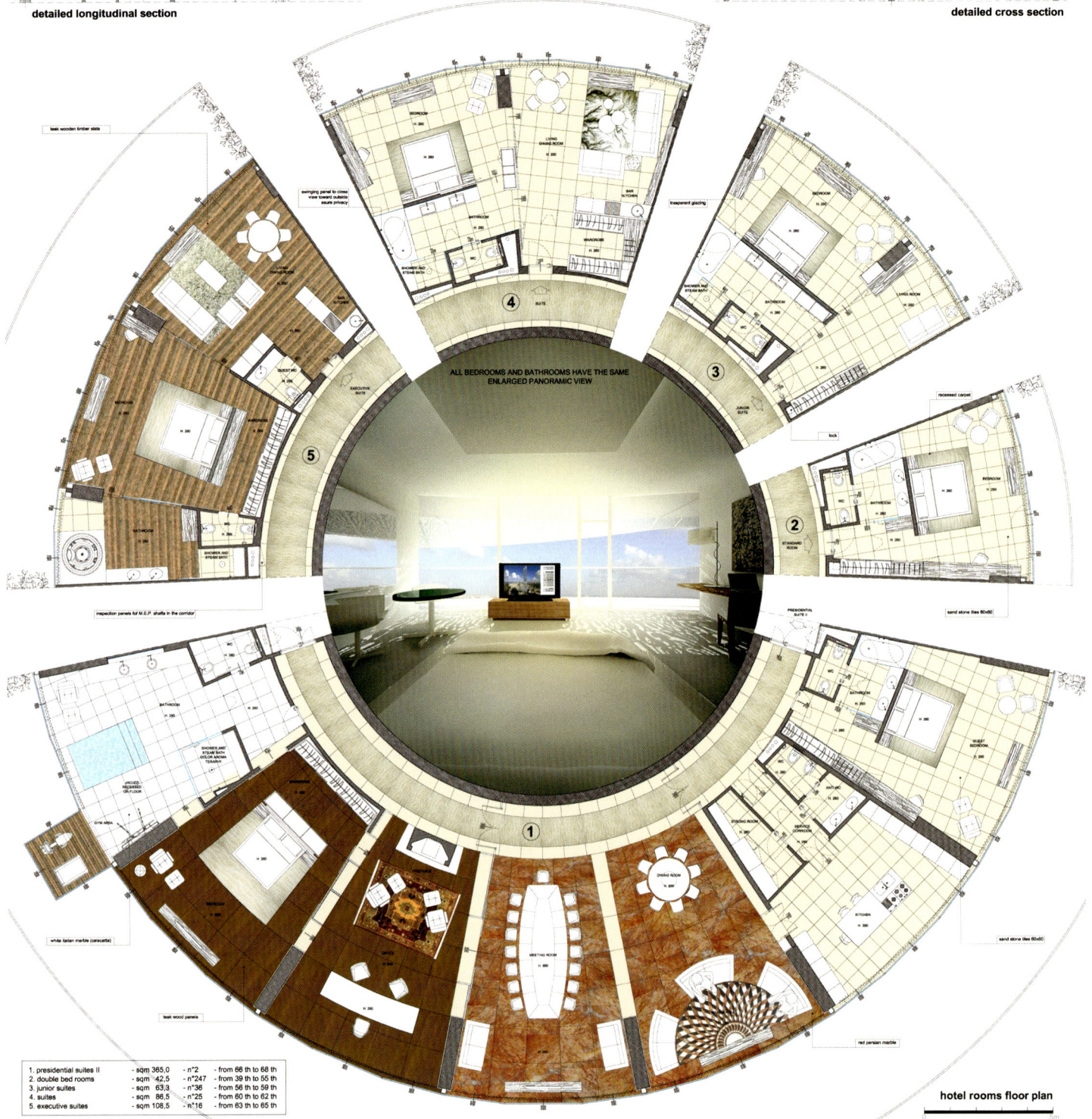

finishing 8 mm ceramic porcelain tiles,5cm screed
8 mm impact-sound insulation, 30 cm lightweight
reinforced concrete, 60-120 cm service space
12+12 cm plasterboard ceiling

double glazing (4+4/12 argon cavity/6+5) and low
emission (face2) internal coating

stainless steel tie bar 30mm to support
cantilever slab over 3 meter

air conditioning silent duct

concrete external ring beam 300x800mm

aluminium sheet with micro perforated decoration
thickness 2mm and maximum lenght 2400mm
angle 60° - east and west side

aluminium sand trap louver to eliminate
sand admittance in sandy weather (TYP)
and for individual fresh air intake

aluminium profile pivoting bracket 40mm
resin floor finishing 30mm max screed and 30mm
extruded polistirene thermal insulation

angle 45° - south east / south side

concrete radial beam 500x800mm

finishing and screed 125mm
acoustic insulation 10mm, concrete slab 300mm

concrete ring beam 1000x300mm

silk curtain and blackout curtain and double railing

stainless steel handrail 60mm diam
balustrade 12+12 with tempered extraclear safety
glass and heat-strenghtend glass (2mm PVB)

angle 30° - south side
eventual photovoltaic panel on south side

cyma reverse with recessed light and air diffuser
linear grids around the bed plasterboard 15 mm

HAUC / AC with motorized damp to control
the quantity of admitted fresh air (control by BMS)

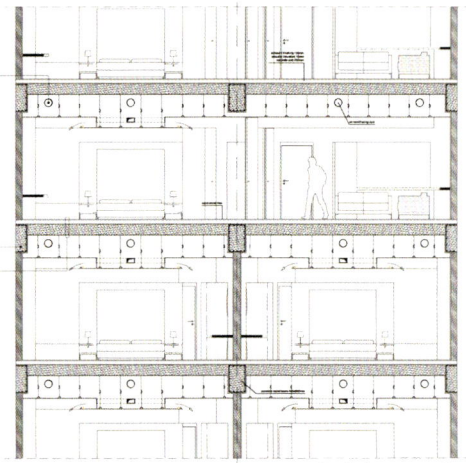

ALL BEDROOMS AND BATHROOMS HAVE THE SAME
ENLARGED PANORAMIC VIEW

1. presidential suites II - sqm 365,0 - n°2 - from 66 th to 68 th
2. double bed rooms - sqm 42,5 - n°247 - from 39 th to 55 th
3. junior suites - sqm 63,3 - n°36 - from 56 th to 59 th
4. suites - sqm 86,5 - n°25 - from 60 th to 62 th
5. executive suites - sqm 108,5 - n°16 - from 63 th to 65 th

hotel rooms floor plan

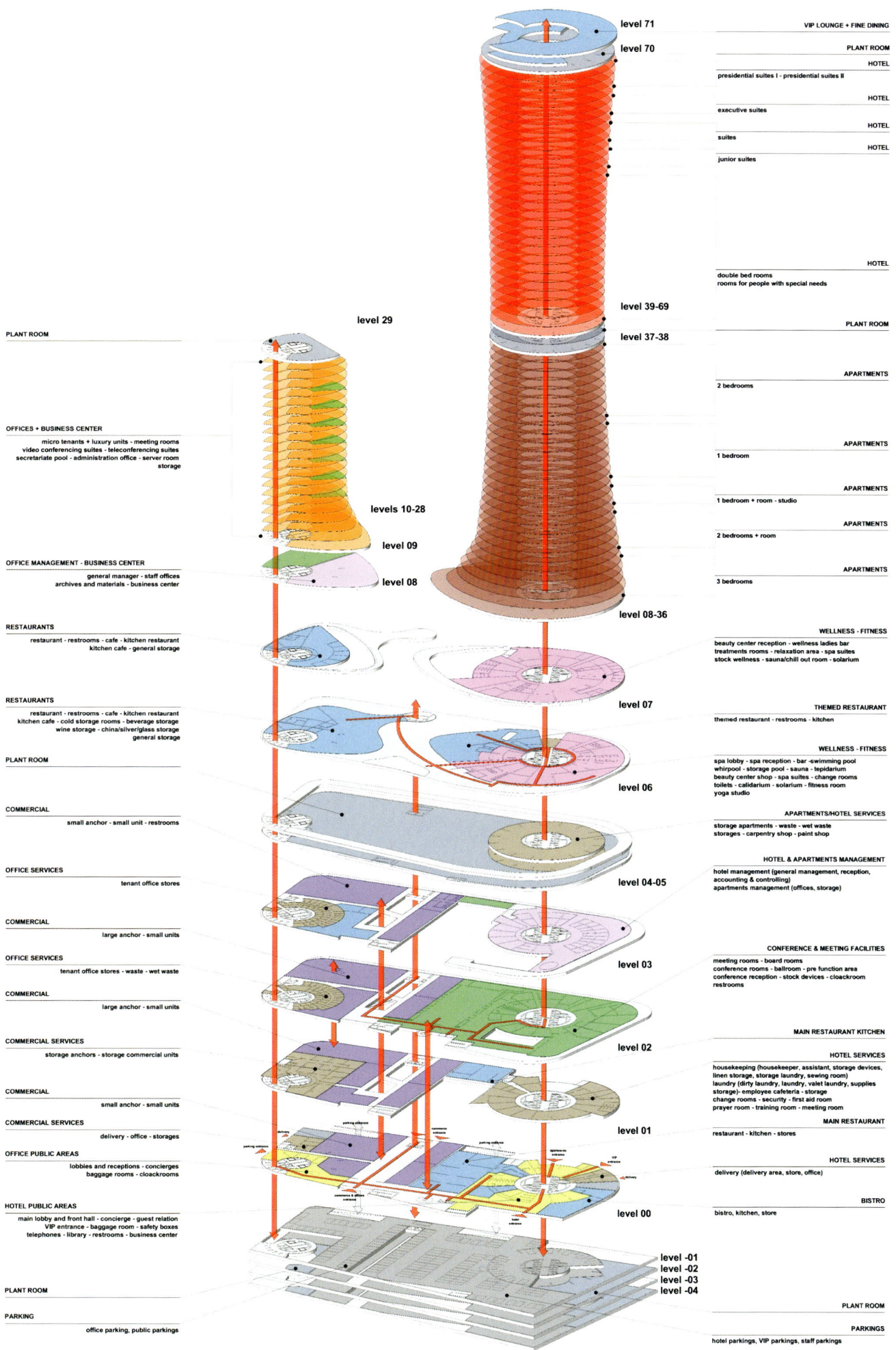

Tower at Suk Al Thalath Al Gadeem, Tripoli

level 71 — VIP LOUNGE + FINE DINING

level 70 — PLANT ROOM

HOTEL
presidential suites I - presidential suites II

HOTEL
executive suites

HOTEL
suites

HOTEL
junior suites

HOTEL
double bed rooms
rooms for people with special needs

level 39-69

level 37-38 — PLANT ROOM

APARTMENTS
2 bedrooms

APARTMENTS
1 bedroom

APARTMENTS
1 bedroom + room - studio

APARTMENTS
2 bedrooms + room

APARTMENTS
3 bedrooms

level 08-36

WELLNESS - FITNESS
beauty center reception - wellness ladies bar
treatments rooms - relaxation area - spa suites
stock wellness - sauna/chill out room - solarium

level 07 — THEMED RESTAURANT
themed restaurant - restrooms - kitchen

WELLNESS - FITNESS
spa lobby - spa reception - bar -swimming pool
whirpool - storage pool - sauna - tepidarium
beauty center shop - spa suites - change rooms
toilets - calidarium - solarium - fitness room
yoga studio

level 06

APARTMENTS/HOTEL SERVICES
storage apartments - waste - wet waste
storages - carpentry shop - paint shop

HOTEL & APARTMENTS MANAGEMENT
hotel management (general management, reception,
accounting & controlling)
apartments management (offices, storage)

level 04-05

CONFERENCE & MEETING FACILITIES
meeting rooms - board rooms
conference rooms - ballroom - pre function area
conference reception - stock devices - cloackroom
restrooms

level 03

MAIN RESTAURANT KITCHEN

level 02 — HOTEL SERVICES
housekeeping (housekeeper, assistant, storage devices,
linen storage, storage laundry, sewing room)
laundry (dirty laundry, laundry, valet laundry, supplies
storage)- employee cafeteria - storage
change rooms - security - first aid room
prayer room - training room - meeting room

MAIN RESTAURANT
restaurant - kitchen - stores

level 01 — HOTEL SERVICES
delivery (delivery area, store, office)

BISTRO
bistro, kitchen, store

level 00

level -01
level -02
level -03
level -04

PLANT ROOM

PARKINGS
hotel parkings, VIP parkings, staff parkings

PLANT ROOM

OFFICES + BUSINESS CENTER
micro tenants + luxury units - meeting rooms
video conferencing suites - teleconferencing suites
secretariate pool - administration office - server room
storage

level 29

levels 10-28

level 09

OFFICE MANAGEMENT - BUSINESS CENTER
general manager - staff offices
archives and materials - business center

level 08

RESTAURANTS
restaurant - restrooms - cafe - kitchen restaurant
kitchen cafe - general storage

RESTAURANTS
restaurant - restrooms - cafe - kitchen restaurant
kitchen cafe - cold storage rooms - beverage storage
wine storage - china/silver/glass storage
general storage

PLANT ROOM

COMMERCIAL
small anchor - small unit - restrooms

OFFICE SERVICES
tenant office stores

COMMERCIAL
large anchor - small units

OFFICE SERVICES
tenant office stores - waste - wet waste

COMMERCIAL
large anchor - small units

COMMERCIAL SERVICES
storage anchors - storage commercial units

COMMERCIAL
small anchor - small units

COMMERCIAL SERVICES
delivery - office - storages

OFFICE PUBLIC AREAS
lobbies and receptions - concierges
baggage rooms - cloackrooms

HOTEL PUBLIC AREAS
main lobby and front hall - concierge - guest relation
VIP entrance - baggage room - safety boxes
telephones - library - restrooms - business center

PLANT ROOM

PARKING
office parking, public parkings

Massimiliano Fuksas architetto Rome
Massimiliano Fuksas建筑事务所 罗马

"... a dramatic and elegant tower that can become a symbol of both the country's proud and deep heritage and of its position as a progressive cultural leader in the 21st century."

"一个引人注目的和优美文雅的塔楼可以成为国家引人为豪的深厚传统的象征，同时也是作为21世纪先进文化的领袖。"

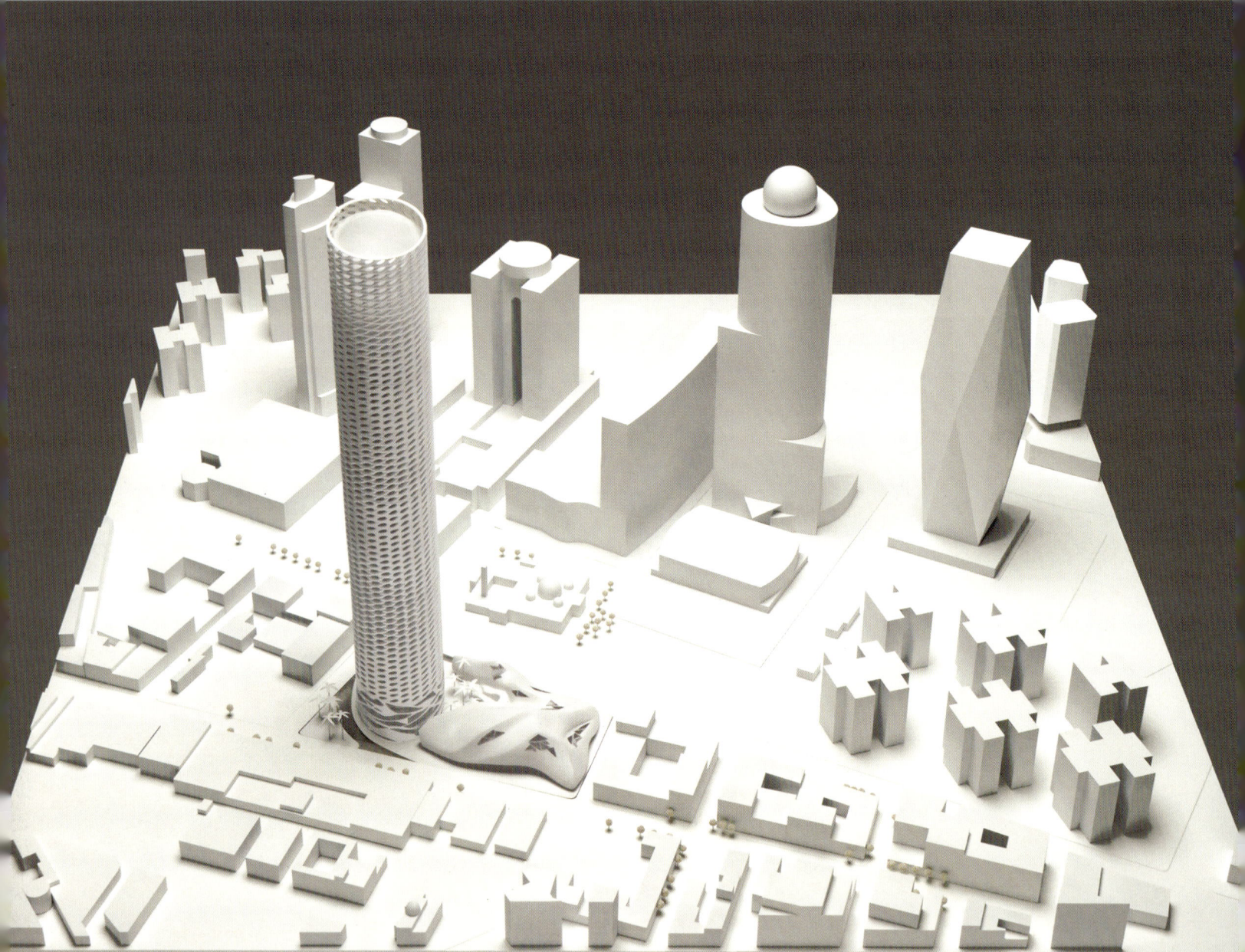

作者 Massimiliano Fuksas, Doriana Mandrelli　合作伙伴及建筑系在校生 Grazia Patruno, Serena Mignatti, Joshua Mackley, Tommaso Villa, Nicola Cabiati, Luca Vernocchi　专家 structural analysis: Knippers und Helbig Beratende Ingenieure, Stuttgart

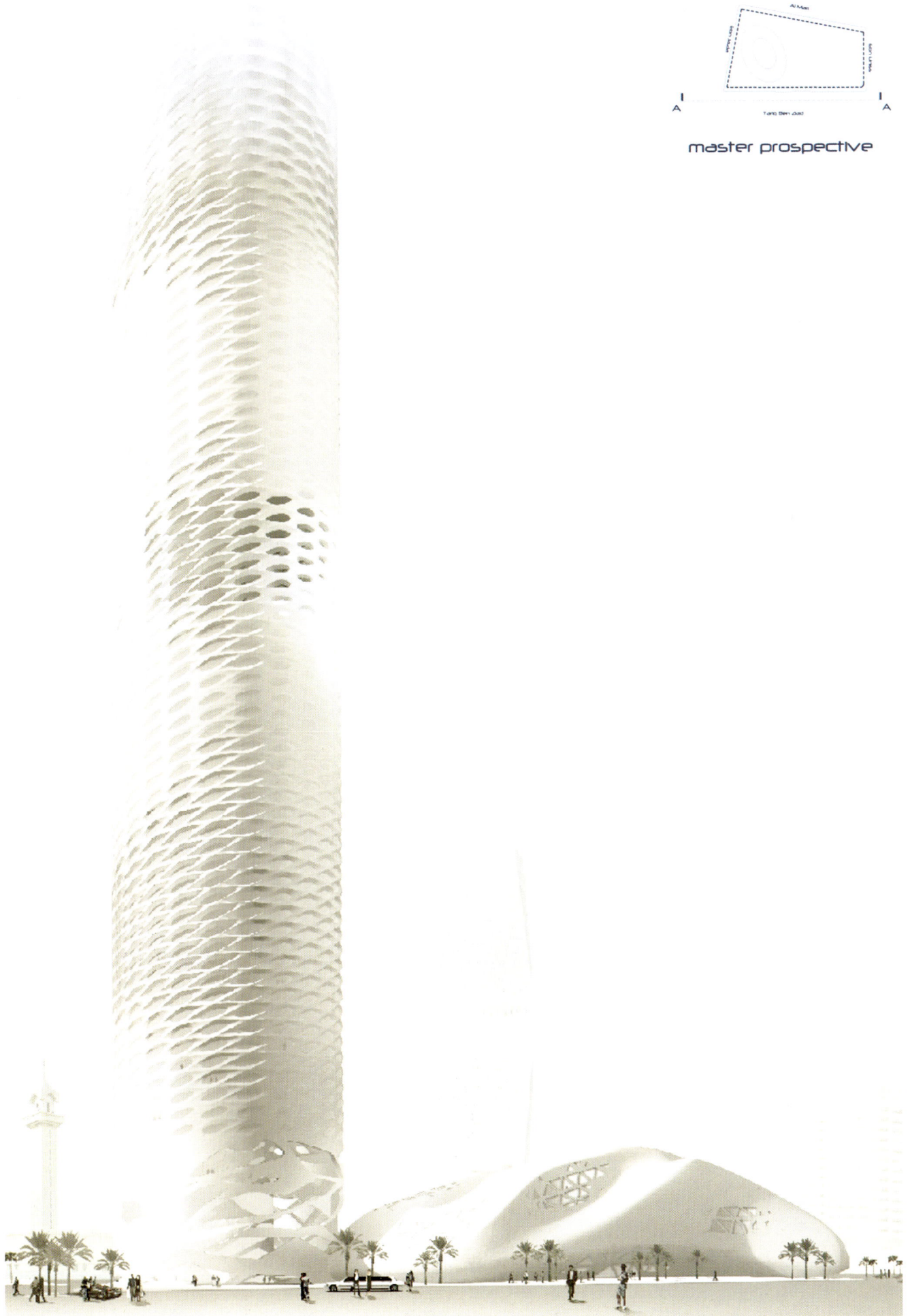

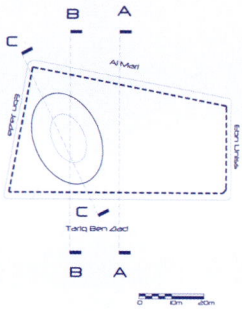

further elevations
scale 1:500

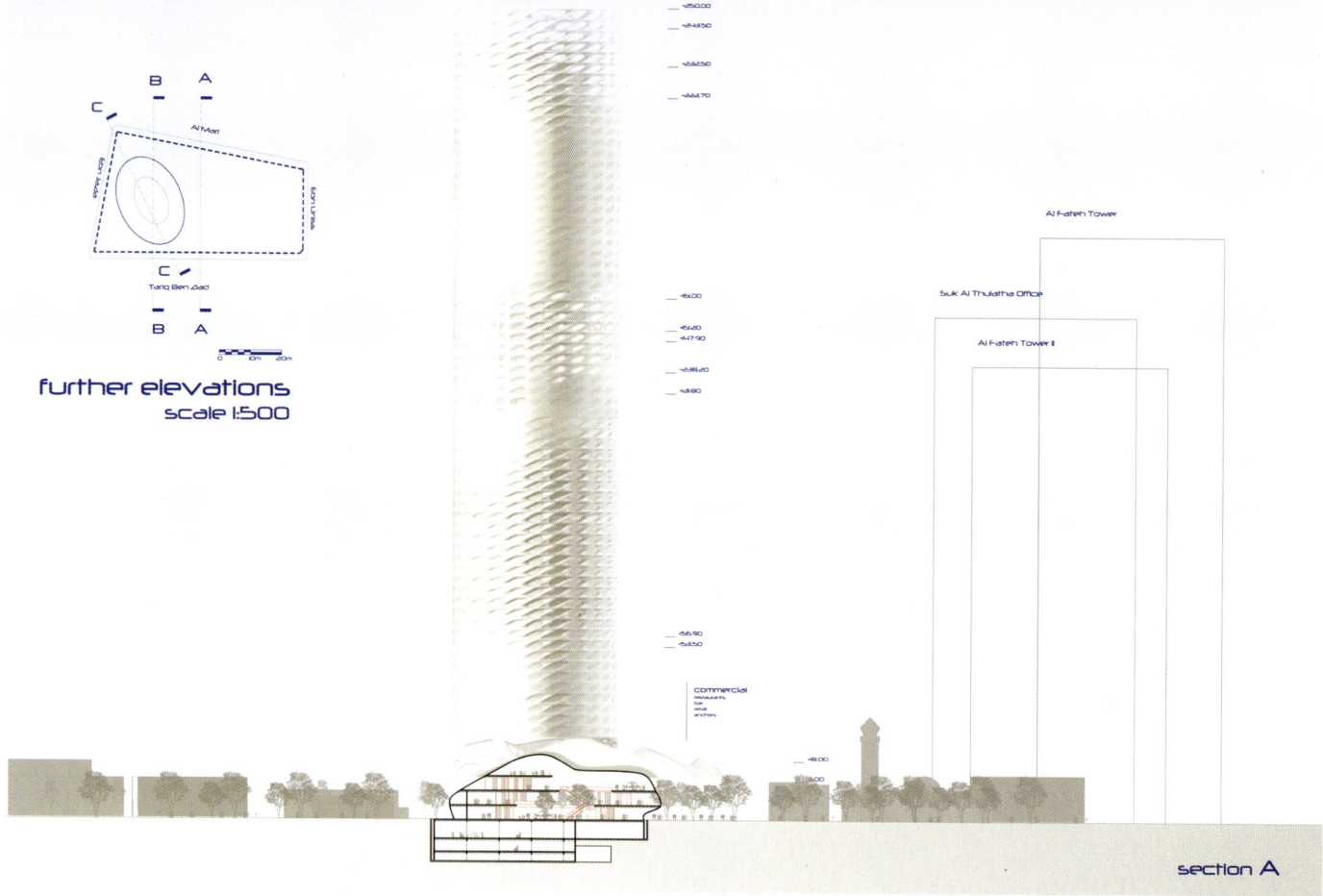

Al-Fateh Tower

Suk Al Thulatha Office

Al-Fateh Tower II

commercial
restaurants
bar
retail
anchor

+50.00
+44.50
+38.50
+32.70

+6.00
+5.20
+4.7.90
+38.20
+3.80

+56.90
+53.50

+8.00
+5.00

section A

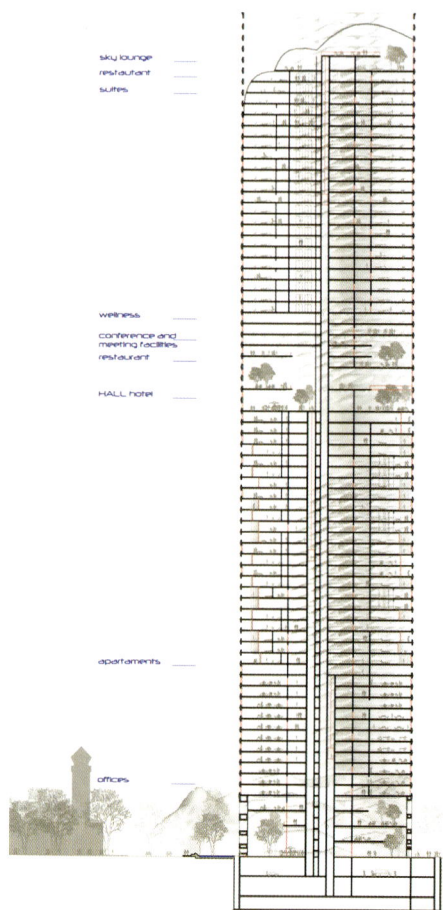

sky lounge
restaurant
suites

wellness

conference and
meeting facilities
restaurant

HALL hotel

apartments

offices

+50.00
+44.50
+38.50
+32.70

+6.00
+5.20
+4.7.90
+38.20
+3.80

+56.90
+53.50

+8.00
+4.00

section B

Al-Fateh Tower

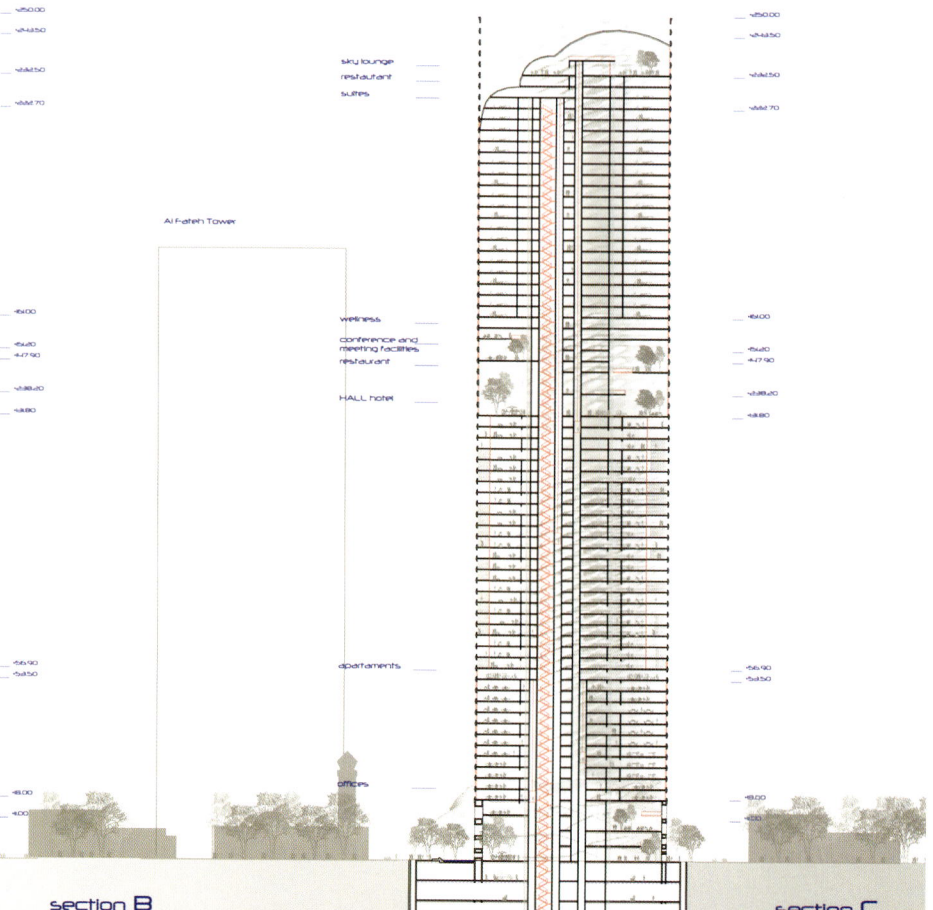

sky lounge
restaurant
suites

wellness

conference and
meeting facilities
restaurant

HALL hotel

apartments

offices

+50.00
+44.50
+38.50
+32.70

+6.00
+5.20
+4.7.90
+38.20
+3.80

+56.90
+53.50

+8.00
+5.00

section C

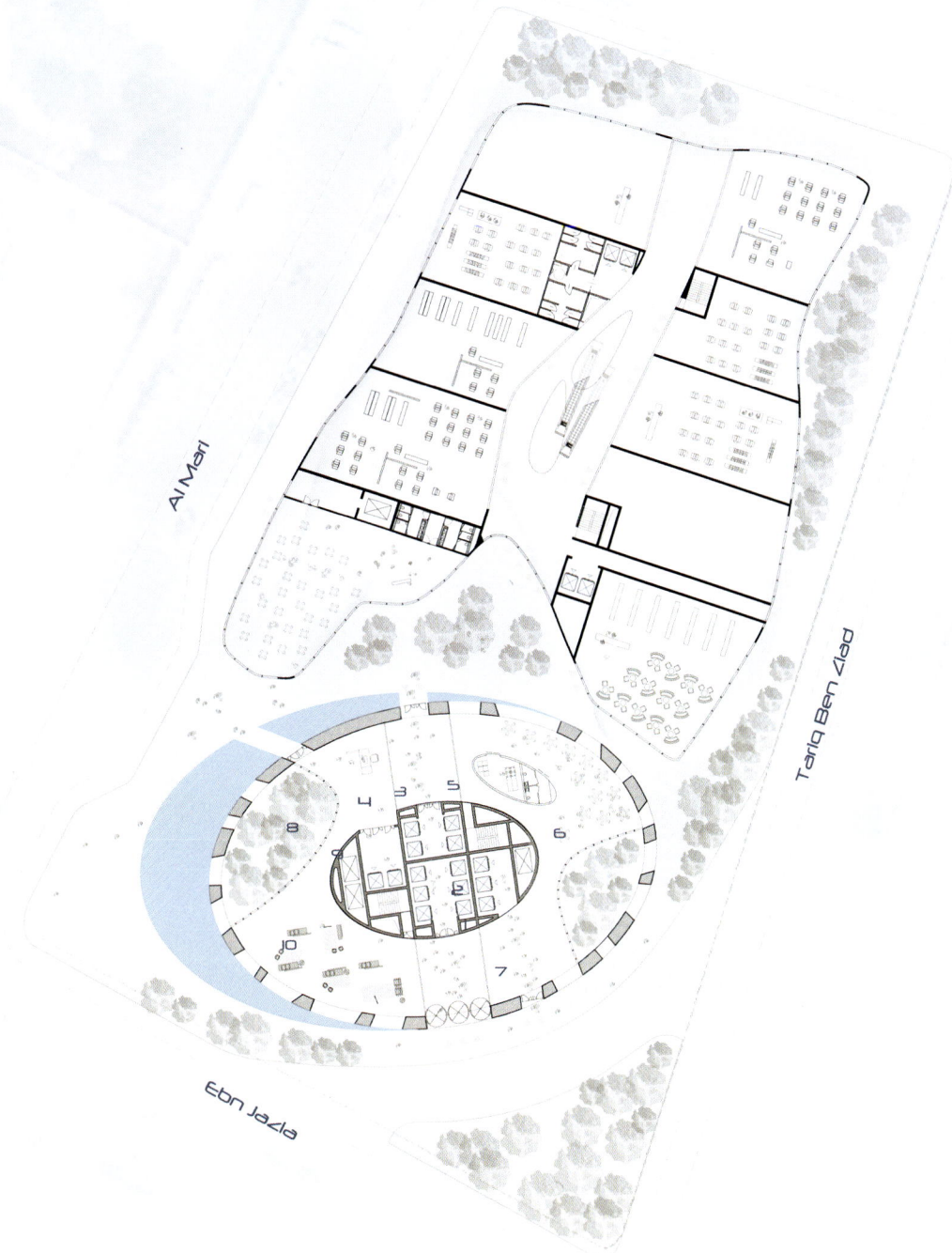

Al Mari

Tariq Ben Ziad

Ebn Jazla

1 hotel public entrance
2 hotel vip entrance
3 office entrance
4 apartment entrance
5 cafè
6 winter garden
7 lounge
8 bistrot
9 kitchen
10 winter garden
11 vip lounge

groundfloor scale 1:200

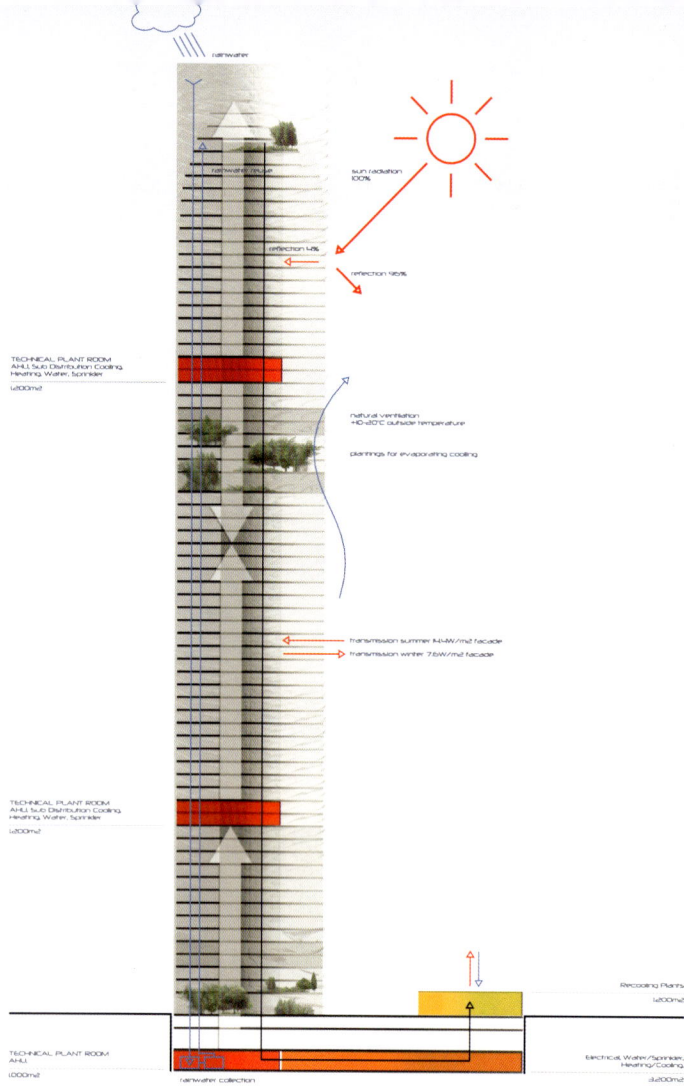

technical and ecological concept

sun radiation
100%

reflection 1-4%

reflection 96%

natural ventilation
+10-20°C outside temperature

plantings for evaporating cooling

transmission summer 14-14W/m2 facade
transmission winter 7.6W/m2 facade

TECHNICAL PLANT ROOM
AHU, Sub Distribution Cooling
Heating, Water, Sprinkler
1x200m2

TECHNICAL PLANT ROOM
AHU, Sub Distribution Cooling
Heating, Water, Sprinkler
1x200m2

TECHNICAL PLANT ROOM
AHU
1.000m2

rainwater collection

Recooling Plants
1x200m2

Electrical Water/Sprinkler,
Heating/Cooling
4x200m2

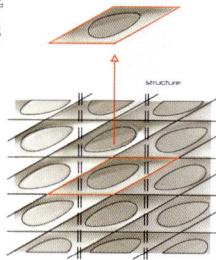

facade concept

prefabricated panels including the whole package
concrete, glass, sun blades etc.

structure

Windows   U-value 1.80   G-value 0.10
Massive elements U-value 0.40   U-value -
Facade total   U-value 0.78   U-value 0.04

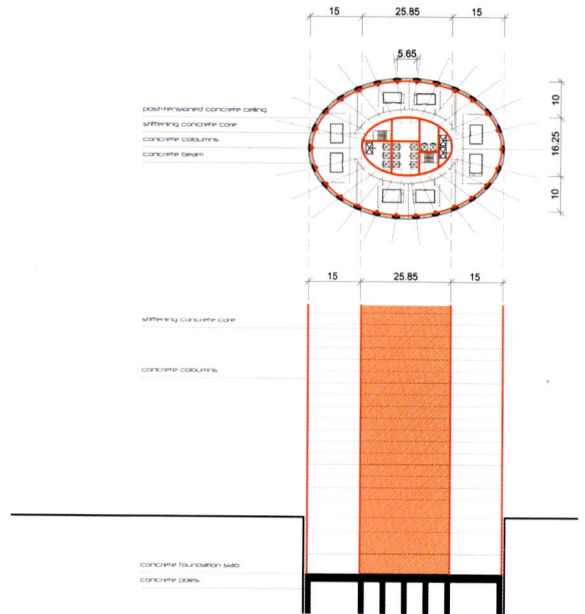

structural concept

15   25.85   15
5.65
16.25   10
10

post-tensioned concrete ceiling
stiffening concrete core
concrete columns
concrete beam

15   25.85   15

stiffening concrete core
concrete columns

concrete foundation slab
concrete piles

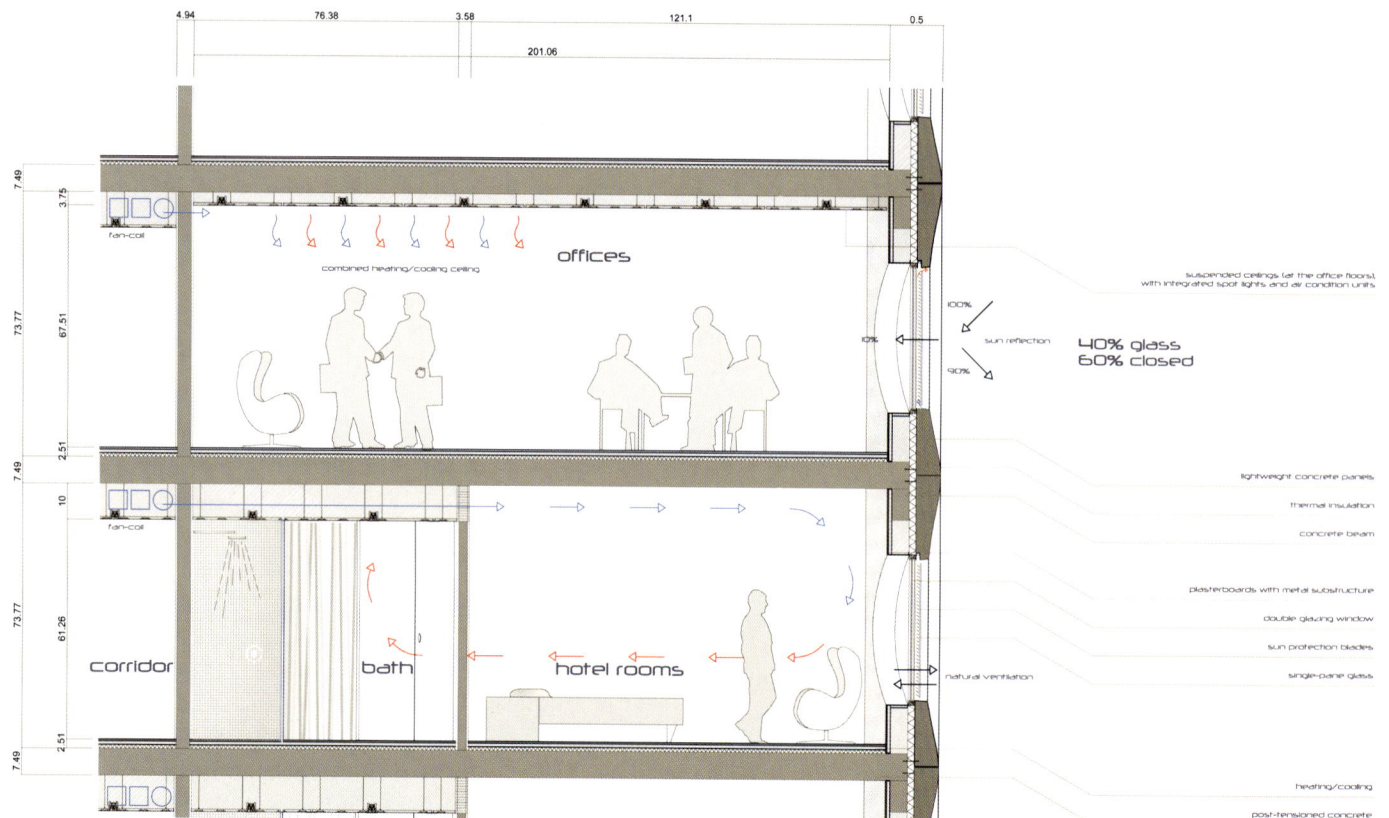

section scale 1:200

4.94   76.38   3.58   121.1   0.5
201.06

fan-coil

combined heating/cooling ceiling

offices

suspended ceilings (at the office floors)
with integrated spot lights and air condition units

100%
90%

sun reflection

40% glass
60% closed

lightweight concrete panels
thermal insulation
concrete beam
plasterboards with metal substructure
double glazing window
sun protection blades
single-pane glass

fan-coil

corridor   bath   hotel rooms

natural ventilation

heating/cooling
post-tensioned concrete

3.75   7.49
67.51   73.77
2.51   7.49
10
61.26   73.77
2.51   7.49

interior of the sky lounge

interior of the hotel foyer

Harry Seidler and Associates Sydney
Harry Seidler and Associates建筑事务所 悉尼

"Hotel Concourse as a kind of Libyan urbanism which is characterised by a strict distinction between public and private use of space."
"酒店广场作为一种利比亚城市化的表现，特点是在公众和私人空间使用之间进行严格的区分。"

作者 Penelope Seidler, Peter Hirst, Hiromi Shiraishi, Henry Feiner   合作伙伴及建筑系在校生 John Curro, Dirk Meinecke, Katrin Schmidt-Dengler   专家 structural analysis: Birzulis Associates, Sydney; Mechanical, Electrical, Hydraulic Engineers, Sydney

View along Al''Mari Street

3

Harry Seidler and Associates, Sydney

Tower at Suk Al Thalath Al Gadeem, Tripoli

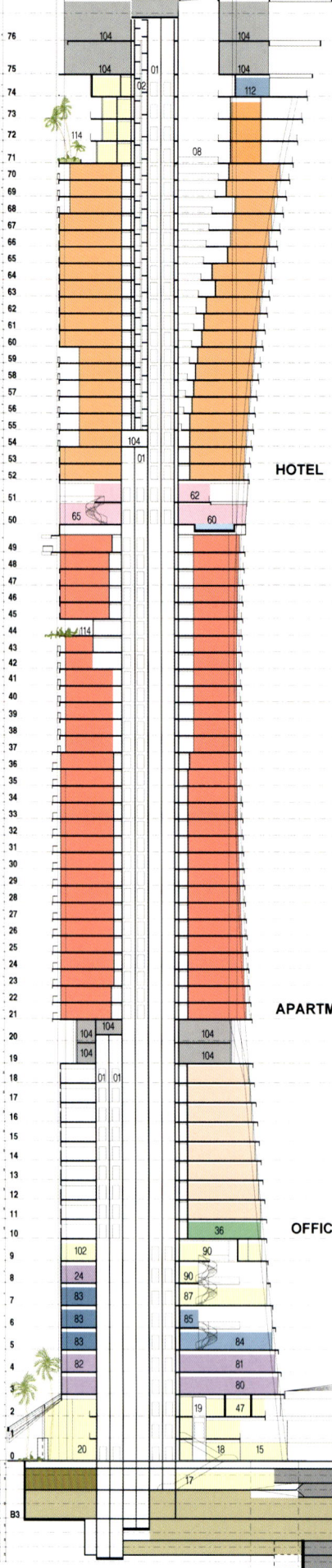

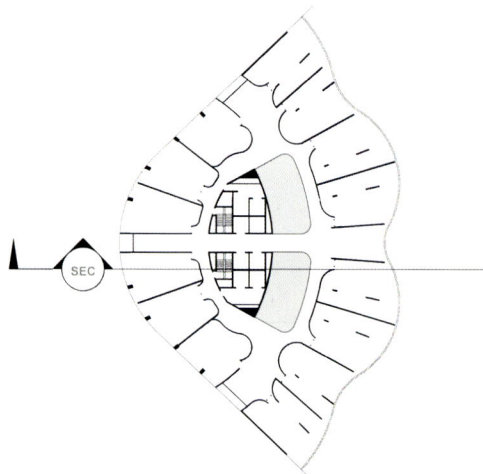

SEC

HOTEL

APARTMENTS

OFFICES

RETAIL

01	ELEVATORS	60	SWIMMING POOL
02	FIRE STAIR	61	WHIRLPOOL
03	TOILET (MALE)	62	BAR
04	TOILET (FEMALE)	63	SAUNA
05	TERRACE	64	TEPIDARIUM
06	SERVICE DUCT	65	SPA LOBBY /BEAUTY CENTER
07	A/C PLANT	66	SPA SUITE
08	VOID	67	YOGA STUDIO
09	STORAGE	68	TREATMENT ROOMS
10	SERVICE LIFT	69	MALE FITNESS ROOM
		70	SPA
		71	FITNESS ROOM
		72	STOCK WELLNESS
		73	SOLARIUM
		74	CALDARIUM
		75	SPA RECEPTION

80	HOTEL MANAGEMENT	15	MAIN LOBBY AND FRONT HALL
81	STAFF AND HOUSEKEEPING	16	RECEPTION
82	KITCHEN /STORES	17	ESCALATORS DOWN TO VIP
83	KITCHEN	18	ESCALATORS UP TO APARTMENT LOBBY
84	ALL DAY DINER	19	GLASS LIFTS
85	THEMED RESTAURANT	20	OFFICE LOBBY
86	BISTRO	21	HOTEL CONCOURSE
87	LARGE BALLROOM	22	RESTROOMS HANDICAPED TOILETS
88	WAITER LIFT	23	PORTERS
89	PRE FUNCTION AREAS	24	CLOAKROOM
90	CONFERENCE RECEPTION	25	BAGGAGE ROOM
91	MEETING ROOM 1	26	SKYLIGHT TO DELIVERY BELOW
92	MEETING ROOM 2	27	CAR/TRACK RAMP DOWN TO BASEMENT
93	MEETING ROOM 3	28	CAR/TRACK RAMP UP FROM BASEMENT
94	MEETING ROOM 4	29	PALM TREES
95	TEA ROOM	30	REFLECTING POOL
96	CONFERENCE ROOM 1	31	WATERFALL TO VIP LOBBY BELOW
97	CONFERENCE ROOM 2	32	SCULPTURE
98	CONFERENCE ROOM 3	33	LUXURY RETAIL SHOPS
99	CONFERENCE ROOM 4	34	LOBBY BAR
100	BOARDROOM 1	35	APARTMENT LOBBY
101	BOARDROOM 2	36	BUSINESS CENTRE
102	BOARDROOM 3	37	RECEPTION MANAGEMENT
103	OFFICE TENANCIES	38	CAFE
104	PLANTROOM	39	GOODS LIFT AND GARBAGE ROOM
		40	COURTYARD
		41	GLASS AWNING DOTTED OVER
		42	ANCHOR TENANT
		43	WINTER GARDEN
		44	RESTAURANT
		45	LIBRARY
		46	PRAYER ROOM
		47	CIGAR LOUNGE
		48	CLUB LOUNGE
		49	OFFICE MANAGEMENT
		50	SECURITY GUARD
		51	ELEVATING SECURITY BARRIER
		52	FIXED SECURITY BARRIER
		110	RECEPTION EXECUTIVE FLOOR
		111	HOUSEKEEPING ROOM
		112	FINE DINING
		113	SKY LOUNGE
		114	SKY GARDEN

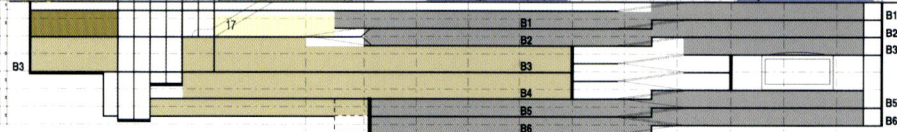

Aerial Perspective

2

250623

TYPICAL SUITE PLAN

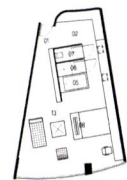

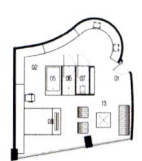

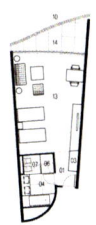

TYPICAL JUNIOR SUITE PLANS

TYPICAL SUITE PLAN

TYPICAL HOTEL ROOM PLANS

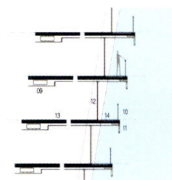

LEGEND

01	ENTRANCE
02	DRESSING
03	WARDROBE
04	BATHROOM (POLISHED GRANITE FLOOR SLABS)
05	SOK BATH
06	SHOWER (GLASS DOOR & WALL)
07	WC (FROSTED GLASS DOOR)
08	BED BASE (WITH ELEVATING
	BACK-TO-BACK PLASMA TV SCREEN)
09	A/C UNITS IN CEILING SPACE
10	CANTILEVERED TOUGHENED GLASS
	BALUSTRADE WITH CURVED S.S. HANDRAIL
11	S.S. ALUCABOND CURVED PANEL DOWN-TURNS
12	SOLAR GREY TINTED TOUGHENED FULL-HT GLASS
13	CARPET ON CONCRETE SLAB
14	FLAMED GRANITE PAVING TO TERRACES

TYPICAL FACADE SECTION

TYPICAL HOTEL ROOMS   S 1 : 50   0   2,5   5   10

COOP HIMMELB(L)AU, Prix/Dreibholz & Partner Vienna
COOP HIMMELB(L)AU, Prix/Dreibholz & Partner建筑事务所 维也纳

"... shaped by urban, cultural, climatic, and view considerations, as well as economies of structure, material, and inner functions, creating a memorable and unmistakable icon in the city-shape ..."
"通过城市、文化、气候和观点，以及结构经济性、材料和内部功能等方面的考虑，塑造出一个城市形状的令人难忘的显而易见的地标性建筑。"

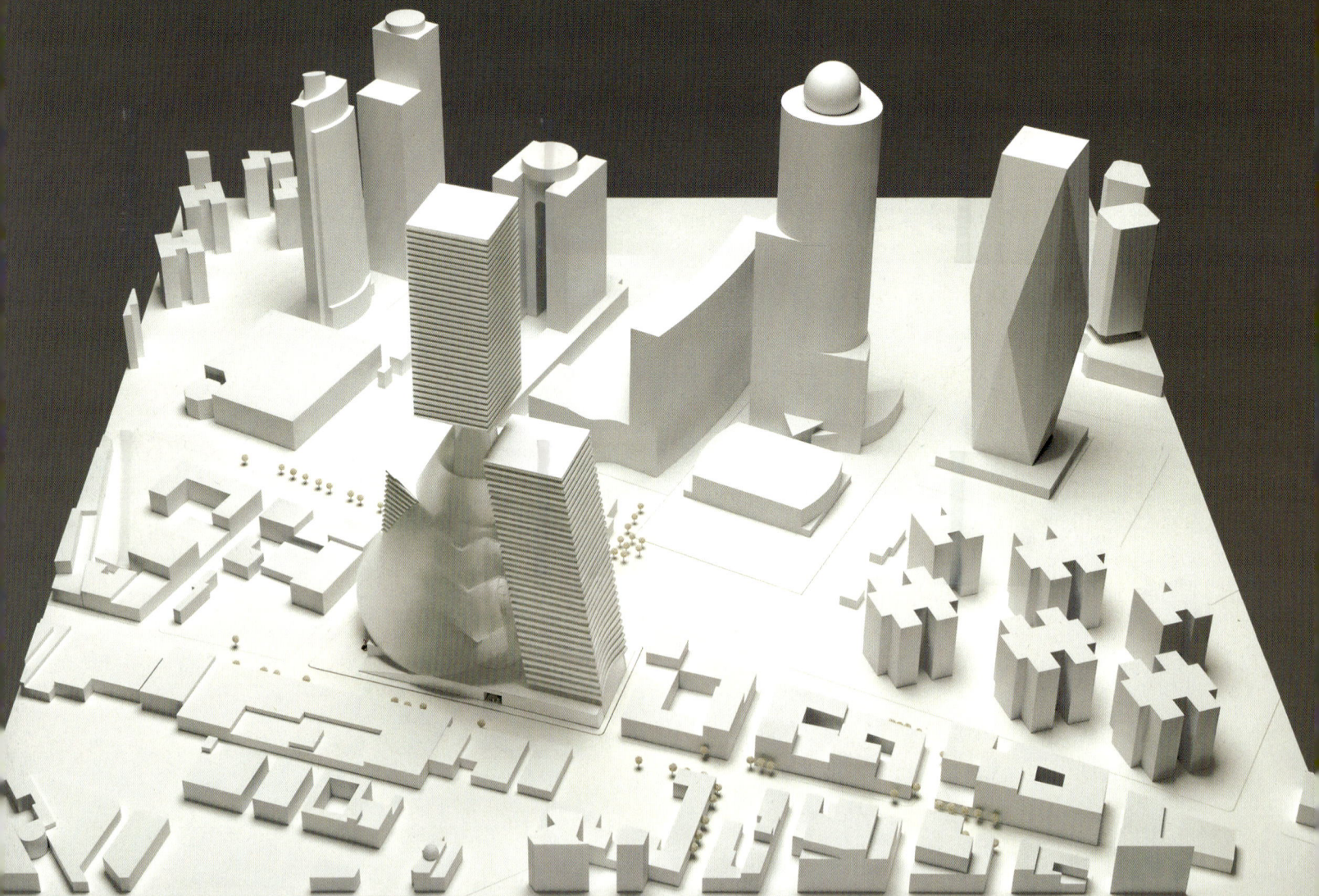

作者 Univ. Prof. Arch. DI Dr. Hc Wolf D. Prix   合作伙伴及建筑系在校生 Michael Volk, Andrea Graser, Pete Rose, Victoria Coaloa, Vinnento Possenti, Juhong Park, Anja Sorger, Guiseppe Zagaria, Elisabeth Swiss, Frank Hildebrandt, Daniela Comito, Chih-Bin Tseng, Paul Hoszowski, Markus Pillhofer   专家 structural analysis: B+G Ingenieure/Bollinger und Grohmann, Frankfurt/Main; building services: ARUP/Brian Cody, London/Frankfurt/Main

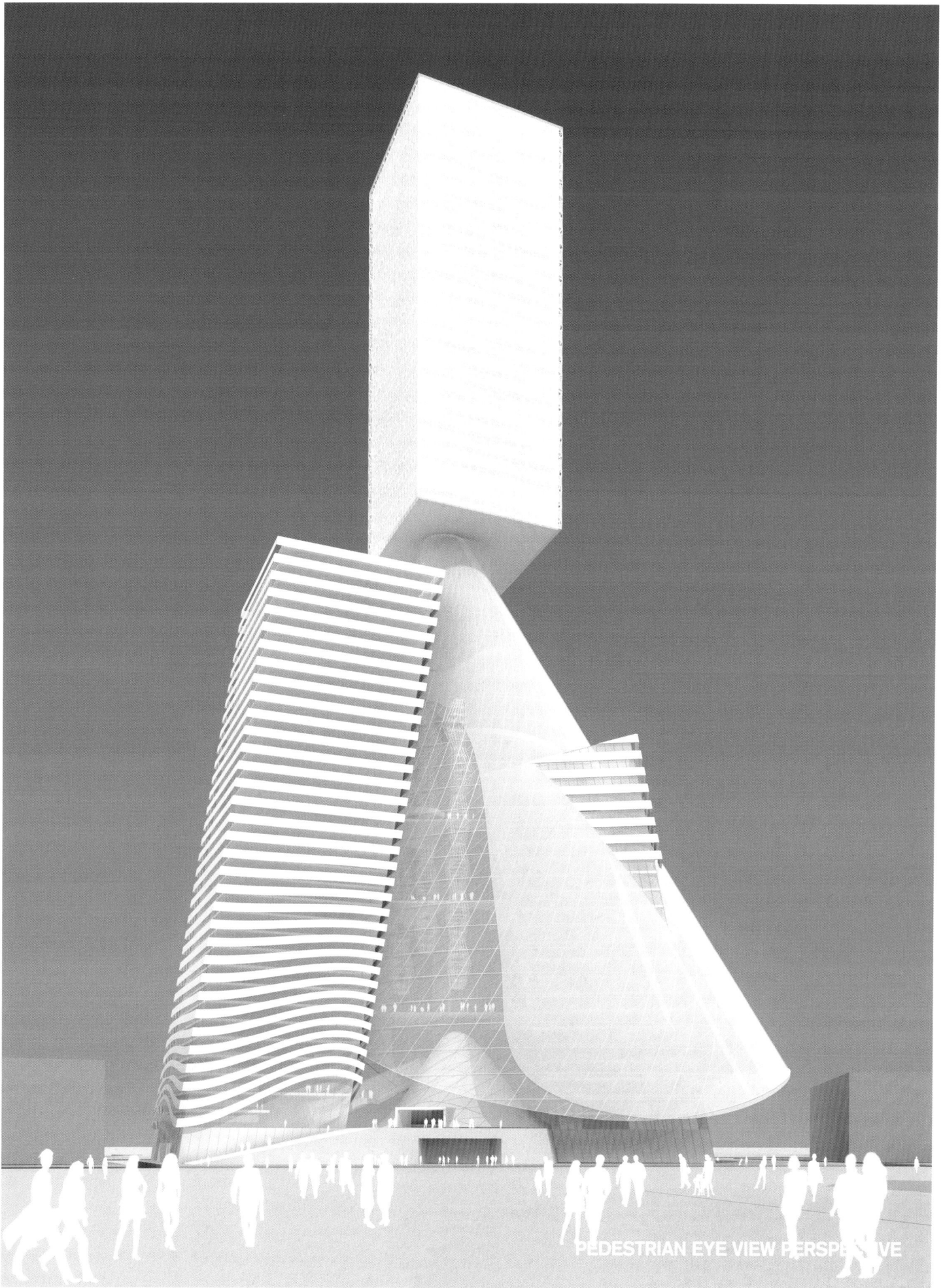

PEDESTRIAN EYE VIEW PERSPECTIVE

COOP HIMMELB(L)AU, Prix/Dreibholz & Partner, Vienna

Tower at Suk Al Thalath Al Gadeem, Tripoli

SHEET #6.1

+225.00m ▼
+218.00m ▼ SKYBAR

SECTION A-A

0 4 8 12 16 20
SCALE | 1:200

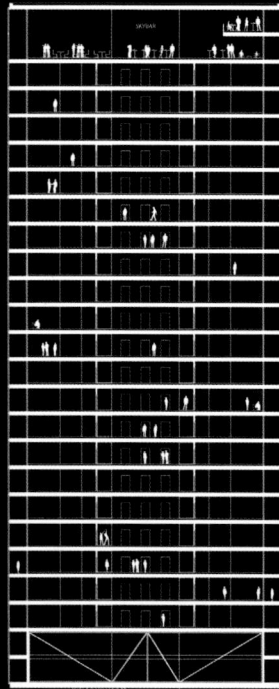

+190.00m ▼ HOTEL

+169.00m ▼ HOTEL

HOTEL TOWER

+137.00m ▼ TECHNICAL AREA

APARTMENT TOWER

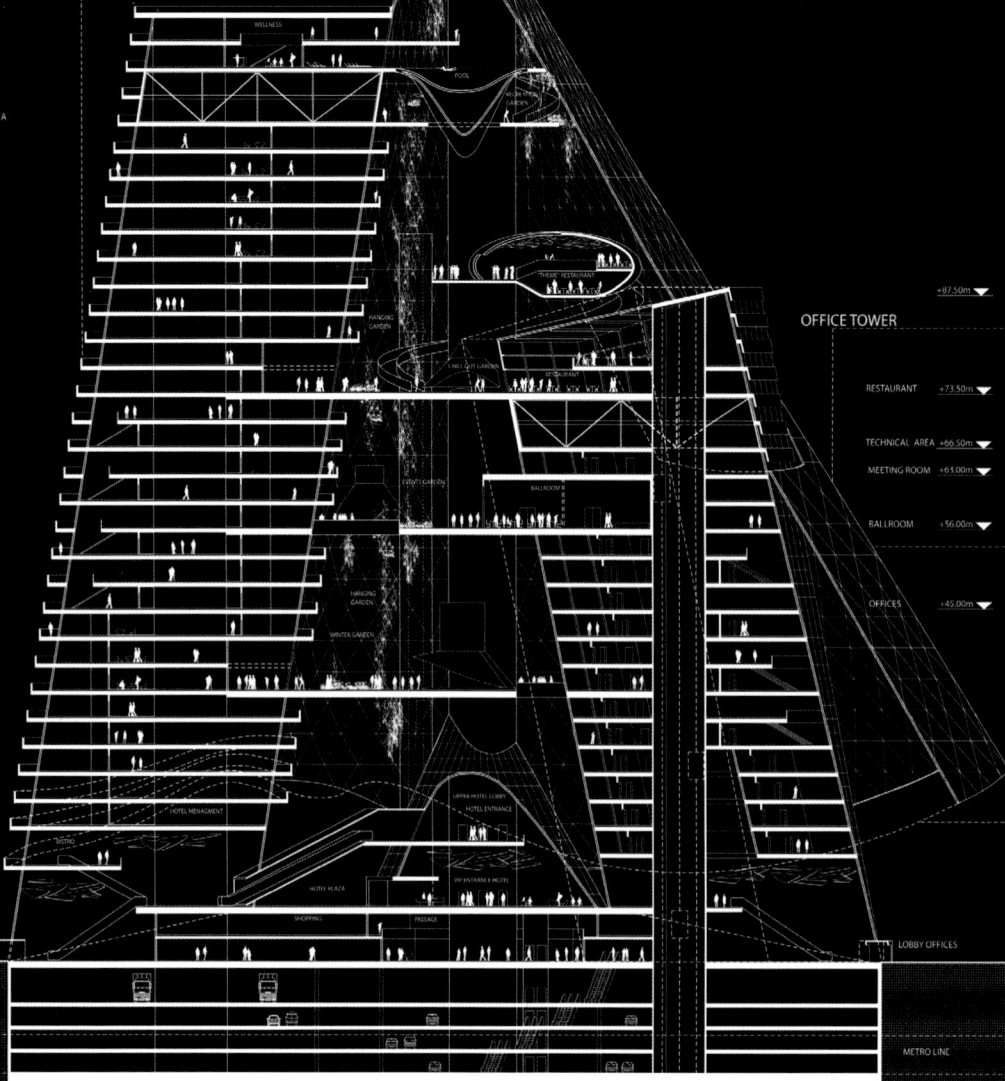

+123.00m ▼
+115.50m ▼ WELLNESS
+115.50m ▼ WELLNESS

POOL

+108.50m ▼ TECHNICAL AREA

+87.50m ▼

"HEART" RESTAURANT

OFFICE TOWER

+73.50m ▼ APARTMENTS

RESTAURANT

RESTAURANT +73.50m ▼

TECHNICAL AREA +66.50m ▼
MEETING ROOM +65.00m ▼

+63.00m ▼ APARTMENTS

BALLROOM B

+56.00m ▼ APARTMENTS

ESTATE GARDEN

BALLROOM +56.00m ▼

+45.50m ▼ APARTMENTS

HANGING GARDEN

OFFICES +45.00m ▼

WINTER GARDEN

UPPER HOTEL LOBBY

HOTEL MENAGMENT

HOTEL ENTRANCE

BISTRO

HOTEL PLAZA

VIP ENTRANCE HOTEL

+7.50m ▼

SHOPPING

PASSAGE

+0.00m ▼ LOBBY HOTEL

LOBBY OFFICES

SERVICE HOTEL

METRO LINE

ENVIRONMENTAL, ENERGY AND BUILDING SERVICES CONCEPTS

Strategies employing the form of the building to assist natural ventilation together with the use of renewable energy sources (wind and solar power) assure an energy efficient design and reduce energy consumption and reliance on fossil fuel energy sources.

Photovoltaic 'shield'
A transparent "shield" in the form of a second skin made from thin, semi-transparent photovoltaic film, wraps itself around the south facing office tower providing effective solar shading for the south facing offices, enclosing the atrium volume which in turns acts as a buffer zone using gardens to improve air quality, and generating electrical energy via the photovoltaic cells. The density of these varies according to orientation and tilt angle, thus creating a visually interesting pattern in the building skin. Tripoli enjoys some of the best climatic conditions in the world for the generation of electricity employing photovoltaic cells.

Wind energy
The building design also enables wind energy to be captured and employed via wind generation plant to generate renewable electrical energy. The building form is used to accelerate the predominantly north-westerly winds, increasing the local wind speed via the Venturi effect and the energy output of the generation plant.

Facade
The design of the building skin allows the use of natural ventilation throughout the entire winter period when temperatures are mild. Excessive wind pressures are reduced via an additional outer façade construction, which is also a solar screen formed in a pattern optimized to orientation.

Alternative energy sources
Biomass or possibly gas fueled combined heat and power generators provide the building with both heat and electrical power. This solution has both ecological and economic advantages compared to more conventional alternatives (c. 60% less CO2 emissions), and also provides a major advantage with regard to security of supply. In warm weather the heat is used to drive an absorption chiller that supplies chilled water to cool the building. All rainwater is collected and used for irrigation and toilet flushing.

Plant rooms
Plant rooms for technical equipment are located as shown in the diagrams and are connected via vertical shafts and risers to the individual floors to provide an efficient technical infrastructure. An integrated security system is provided including CCTV (closed circuit television) surveillance of public areas, full function access control at selected entrances and lifts and central monitoring equipment within a main lobby security/ reception desk. A complete Building Management System (BMS) is provided consisting of multiple Direct Digital Control (DDC) data processing outstations and a central management system.

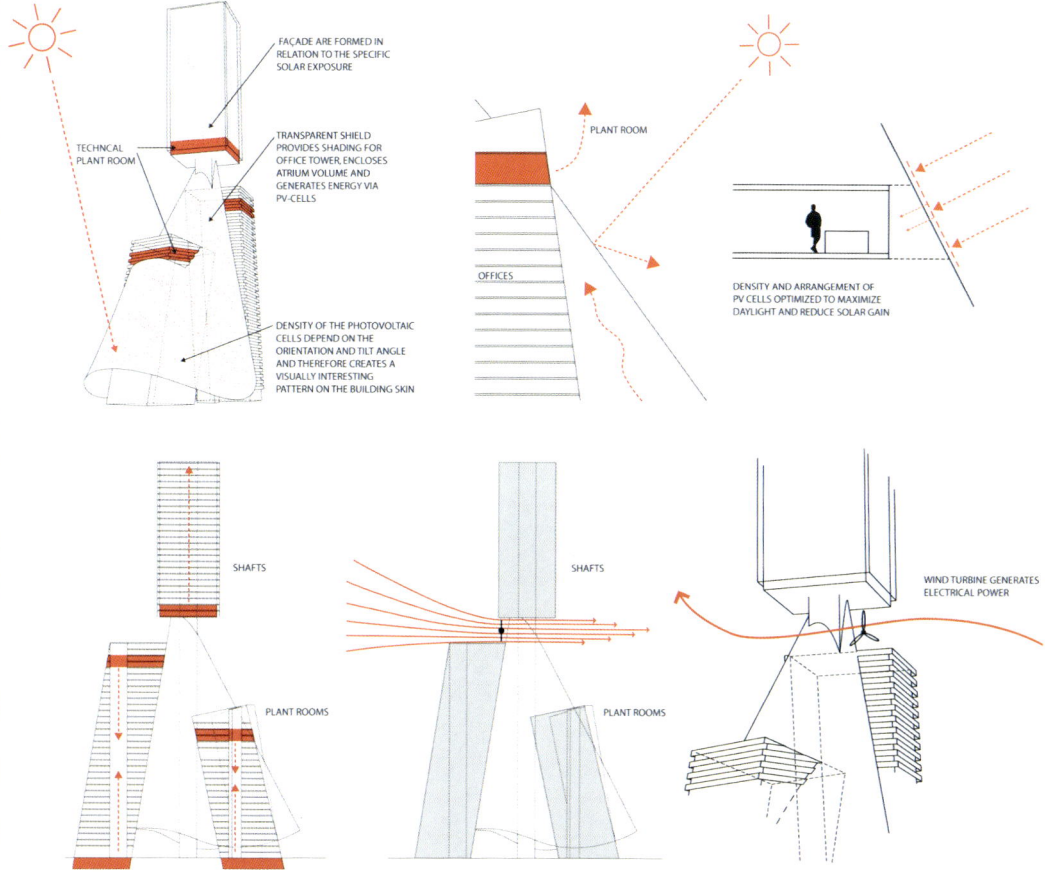

FAÇADE ARE FORMED IN RELATION TO THE SPECIFIC SOLAR EXPOSURE

TECHNCAL PLANT ROOM

TRANSPARENT SHIELD PROVIDES SHADING FOR OFFICE TOWER, ENCLOSES ATRIUM VOLUME AND GENERATES ENERGY VIA PV-CELLS

DENSITY OF THE PHOTOVOLTAIC CELLS DEPEND ON THE ORIENTATION AND TILT ANGLE AND THEREFORE CREATES A VISUALLY INTERESTING PATTERN ON THE BUILDING SKIN

PLANT ROOM

OFFICES

DENSITY AND ARRANGEMENT OF PV CELLS OPTIMIZED TO MAXIMIZE DAYLIGHT AND REDUCE SOLAR GAIN

SHAFTS

SHAFTS

WIND TURBINE GENERATES ELECTRICAL POWER

PLANT ROOMS

PLANT ROOMS

THERMAL COLLECTORS — HEATING LOADS

USE OF HEAT REJECTION FOR WELLNESS AND HOTEL DHWS

FUEL GASS OR BIOMASS — COMBINED HEAT AND POWER COGENERATION PLANT — ABSORPTION CHILLER — COOLING LOADS

COMPRESSION CHILLER

ELECTRICAL UTILITY — ELECTRICAL LOADS

PHOTOVOLTAIC | WIND GENERATORS

STRUCTURAL DESIGN

The Suk Al Thalath building complex consists of three rectilinear high-rise towers and a cone-shaped atrium that interact with each other structurally to form a cohesive single structural system.

Earthquake
The building site belongs to a seismic zone with high requirements. As the seismic and wind loads are the governing parameters affecting the stability of the structure, these loads are mainly considered by the structural design of the building complex. Sufficient ductility is achieved by designing the structures as highly integrated and developed total systems, and by accurately detailing the individual elements with special respect to connections.

Apartment and Office Towers
The 33-storey Apartment tower and 22 Storey Office tower are both rectangular shaped concrete structures of 123 meters and 90 meters tall respectively.
The main structure of the Apartment and Office Towers are comprised of the inner core of concrete walls and outrigger systems situated at their top levels as service storeys. These outriggers provide necessary stiffness against horizontal forces for the Apartment and Office Towers, and are further utilized to make a robust connection to stabilize the upper Hotel Tower.
Vertical forces are carried through concrete columns in a grid of 8.75 by 4.60 m along the facades, and through the walls of the concrete core. The floor slabs are designed as flat slabs with a maximum span of 10.00m.

Hotel Tower
The hotel tower is a 24-storey building located atop the Cone structure of the Atrium. As the first floor is situated at a height of about 137m, the concrete core below supports the building vertically. To carry the vertical forces from façade and inner columns into the core, the two lower floors are designed as service storeys in a steel framework. Unlike the apartment and office towers, floors and columns are provided as a composite structure. This has advantages for construction sequencing and the reduced mass in the design is favourable for load bearing behaviour in relation to earthquake forces.

Cone
The cone is made of a triangulated steel grid with the concrete core of the hotel tower in its centre. As a lightweight structure it is stabilised vertically by the apartment and office tower as well as at its top by the core and horizontally by the platforms placed between both. The platforms are again vertically supported by the steel structure of the cone.

Subsurface Building Complex
The subsurface tract includes four storeys. The grid of 8.20 by 8.20 m allows that the floors be constructed as flat concrete slabs. The Horizontal stiffness of the lower floors is provided through concrete walls and concrete cores.

Foundation
A combined pile-raft-foundation is recommended. It is a geotechnical composite structure comprising piles, a thick foundation slab and the soil below. The building bears on the ground and transfers its vertical loads through compression underneath the foundation slab and by surface friction and base pressure of the piles. The slab thickness is adapted to the individual stress at each area and varies between 1.00m and 3.00m. Piles are arrayed below the cores and columns of the high-rise buildings.

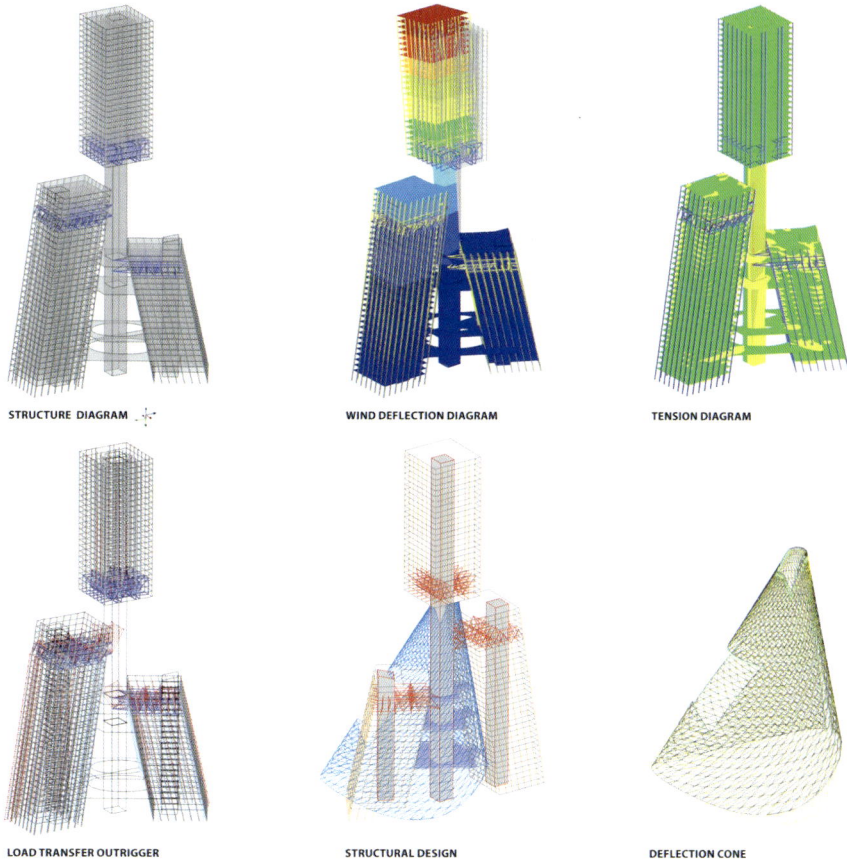

STRUCTURE DIAGRAM

WIND DEFLECTION DIAGRAM

TENSION DIAGRAM

LOAD TRANSFER OUTRIGGER

STRUCTURAL DESIGN

DEFLECTION CONE

Aedas Dubai

Aedas建筑事务所 迪拜

"The 5 elements which graciously form 'the hand of Fatima' are transfered into a landmark to commemorate the people's past adversities, preserve their present way of life, and enhance future prosperity."

"优雅的形成'法帝玛之手'的五个元素被转化为一个地标性建筑，纪念人们苦难的过去，保留现在人们的生活方式，增强未来的繁荣。"

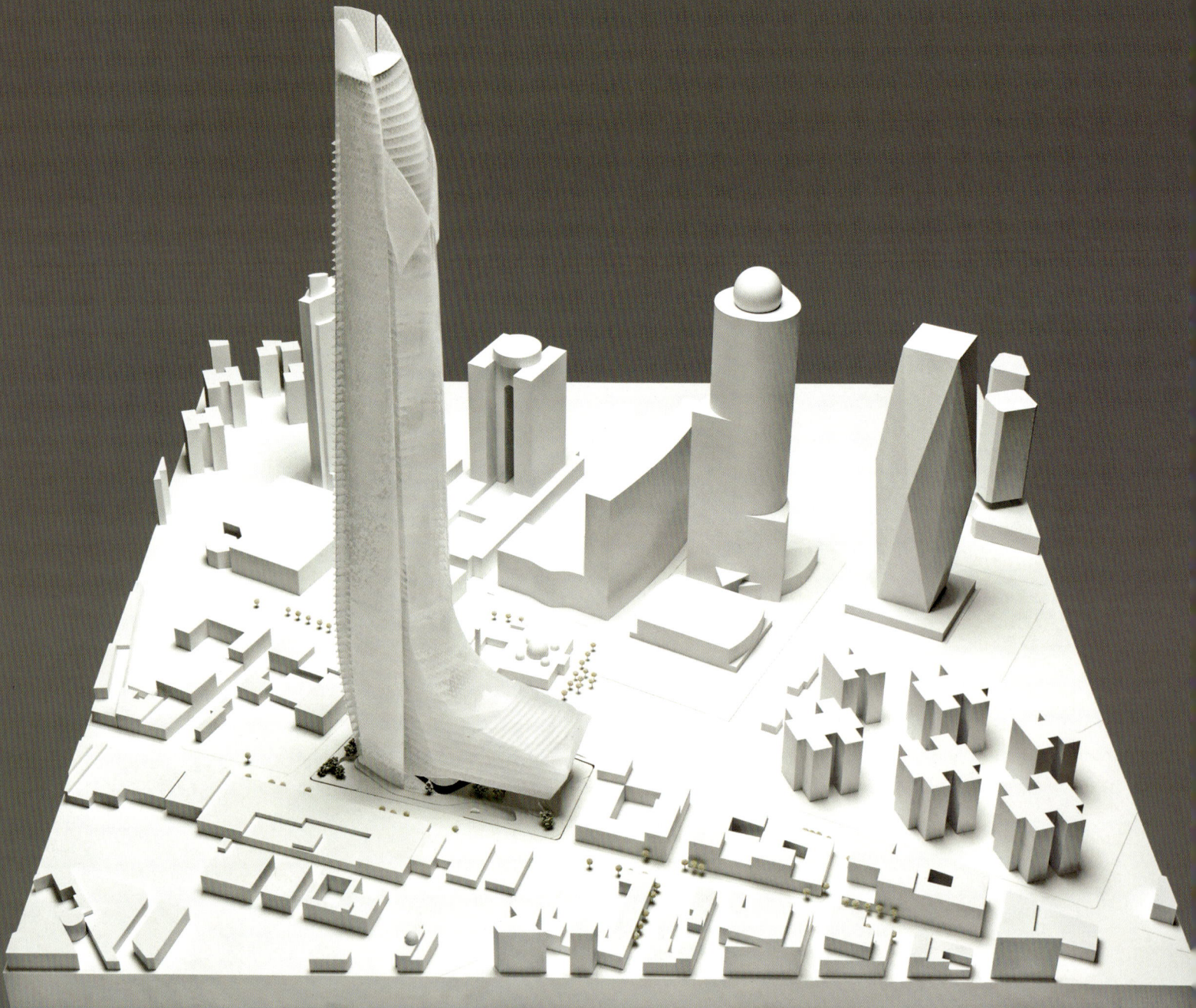

作者 Boran Agoston, Peter Engstrom   合作伙伴及建筑系在校生 Erwin Paul Lota, Byron Emmerson, Jonathan Wong, Rey Majadillas, Reynaldo Cleofe, Dan Albert Formalejo, June Marzan, Eric Chan, Raymond, RJ Models, Judith Kimpran (Aedas - Advanced Modeling Group)
专家 structural analysis: Predrag Eror, Meinhardt Pte Ltd., Singapore; sustainability: Sinsia Stankovic, BDSP Partnership, London; Alan Harles, BDPS Partnership, London; landscape planning: Sinsia Stankovic, BDSP Partnership, London; Alan Harles, BDPS Partnership, London

master perspective

3.0

tower at suk al thalath al gadeem

PHASE 1

PHASE 2

March 22 | 08:00 | 10:00 | 12:00 | 14:00 | 16:00

June 21 | 08:00 | 10:00 | 12:00 | 14:00 | 16:00

December 21 | 08:00 | 10:00 | 12:00 | 14:00 | 16:00

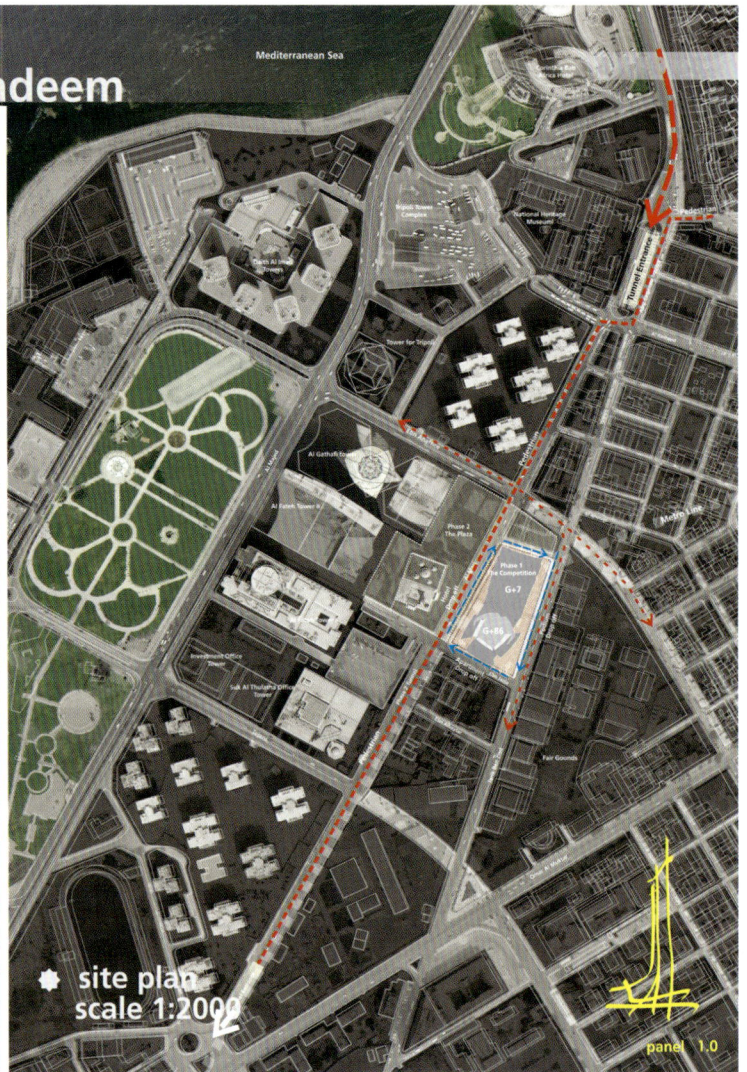

site plan
scale 1:2000

Mediterranean Sea

panel 1.0

elevation from al mari street   scale 1:200

4.0

longitudinal section   scale 1:200

6.1

Henning Lars

Kerry Hill Architects

Zaha

Snøhetta

DELUGAN MEISSL ASS

Atelier Christian de

en Architects

Hadid

OCIATED ARCHITECTS

Portzamparc

Darat King Abdullah II
in Amman

Darat King Abdullah Ⅱ
约旦安曼

Darat King Abdullah II for Culture and Arts Amman
Darat King Abdullah II 文化和艺术中心，约旦安曼

Restricted project competition preceded by an application procedure
限制严格的项目竞赛，开始前有申请程序。

地点 Amman   时间 10/2007–04/2008   主办方 Greater Amman Municipality (GAM)   参赛者 6   面积 12,000 sq m
竞赛费用 110,000 EUR   专业评奖委员会 Prof. Dr. Gulzar Haider, chairman , Lahore; Prof. Klaus Kada, Graz; Prof. Hilde Léon, Berlin; Jafar Tukan, architect, Amman   其他专业评奖委员会 Dominique Lyon, architect, Paris   专家委员会 Omar Maani, mayor of Amman; Dr. Kifah Fakhoury, head of "National Music Conservatory", Amman; Michael Schindhelm, head of culture department, Dubai
其他专家委员会 Amer Bashir, architect and vice mayor of Amman; Lina Attel, general director "Performing Arts Center" (PAC), Amman

With a population of over 2 million, Amman is one of the most vibrant and modern cities in the Middle East. The task set for this competition was a design for Amman's new culture and art center, the "Darat King Abdullah II". The project's principal, the Greater Amman Municipality, or GAM, envisions the center as the hot spot of the regional art scene. It will accommodate the performing art center and host events like concerts, music, and theatre performed by foreign and Jordanian artists, among them the Amman Symphony Orchestra, the State Music Conservatory and various other sources. The "Darat King Abdullah II" is one of several projects meant to liven up the city center and bridge social barriers separating its residents. To this end, a number of public buildings will be constructed in what is called the "GAM Strip", a site at the heart of the city, and a point of contact between two hugely different neighborhoods. The "Darat King Abdullah II" aside, the city administration building, the Hussein Culture Center, and the National Museum (due to open in 2008) are already sited along a stretch of valley at the foot of Jabal Al Akhdar. The 19,000 square metres site of the new culture and art center is at the western end of the strip. The spatial program comprises 12,000 square metres of ancillary usable area, a concert theater with 1,600 seats, a smaller theater with 400 seats, rehearsal rooms and lecture halls, top end stage equipment, a sweeping foyer, a restaurant, a café and club combination, as well as administrative offices.

安曼人口过200万，是中东最有活力和最现代的城市之一。竞赛任务是为安曼设计一个的新文化和艺术中心"Darat King Abdullah II"。项目的主办方，大安曼市政当局把这个中心看成是该区艺术界的热点。里边的表演艺术中心可以举办像音乐会、戏剧的活动，由外国和约旦艺术家与安曼交响乐队、国家音乐学院和其他组织一起表演。"Darat King Abdullah II"与其他几个项目的目的一样，都是要增强城市中心的活力，消除离间居民的社会障碍。为了达到这一目的，很多公共建筑物将会在市中心的一个被称为"GAM带"的地点建起。除了"Darat King Abdullah II"以外，还有城市政府大楼、Hussein文化中心和国家博物馆。新文化和艺术中心，占地1.9万平方米，位于"GAM带"的西端。空间规划包括1.2万平方米的辅助使用面积、一个有1600个座位的大电影院、一个有400个座位的小电影院、彩排室、讲座厅、尖端舞台设备、宽敞的休息室、饭店、咖啡厅俱乐部以及行政办公楼。

Aerial view　鸟瞰图

Competition site　竞赛地点

Competition site　竞赛地点

Ancient amphitheater　古圆形剧场

Darat King Abdullah II, Amman

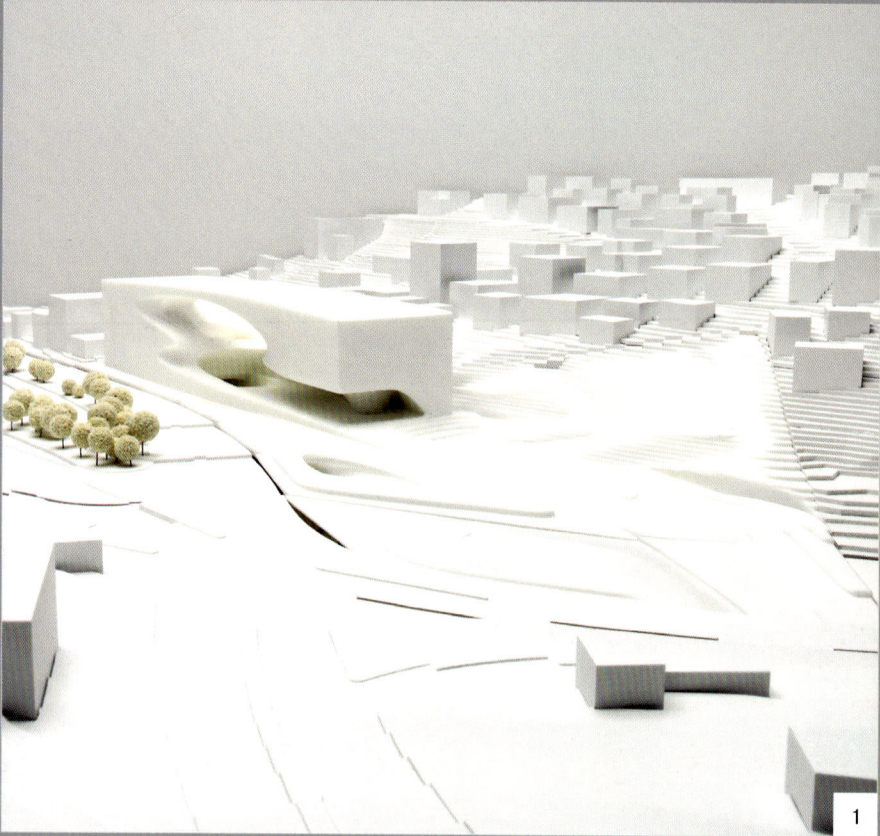
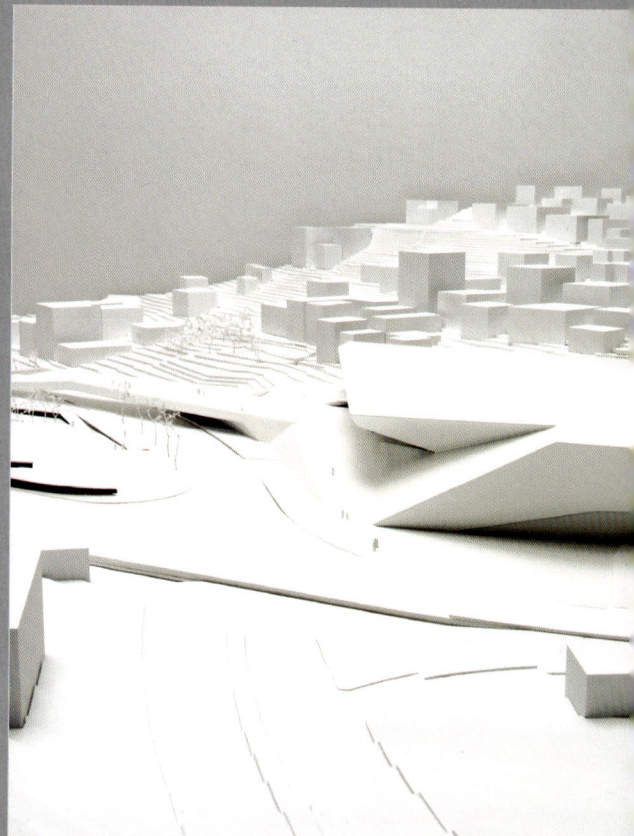
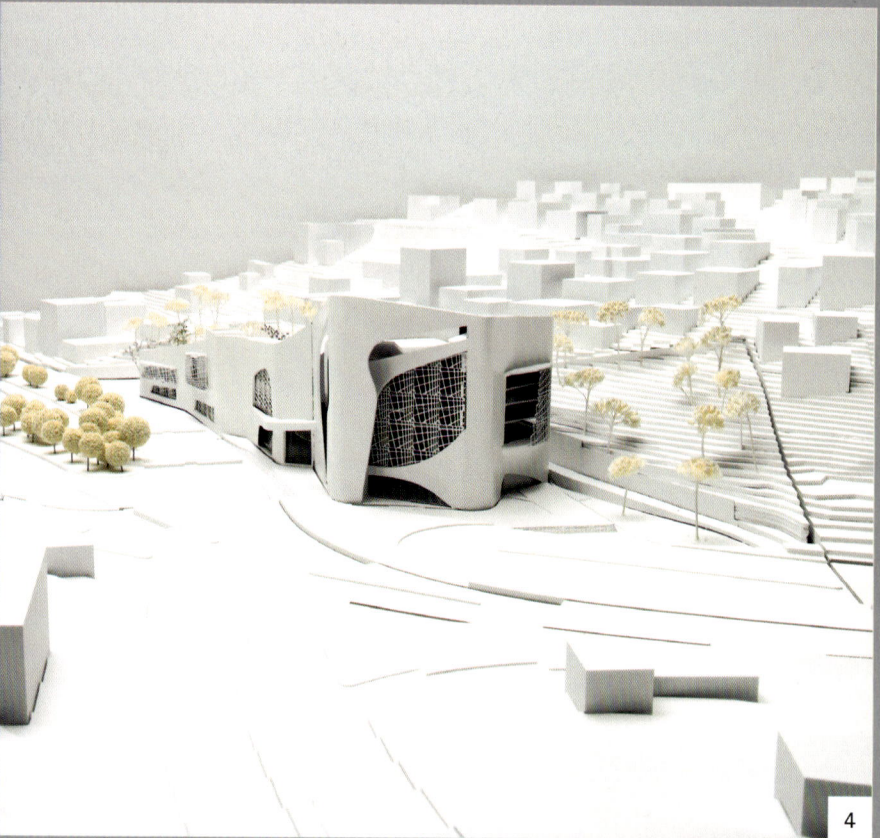
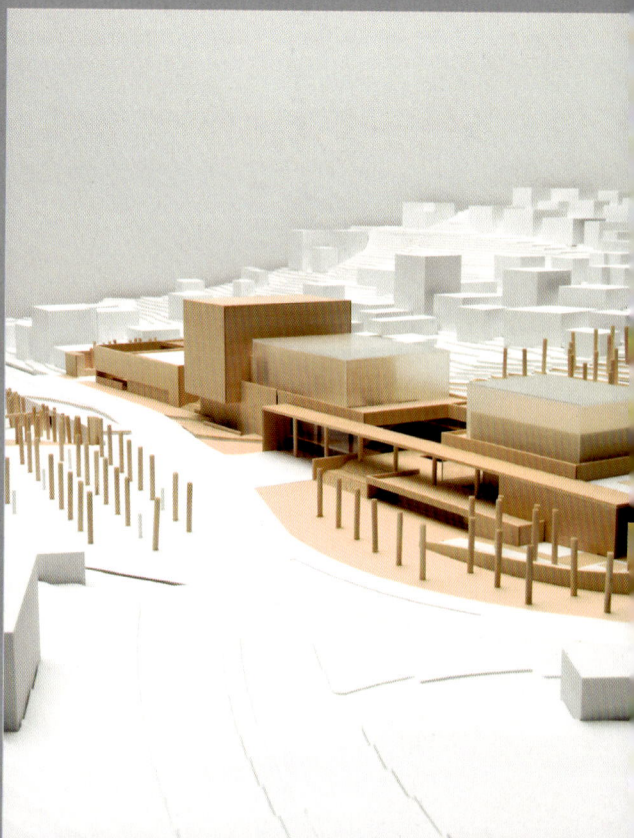

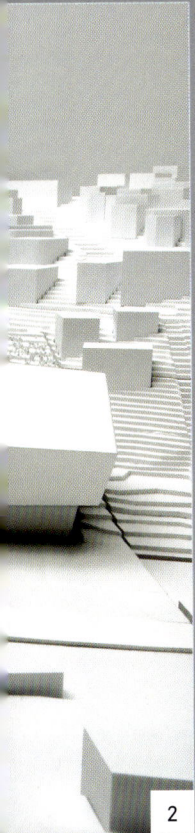

Qualified participants
合格的参赛者

1
1st prize　一等奖
Zaha Hadid Architects,
London

2
1st prize　一等奖
DELUGAN MEISSL ASSOCIATED
ARCHITECTS, Vienna

3
3rd prize　三等奖
Snøhetta, Oslo

4
Further participant　其他参赛者
Atelier Christian de Portzamparc,
Paris

5
Further participant　其他参赛者
Kerry Hill Architects,
Singapore

6
Further participant　其他参赛者
Henning Larsen Architects,
Copenhagen

Zaha Hadid Architects London
Zaha Hadid建筑事务所 伦敦

"... inspired by the uniquely beautiful monument of Petra [...] we are applying the principle of fluid erosion and carving to the mass of the building for the performing arts centre."

"受具有独特美感的Petra遗址的灵感激发，我们在建设表演艺术中心的时候运用流体冲蚀和雕刻的原理。"

作者 Zaha Hadid, Patrick Schumacher   合作伙伴 Christos Passas, Tariq Khayyat, Dominiki Dadatsi, Marya Araya, Sylvia Georgiadou, Bence Pap, Eleni Paviidou, Daniel Santos, Daniel Widrig, Sevil Yazipi

Zaha Hadid Architects, London

Darat King Abdullah II, Amman

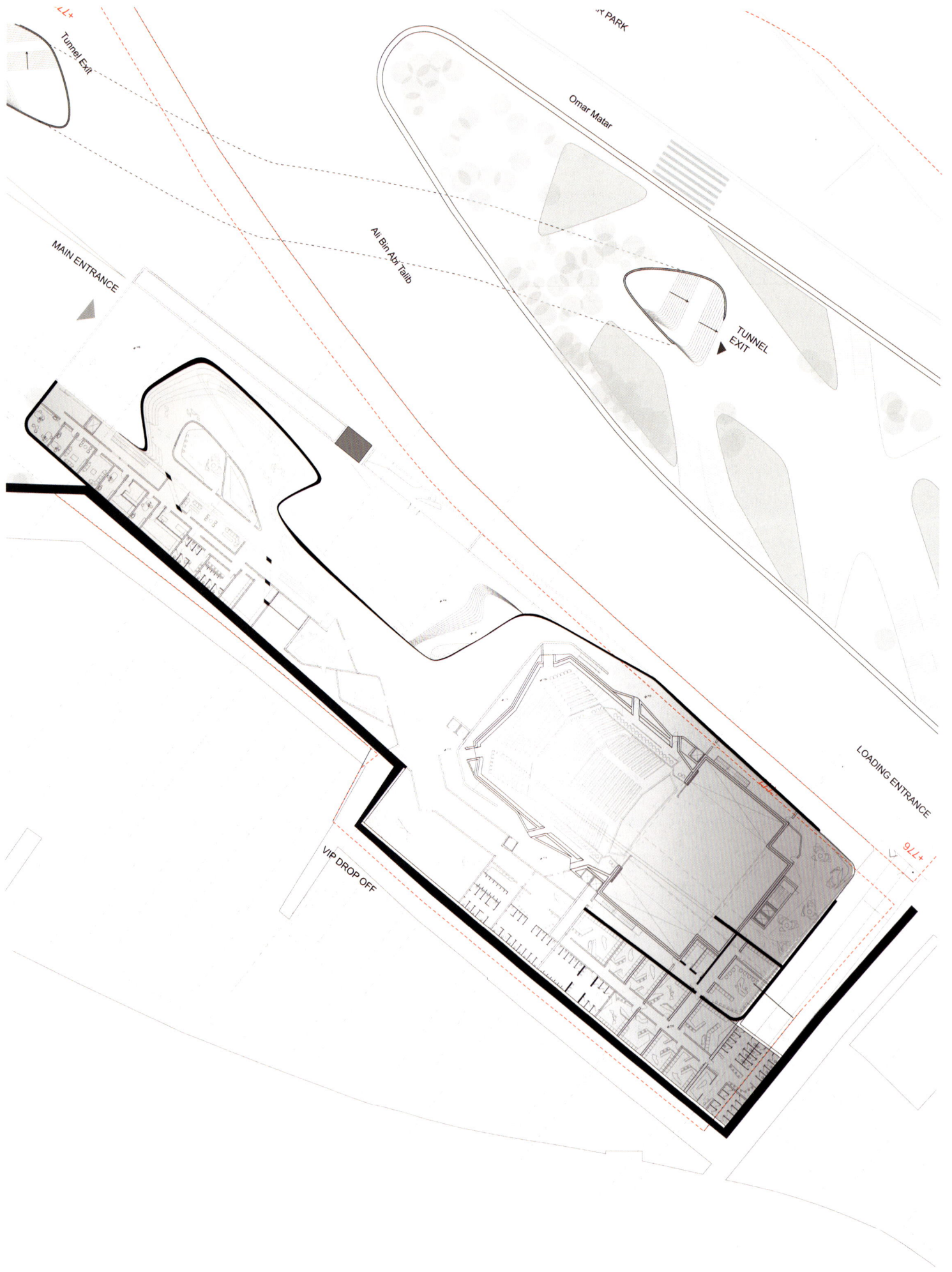

Tunnel Exit

Omar Matar

Ali Bin Abi Talib

TUNNEL
EXIT

MAIN ENTRANCE

LOADING ENTRANCE

VIP DROP OFF

Princess Basma

LOADING EXIT

MAIN DROP OFF

TUNNEL ENTRANCE

CAR PARK

Omar Matar

Ali Bin Abi Talib

TUNNEL EXIT

Stairway A

Subway B

VIP DROP OFF

LOADING ENTRANCE

▮▮▮▮▮ CAR DROP OFF ▮▮▮▮▮ PEDESTRIAN ACCESS ▮▮▮▮▮ LOADING ▮▮▮▮▮ GAM STRIP TUNNEL CONNECTION ▮▮▮▮▮ VIP ACCESS

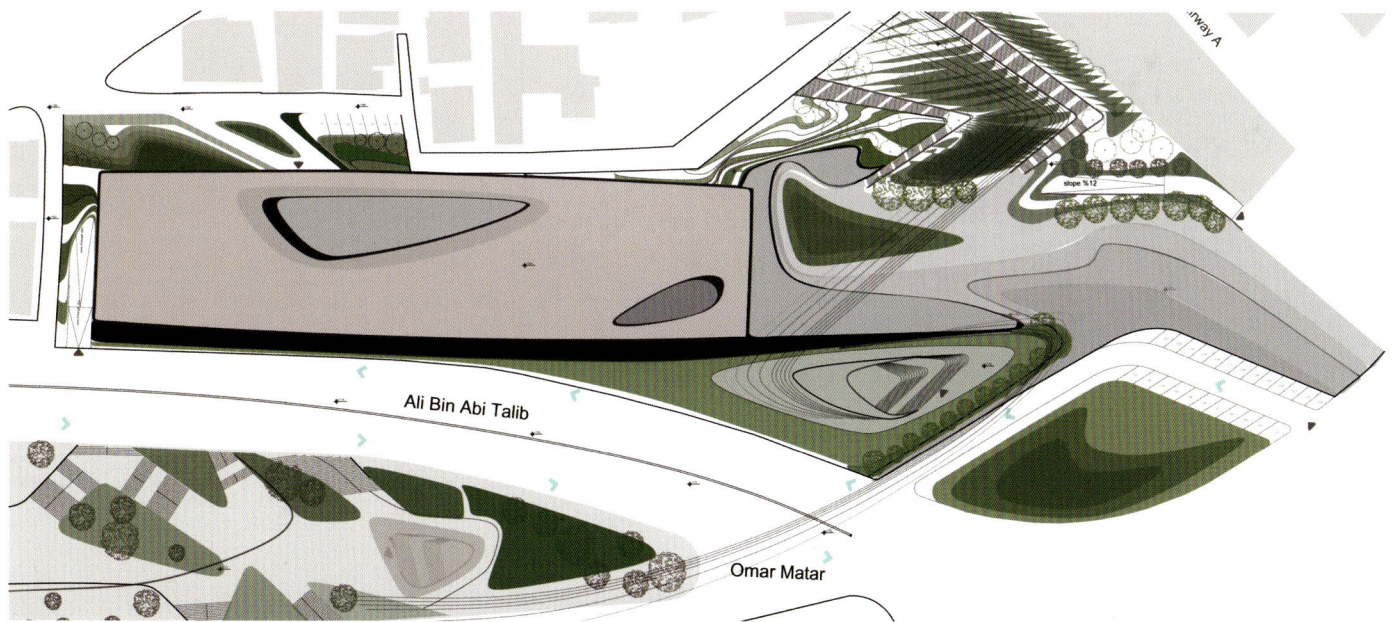

Stairway A

slope %12

Ali Bin Abi Talib

Omar Matar

VIEW FROM THE PARK AT ZONE A

Division of Hard, and Soft landscape

-3.75_LEVEL_MAIN AUDITORIUM (VIP) / REHEARSAL ROOMS   1:500

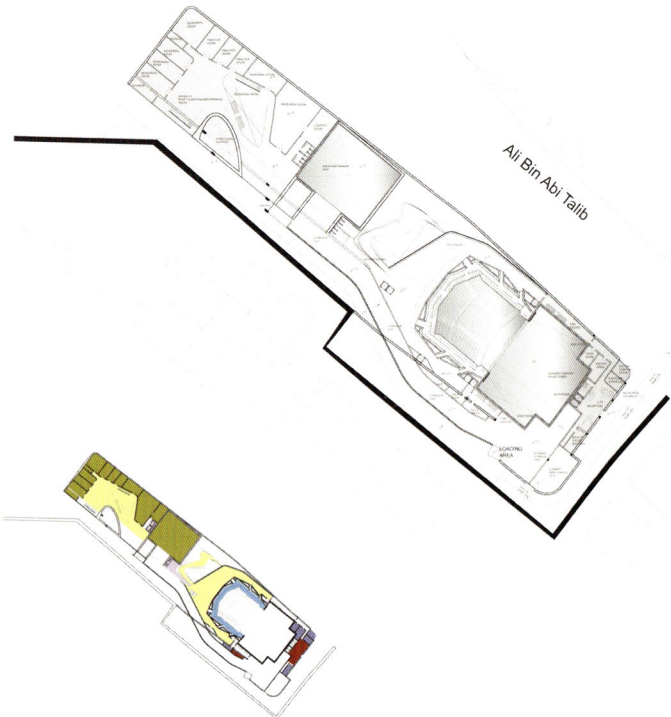

10 m   50 m

Ali Bin Abi Talib

+5m_LEVEL MAIN AUDITORIUM   1:500

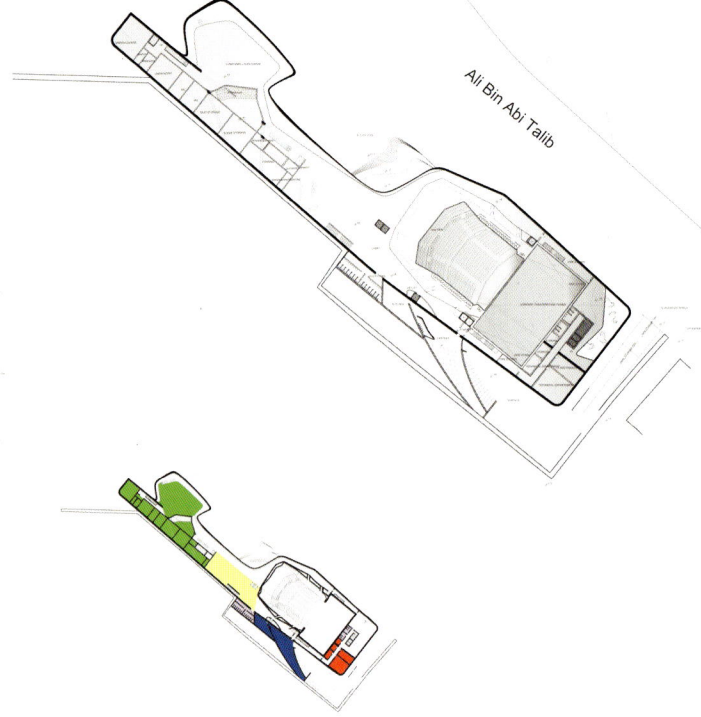

10 m   50 m

Ali Bin Abi Talib

+18m_LEVEL_CAFE AND RESTAURANTS   1:500

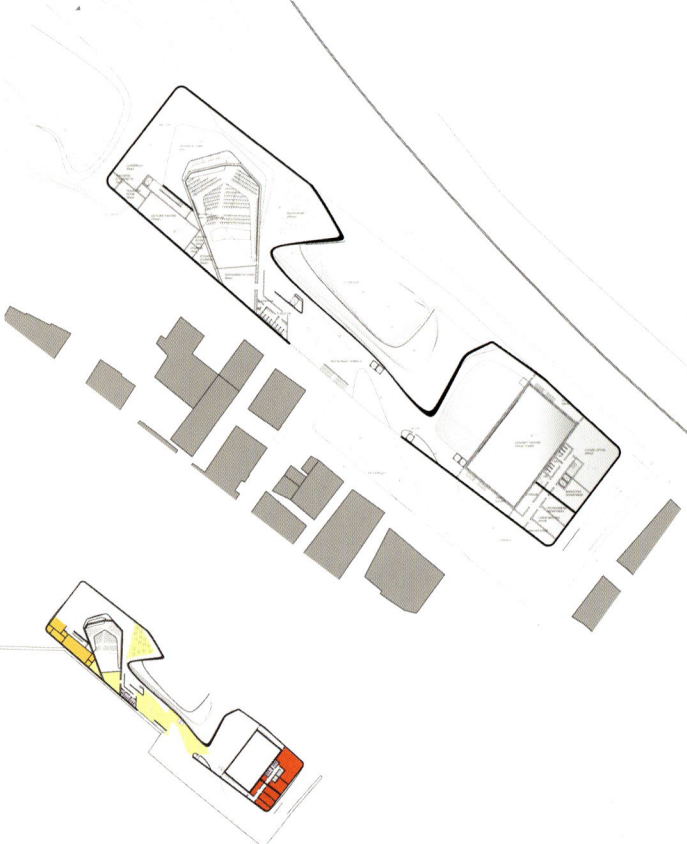

10 m   50 m

+22_LEVEL   1:500

10 m   50 m

Ali Bin Abi Talib

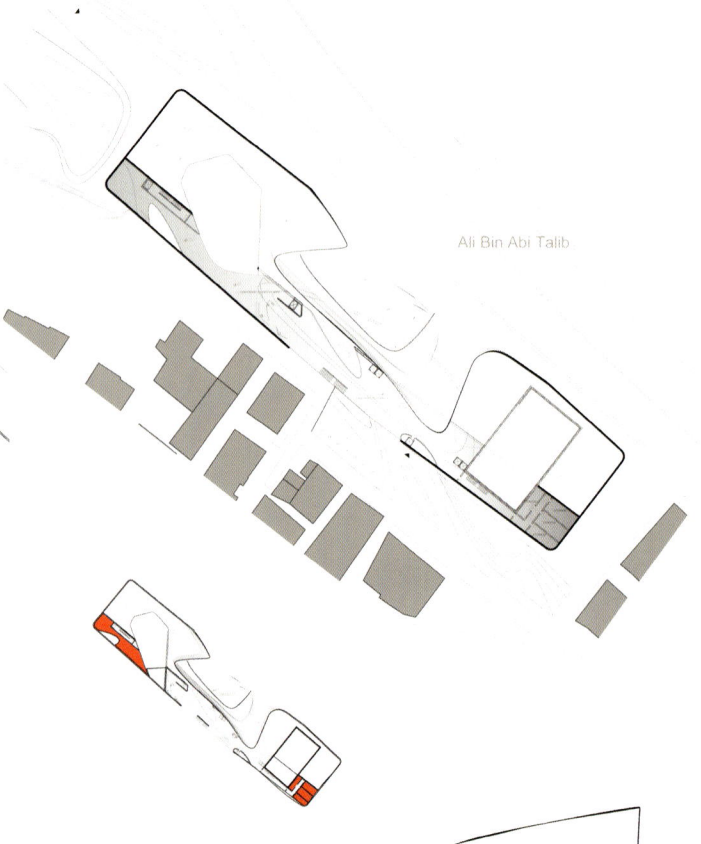

| 1 | 2 | 3 | 4 | 5 | 6 | 7 | 8 | 9 | 10 | 11 | 12 |

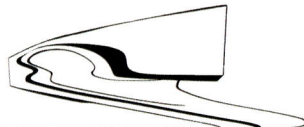

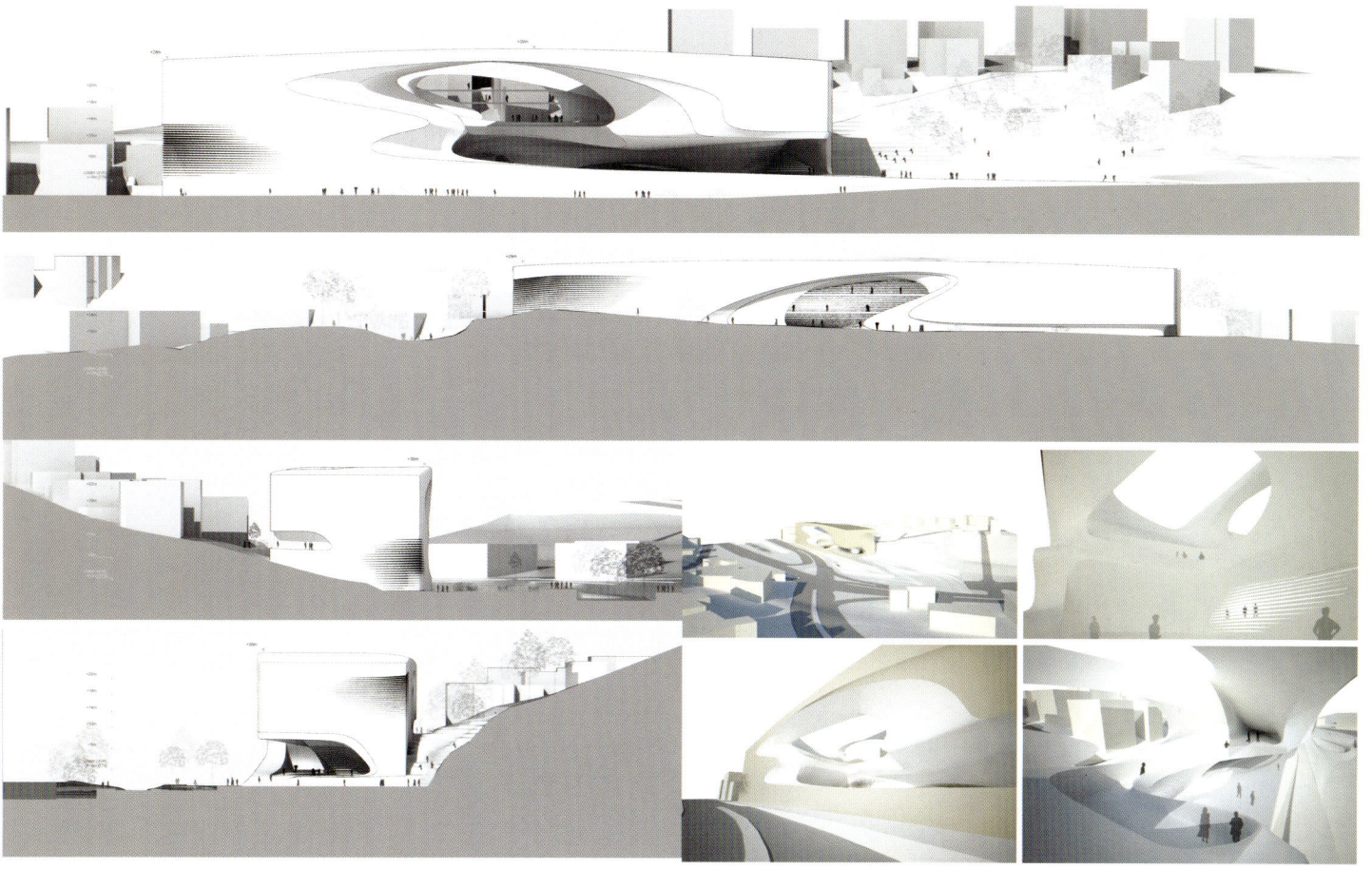

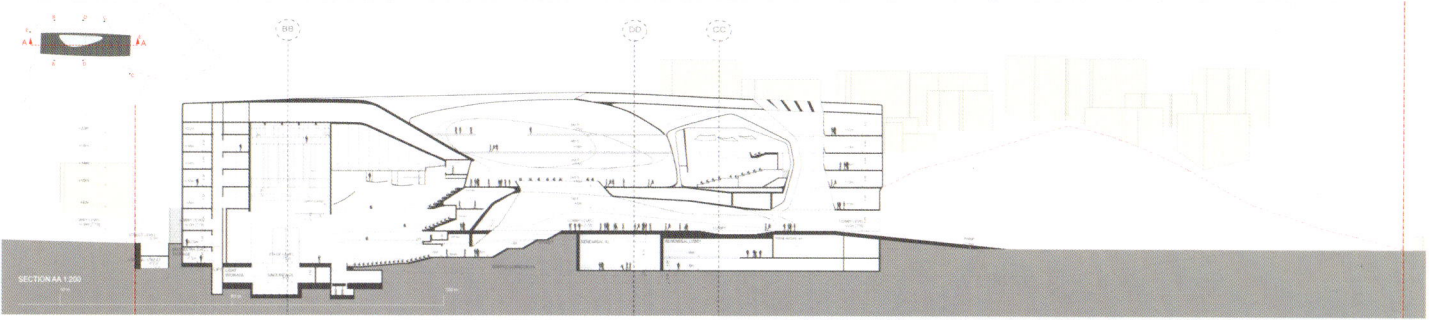

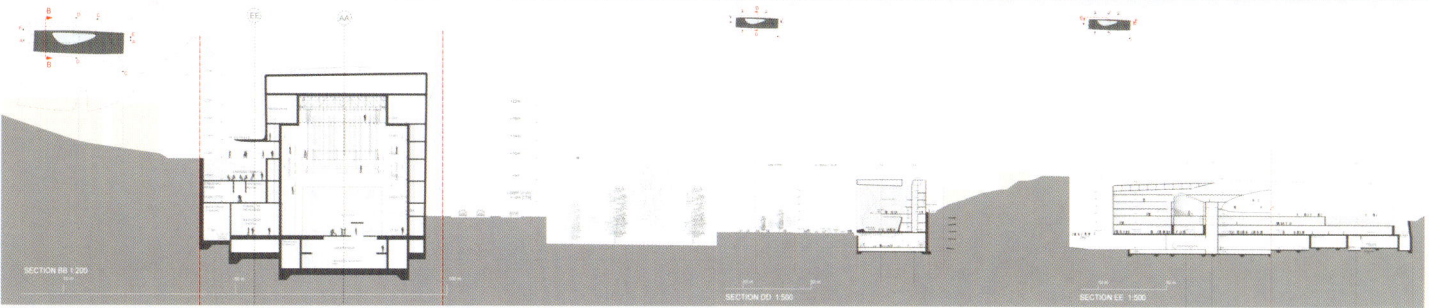

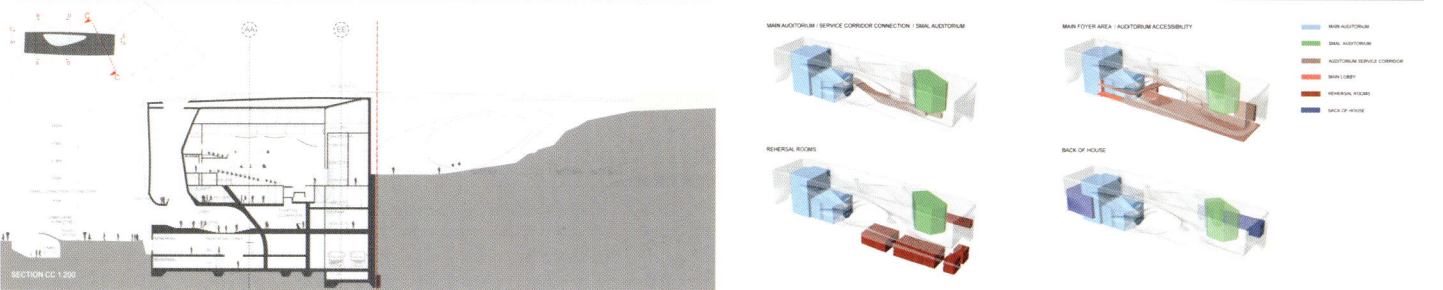

MAIN AUDITORIUM

MAIN AUDITORIUM

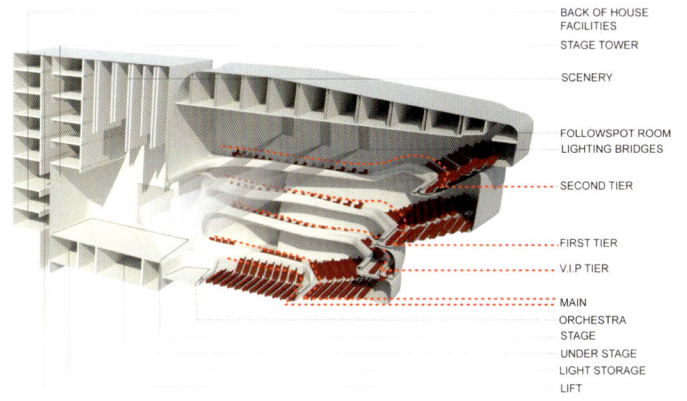

BACK OF HOUSE
FACILITIES

STAGE TOWER

SCENERY

FOLLOWSPOT ROOM
LIGHTING BRIDGES

SECOND TIER

FIRST TIER
V.I.P TIER

MAIN
ORCHESTRA
STAGE
UNDER STAGE
LIGHT STORAGE
LIFT

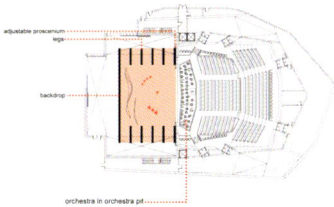

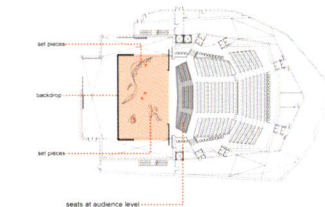

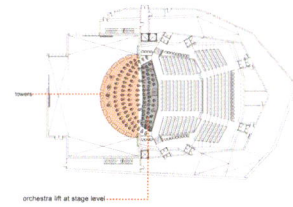

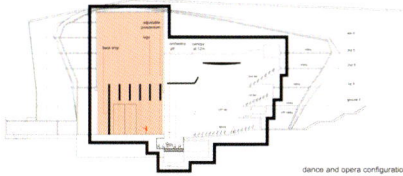

dance and opera configuration

drama configuration

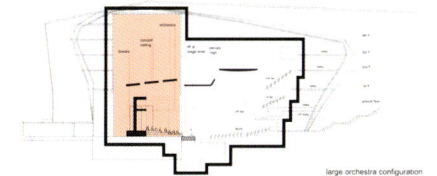

large orchestra configuration

DELUGAN MEISSL ASSOCIATED ARCHITECTS Vienna
DELUGAN MEISSL联合建筑事务所 维也纳

"The differentiated but interconnected spatial sequences of public spaces, foyers, and theater halls turn the Darat King Abdullah II into a lively platform for conversations, performances, and societal action ..."

"互不相同而又彼此连接的公共空间、休息室和剧院大厅等一系列的空间建筑使Darat King Abdullah II成为人们谈话、表演和进行社交活动的场所……"

作者 Elke Delugan-Meissl, Roman Delugan   合作伙伴 Martin Josst, Sebastian Brunke, Jörg Rasmussen, Oana Maria Nituica, Claudiu Barsan-Pipu, Marina Kolloch, Thomas Theilig, Xiaozhen Zhu, Peter Pichler, Jan Saggau   专家 structural analysis: Werkraum Wien; open space planning: Rajek Borosch, Landschaftsarchitektur, Vienna; concert, theatre, and acoustics: Müller BBM Akustik, Munich; building services and air conditioning: Scholzegruppe, Vienna

DARAT KING ABDULLAH II

DELUGAN MEISSL ASSOCIATED ARCHITECTS

PUBLIC STAGE

RESTAREA

CONNECTING

BUFFER ZONE TO TRAFFIC

EXISTING FOREST + NEW FOREST

TERRACES

PUBLIC PLAZA

SITE PLAN 1:500

FOREST   GREEN BUFFER   PUBLIC STAGE   TERRACES

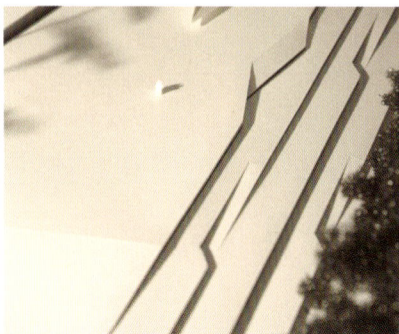

DESIGN OF OPEN-AIR SPACES:

The new landscape of the Darat King Abdullah II is characterized by two basic movements that develop from the existing geological layering and the morphology of the city. The trees are embedded into these horizontal and vertical movements that connect the city districts or are generously compensated by new plantings of trees. The entire landscape is made accessible to the public.

Horizontal layering
The first movement follows the sloping line of the hill chains and picks up the principle of the geological layering in the form of diverse terraces. The urban terrace landscape permeates the concert hall, staggers across the "bridge" and the "city stage" that is on the street level, in order to continue across several levels in the park of the GAM Stripe. The different levels are connected across ramps and stairs and create additional access and connection points in the surrounding cityscape.

With the aid of the park terraces, the level is raised in the direction of the bridge so that the height of the stairs can be reduced. Similar to the geological layering of the sandstone, the walls of the terraces consist of layers of differently colored concrete. The seating areas on these plateaus are in the shade of pergolas and provide small areas of respite. The upper landscape layer is formed by the expanded forest. As far as possible, the existing trees in the north-east are kept and the required reductions on the south-east slope are generously compensated. The entire area will be accessible to the residents.

City and landscape axis
The second movement is developed as an axis of green and open space from the GAM Stripe, flows into the street area of the Princess Basma Road as a tree structure, and continues to follow the traffic axis through the city. The new design of the city park here integrates the existing trees, which becomes a shaping element in that the terraces are adjusted to the trees in terms of levels and locations. The terrace system enables a new sequence of important urban open space functions and becomes a connecting space in the urban landscape. Starting at the Hussein Cultural Center, the park is organized into a sport, play, quiet and connection zone in order to open up as an urban location in the intersection area of the Ali Bin Abi Talib, Princess Basma and the new service road towards the concert hall. The urban "stage", easily visible from the concert hall and the pedestrian bridge, is enveloped by raised botanical terraces. This green buffer towards the Oma Mater Road is interrupted by broad aisles and thus provides interesting views of the square and the concert hall. The square is structured and enlivened with a play of water.

The horizontal and vertical movement melt into a new city and park landscape that connects the city areas, leads to the concert hall and allows the creation of new, adequate open-air stages.

03

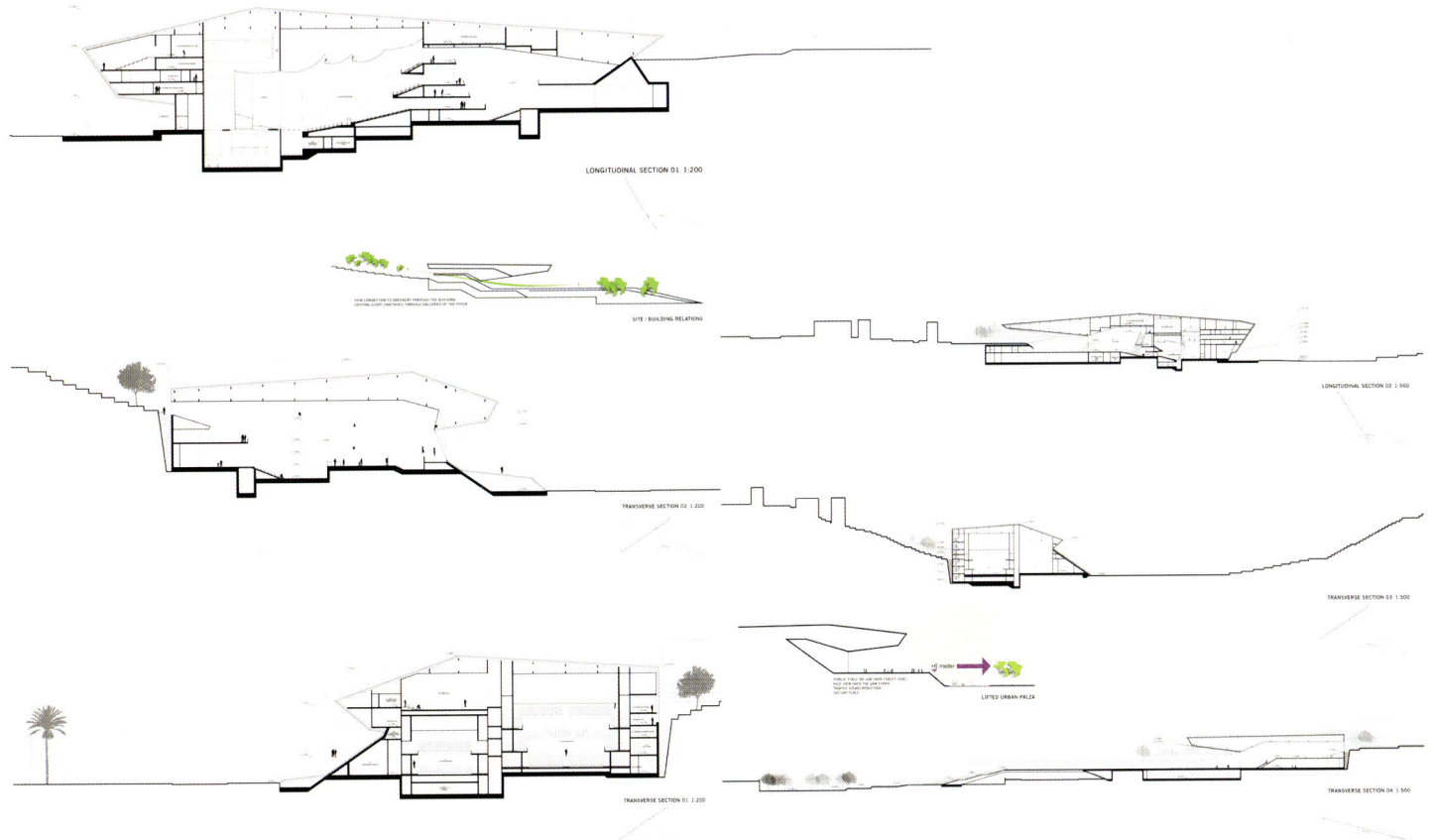

LONGITUDINAL SECTION 01 1:200

SITE : BUILDING RELATIONS

LONGITUDINAL SECTION 02 1:500

TRANSVERSE SECTION 02 1:200

TRANSVERSE SECTION 03 1:500

TRANSVERSE SECTION 01 1:200

LIFTED URBAN PALZA

TRANSVERSE SECTION 04 1:500

FUNCTIONAL OVERVIEW

REHEARSAL, AND PRACTICE ROOMS
EDUCATION AND SEMINARS
ADMINISTRATION
CHANGING AND DRESSING ROOMS
GENERAL BACKSTAGE / LOADING DOCK

CATERING AND RETAIL
CONCERT THEATER
SMALL THEATER
REHEARSAL ROOM
PUBLIC ACCESS
VIP ACCESS / HOSPITALITY

FLOORPLAN LEVEL +01_ENTRANCE LEVEL_1:200

LEVEL 03

LEVEL 04

LEVEL 05

LEVEL 06

LEVEL -01

LEVEL 00

LEVEL 01

LEVEL 02

PUBLIC AREA

EDUCATIONAL PROGRAMS

CONCERT THEATRE

SMALL THEATRE

REHEARSAL

ADMINISTRATION AND BACKSTAGE

TECHNICAL PLANT ROOMS

PARKING SPACES

RESTROOMS

CIRCULATION AREAS

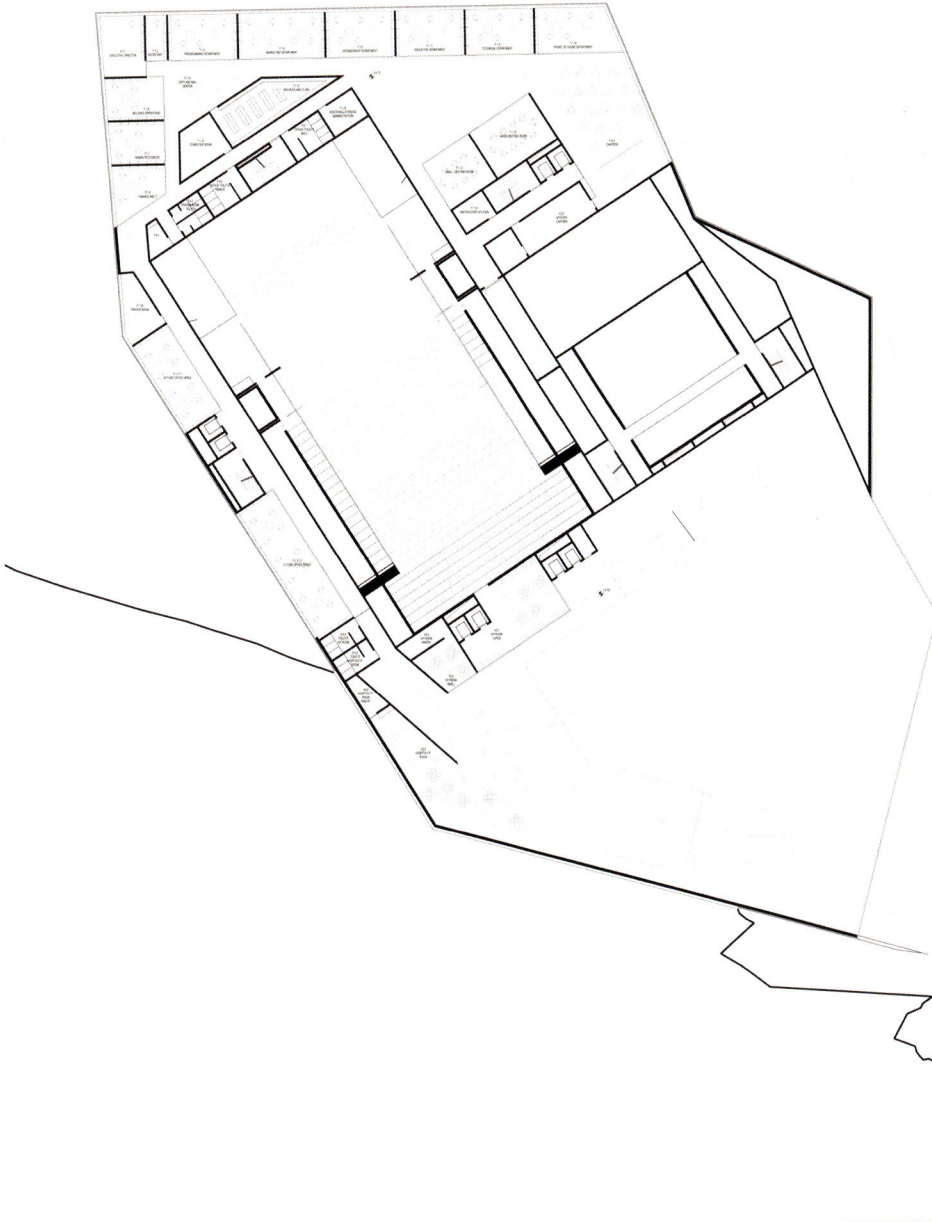

FLOORPLAN LEVEL +03_VIPLEVEL_ADMINISTRATION_1:200

10

PATTERN_01: FLOOR

PATTERN_01: FLOOR

PATTERN_02: WALL

PATTERN_03: ROOF

PATTERN_03: ROOF

PATTERN_02.1: WINDOW

PATTERN_03: SOLAR INTEGRATION ONTHE ROOF

PATTERN_02.1: WINDOW

PATTERNDETAILS

RAYTRACING CONCERT THEATRE_CONCERT MODE

RAYTRACING CONCERT THEATRE_OPERA MODE

RAYTRACING SMALL THEATRE

DISPLACEMENT FLOW FOR THE HALLS

THE MAIN PART OF THE SUPPLY AIR IS LED BENEATH THE FLOOR LEVELS OF THE HALLS AND WILL BE DISTRIBUTED VIA A DOUBLE FLOOR AND FINALLY VIA STEP - OR CHAIR - OUTLETS DIRECTLY TO THE PERSONS, A METHOD WELL ACCEPTED IN MANY OPERA HOUSES. FIG.2 SHOWS THE PRINCIPAL AIR FLOW WITH DISPLACEMENT FLOW PATTERN THAT HAS BEEN PROVED AS OPTIMAL. ON STAGE WE PROPOSE SWIRL NOZZLES IN SIDE WALLS.

FIG.3 IS AN ENLARGEMENT OF FIG.2 AND SHOWS THE DETAIL OF AIR ENTRANCE IN THE LOWER FLOOR PART, THE DISTRIBUTION IN THE WHOLE FLOOR AND THE BLOWING OUT VIA THE STEP-OUTLETS.

VENTILATION CONCEPT THEATRES

3D MODEL OF THE STRUCTURE DESCRIPTION

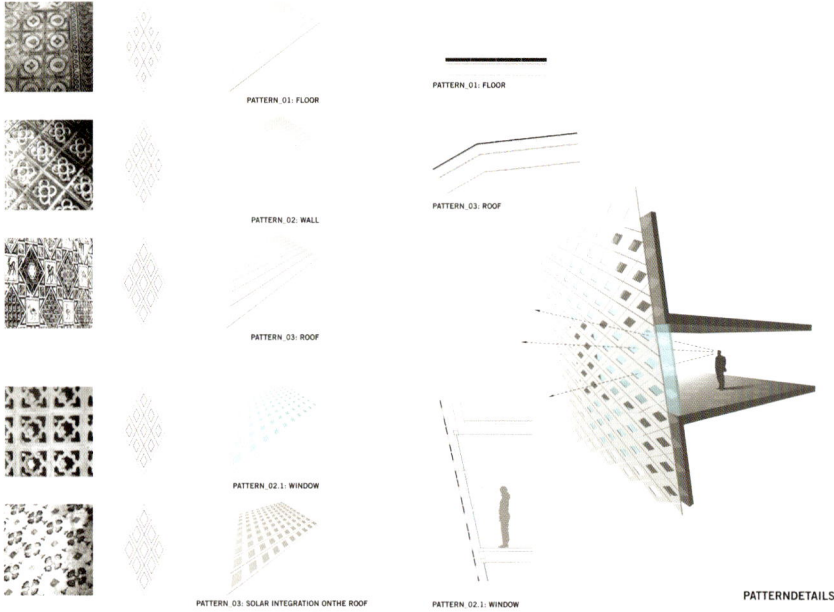

THE DESIGNED CONCERT HALL IS A BUILDING CONSTRUCTED IN STEEL AND CONCRETE, WITH THE LOCATION AMMAN / JORDANIA

DUE TO THE HIGH SPANS AND CANTILEVERS AND THE GEOMETRICAL REQUIREMENTS, THE ROOFCONSTRUCTION IS DESIGNED AS A ORTHOGONAL STEEL TRUSS, UP TO 10M HIGH, WHICH COVERS THE WHOLE BUILDING. THE TRUSS ITSELF STANDS IN ONE PART ON THE WALLS OF THE CONCRETE HALLS, IN THE OTHER PART ON COLUMNS AND CONCRETE WALLS ALONG THE ENTRANCE. FOR THE TWO COCERT HALLS AND THE SURROUNDED STOREYS, WE PROPOSE A CONCRETE CONSTRUCTION. THIS PART WILL WORK AS MASSIVE CORE AND SERVE FOR THE STIFFNESS OF THE BUILDING IN HORIZONTAL DIRECTION IN THE CASE OF WIND OR EARTHQUAKE IMPACT.

THE FOUNDATION OF THE BUILDING IS CONSIDERED TO BE A PILE FOUNDATION. PILES WILL BASICALLY BE NECASSARY BELOW ALL COLUMNS AND WALLS.

PRIMARY TRUSSWORK OF THE ROOF STRUCTURE (MARKED IN RED). THE ROOF CLADDING IS FIXED TO THIS TRUSS.

SECUNDARY TRUSSWORK OF THE ROOF STRUCTURE (MARKED IN RED) ALONG THE BORDER OF THE CANTILEVER AND THE ENTRANCE FACADE.

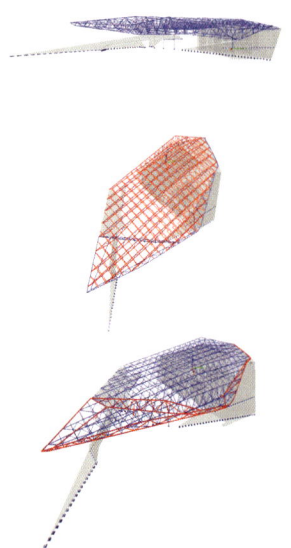

THE SUPPORTING STRUCTURE IN THE AREA OF THE ENTRANCE IS PROPOSED IN CONCRETE.

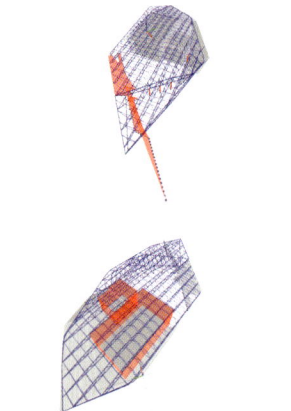

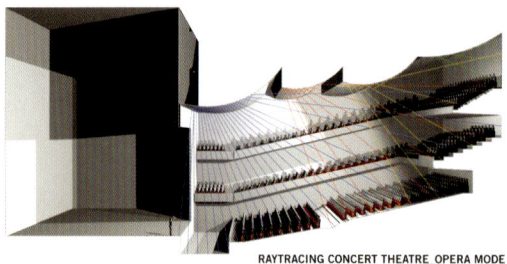

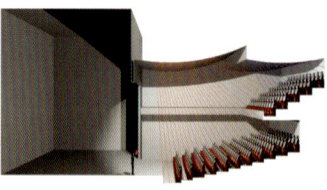

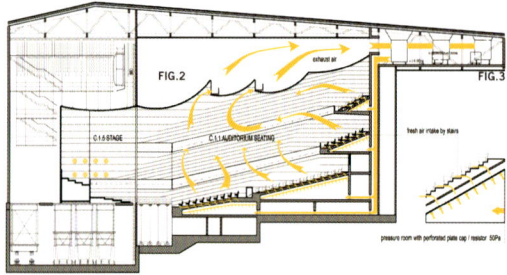

THE TWO CONCERT HALLS AND...

... AND THE SOURROUNDED WALLS ARE PROPOSED IN CONCRETE TOO. THEY SERVE THE STRUCTURE FOR THE STIFNESS AGAINST WIND LOADS AND EARTHQUAKE.

DISCUSSION OF THE RESULTS

DEFLECTION OF THE STRUCTURE UNDER DEAD LOADS. THE CANTILIVER HAVE A SPAN OF 40M RESPECTIVELY 75M.

DEFLECTION OF THE STRUCTURE UNDER WIND LOADS (ASSUMED WIND SPEED: 135KM/H)

BOTH RESULTS SHOW THE EXCELLENT LOAD BEARING BEHAVIOUR OF THE STRUCTURE, IN PARTICULAR THE ROOF STRUCTURE.

CAPACITY OF STEEL STRUCTURE. APART FROM SOME PEAKS THE STEEL STRUCTURE IS FAR FROM BEING AT ITS LIMITS; THERE IS STILL POTENTIAL FOR OPTIMIZATION.

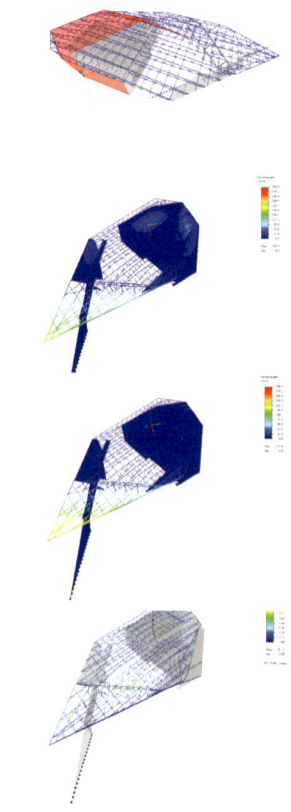

STRUCTURAL DESIGN CONCEPT

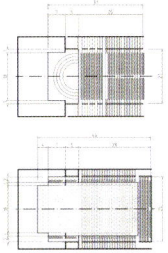

CONCERT THEATRE_CONCERT MODE

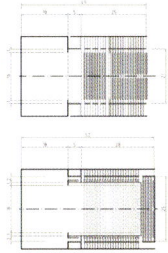

CONCERT THEATRE_OPERA MODE

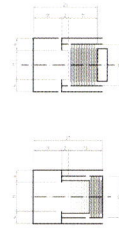

SMALL THEATRE

Snøhetta Oslo
Snohetta建筑事务所 奥斯陆

"A prestigious, contemporary, and stimulating complex for the performing arts, truly integrated into the landscape and urban context."

"一个有声望的、现代的和令人兴奋的表演艺术建筑群真正与周围景观和城市的文脉融合在一起。"

作者 Robert Greenwood, Oslo   合作伙伴 Kjerstin Bjerka, Peter Dang, Peter Girgis, Tine Hegli, Andreas Nypan, Julian Prizes, Erik Vitanza   专家 civil structural services and specialist engineers: Buro Happold Ltd., Glasgow; acoustical consultant: Arup Acoustics, Winchester, Hampshire

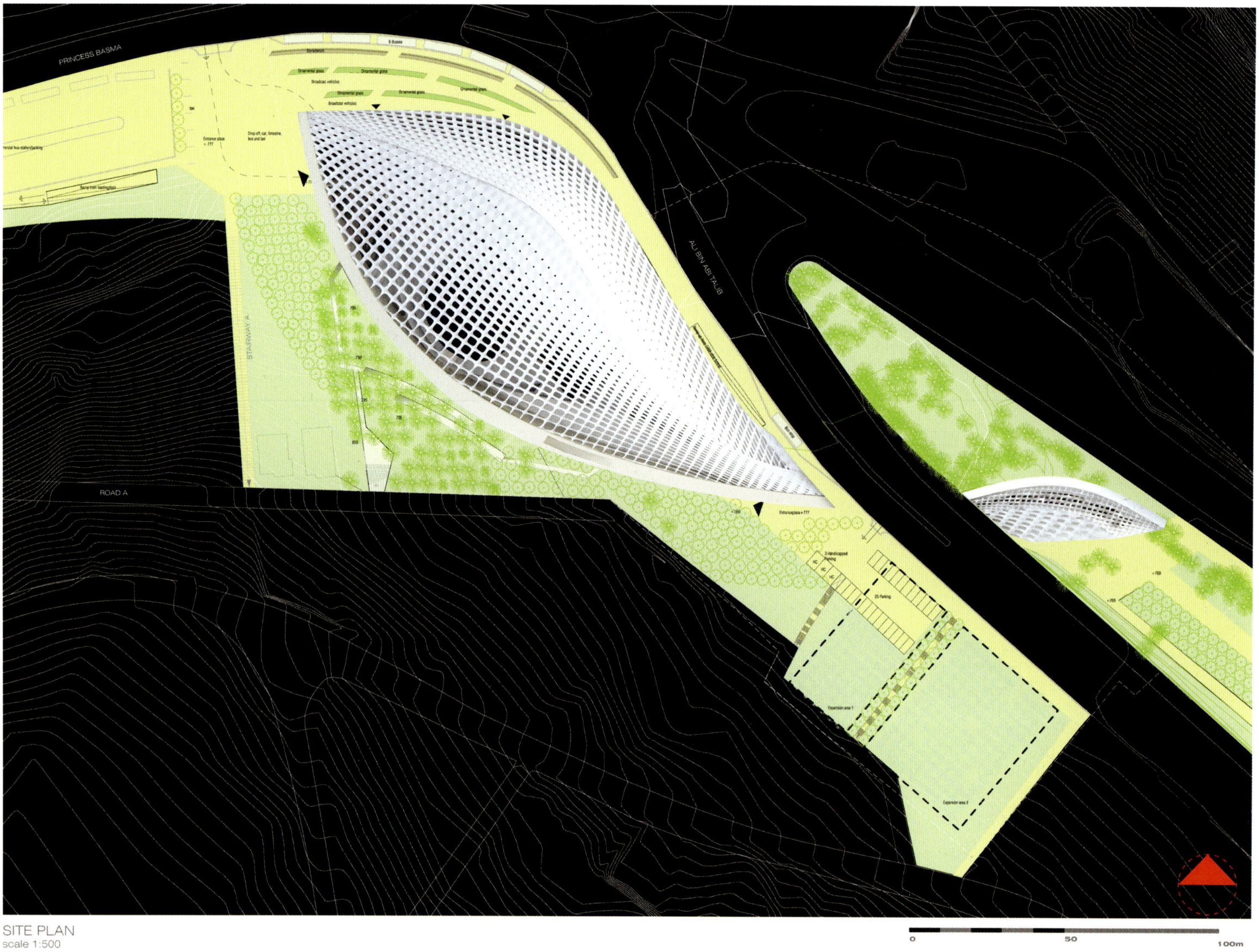
PRINCESS BASMA
ALI BIN ABI TALIB
STAIRWAY A
ROAD A
0
50
100m

SITE PLAN
scale 1:500

Sandstone for the urban plazas and the arch wall

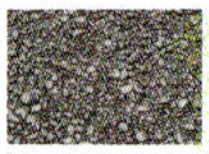

Black gravel in the surrounding park under the trees

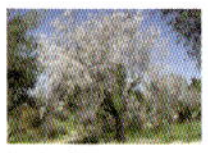

Amygdalis Communis - trees surrounding the centre, white flowers in springtime

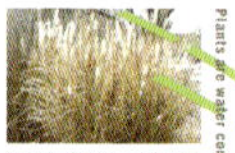

Pennisetum setaceum - Ornamental grass on Entrance Plaza

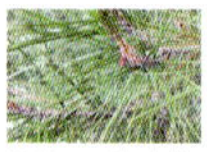

Pinus Helepensis - Visual backdrop for the centre and expansion sites 1 and 2

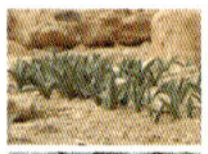

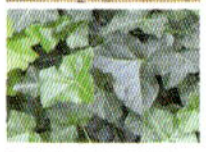

Succulents, gravel an and Hedera Helix covering the surface in the park

Plants are water conserving and will handle reused grey water for irrigation

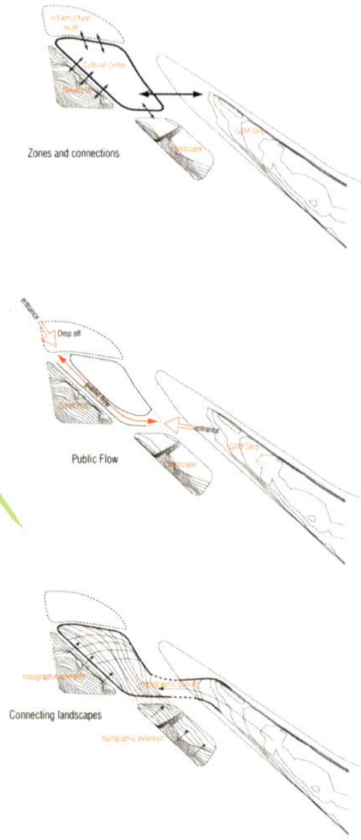
Zones and connections
Public Flow
Connecting landscapes

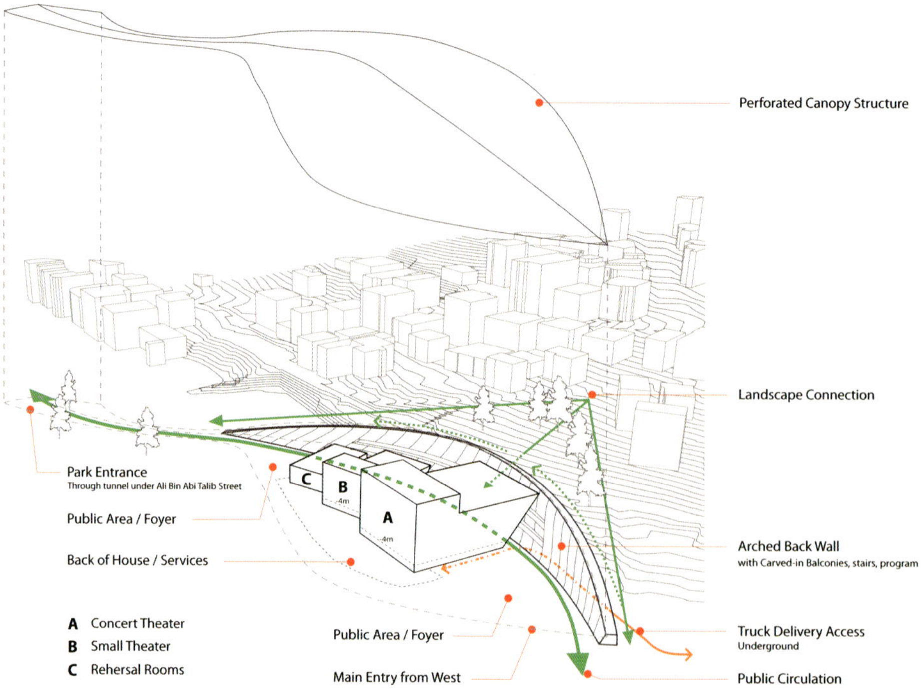
Perforated Canopy Structure
Landscape Connection
Park Entrance
Through tunnel under Ali Bin Abi Talib Street
Public Area / Foyer
Back of House / Services
A Concert Theater
B Small Theater
C Rehearsal Rooms
Public Area / Foyer
Main Entry from West
Arched Back Wall
with Carved-in Balconies, stairs, program
Truck Delivery Access
Underground
Public Circulation

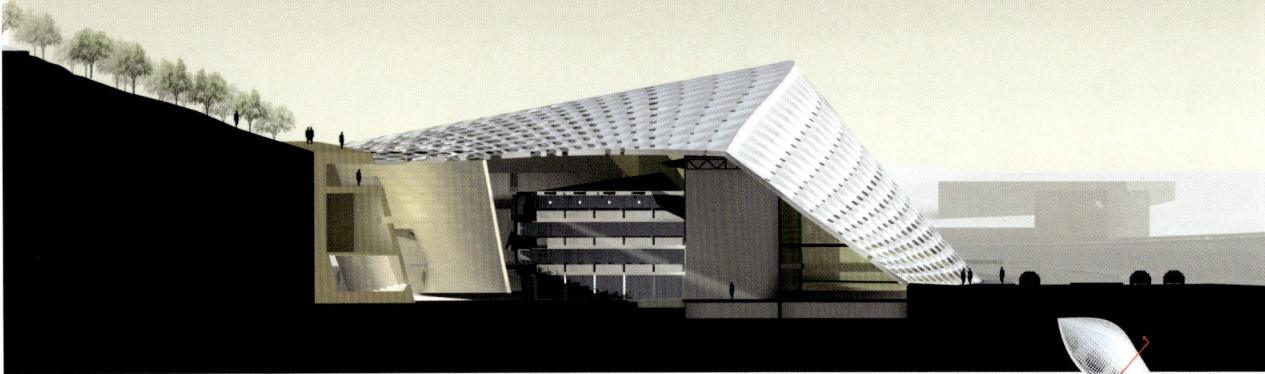
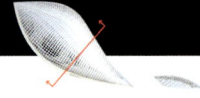

CTION THROUGH SMALL THEATER
e 1:200

NG SECTION THROUGH SMALL THEATER & CONCERT THEATER
e 1:200

NG SECTION - ARCHED WALL
e 1:200

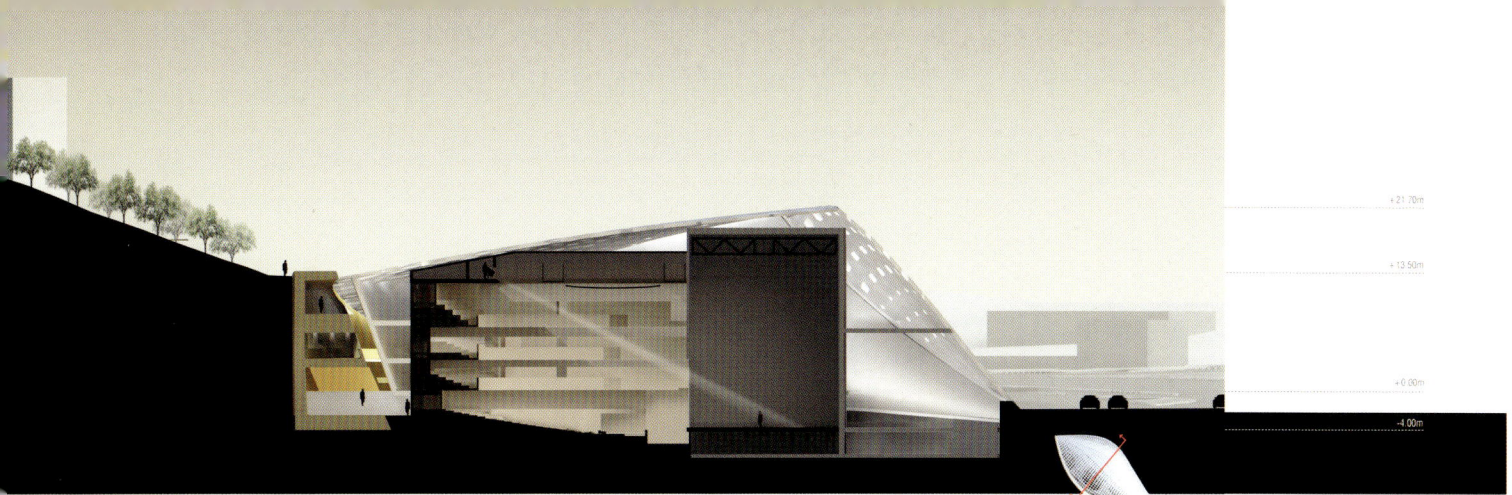

ON THROUGH CONCERT THEATER
'00

+21.70m

+13.50m

+0.00m

-4.00m

CONCEPT COLLAGE REPRESENTING CULTURE AND TEXT ORE INSPIRATION REFERENCING NATURAL LIKE
ADVANTAGES, EMPHASISING SPATIAL DIFFERENCES OF THE OPEN AND MAJESTIC, AS WELL AS INTIMATE
MEETING PLACES FOR THE PEOPLE OF AMMAN.

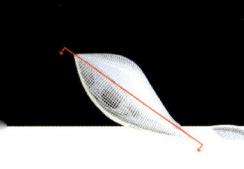

+19.00 m

+0.00 m
-3.00 m

+13.50m

+0.00m

Atelier Christian de Portzamparc Paris
Atelier Christian de Porzamparc建筑事务所 巴黎

"The main component is the desire for a lively building that can become a binding factor between the two halves of the city."

"设计的主要组成部分是想设计出一个充满活力的建筑物，把城市的两个部分加以结合。"

作者 Christian de Portzamparc   合作伙伴 Bertrand Beans, M-E Nicoleau, Andre Terzibachian, Duccio Gardelli, Burkhart Schiller, Bettina Reali, Hyun-Jung Song, Isabella Burck, Paul Chaulet, Fabiana Aravjo, Ricardo Marotta, Veronica Fiorini, Luisa Fonsela
专家 structural analysis: SIDF, Marseilles; acoustics: XU Acoustics, Paris; scenograph: Ducks Slend, Les Pleiades

NORTH-EAST ELEVATION

OPEN AIR-THEATER BY NIGHT

VIEW FROM THE NORTH BY NIGHT

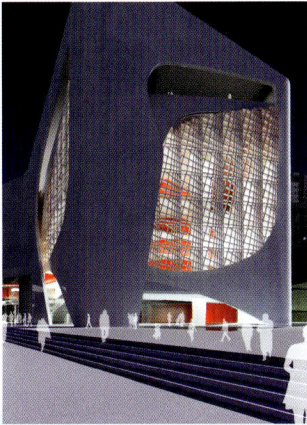

VIEW FROM THE NORTH

THE SPICE TERRACE AND THE RESTAURANT

THE PARFUM GARDEN

VIEW FROM THE WEST

VIEW FROM THE CONNECTING BRIDGE

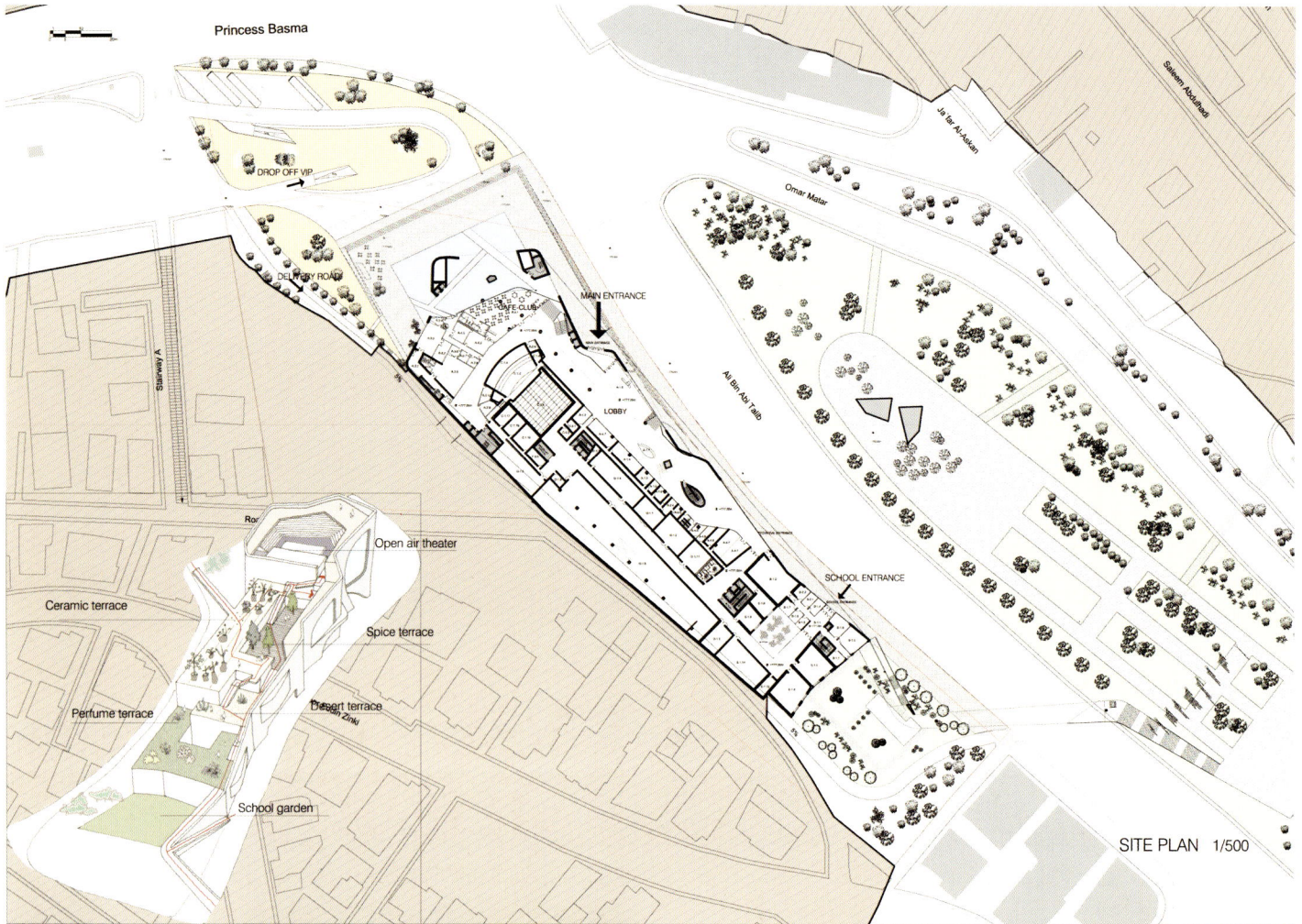

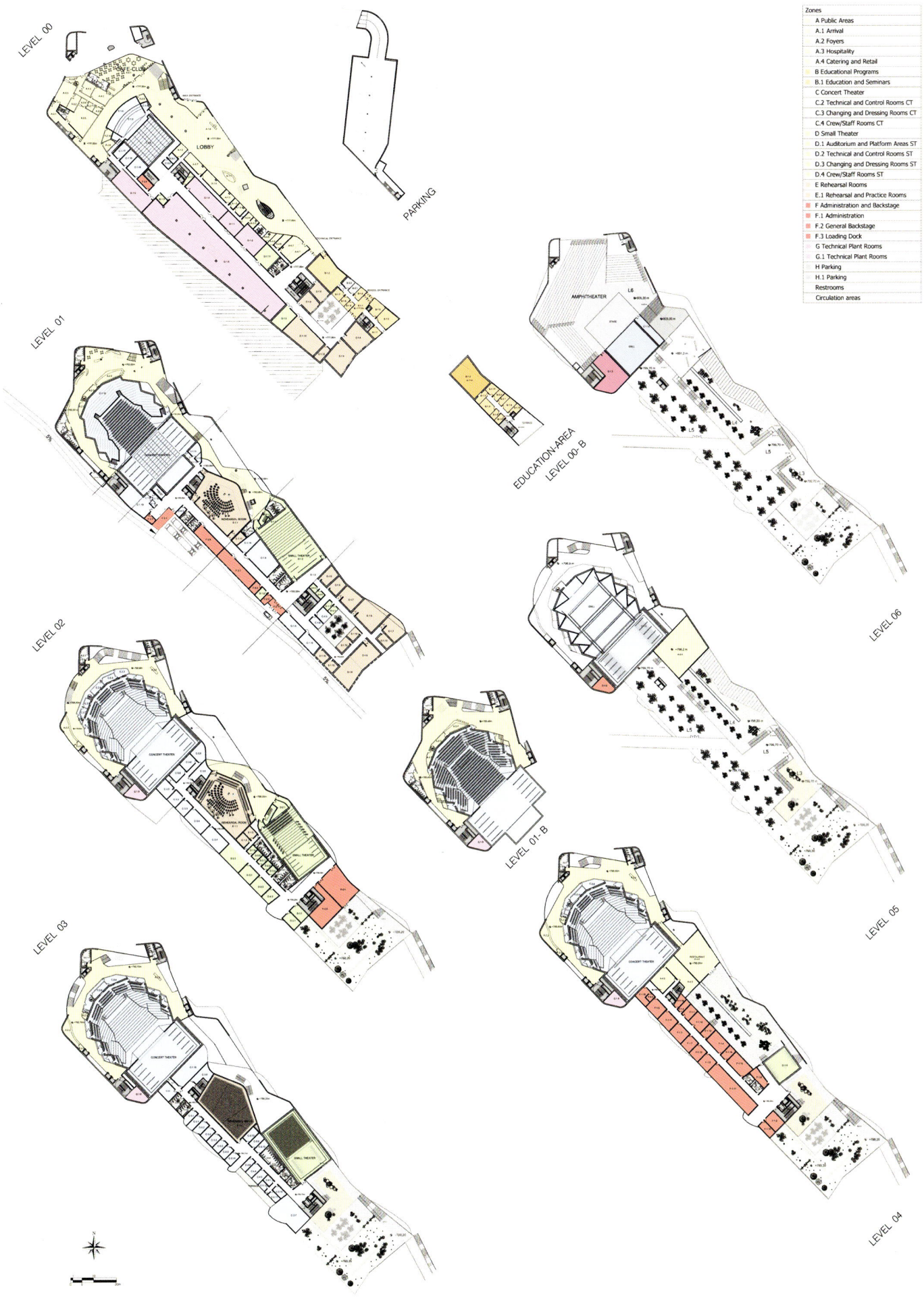

517

Zones
A Public Areas
A.1 Arrival
A.2 Foyers
A.3 Hospitality
A.4 Catering and Retail
B Educational Programs
B.1 Education and Seminars
C Concert Theater
C.2 Technical and Control Rooms CT
C.3 Changing and Dressing Rooms CT
C.4 Crew/Staff Rooms CT
D Small Theater
D.1 Auditorium and Platform Areas ST
D.2 Technical and Control Rooms ST
D.3 Changing and Dressing Rooms ST
D.4 Crew/Staff Rooms ST
E Rehearsal Rooms
E.1 Rehearsal and Practice Rooms
F Administration and Backstage
F.1 Administration
F.2 General Backstage
F.3 Loading Dock
G Technical Plant Rooms
G.1 Technical Plant Rooms
H Parking
H.1 Parking
Restrooms
Circulation areas

LEVEL 00

PARKING

EDUCATION-AREA
LEVEL 00- B

LEVEL 01

LEVEL 01- B

LEVEL 02

LEVEL 03

LEVEL 04

LEVEL 05

LEVEL 06

ACOUSTIC STUDY OF THE CONCERT THEATRE

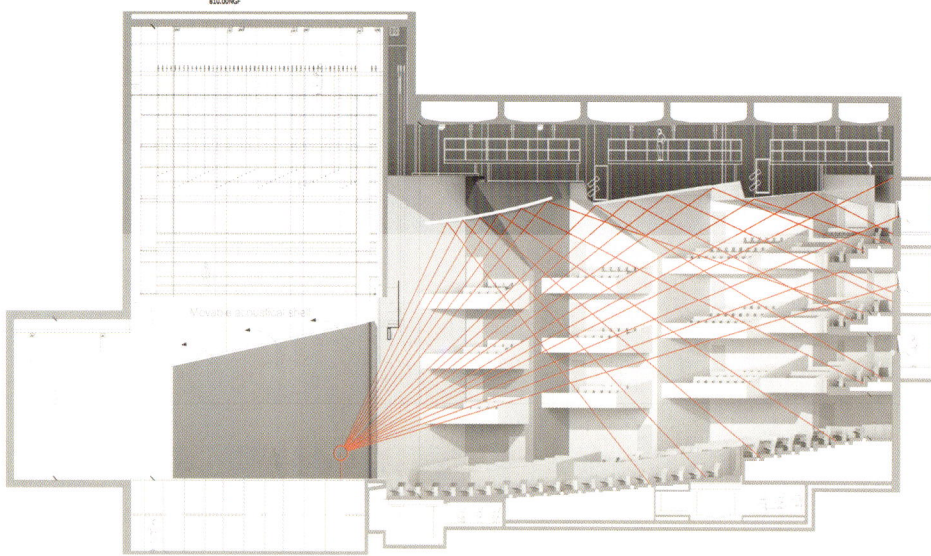

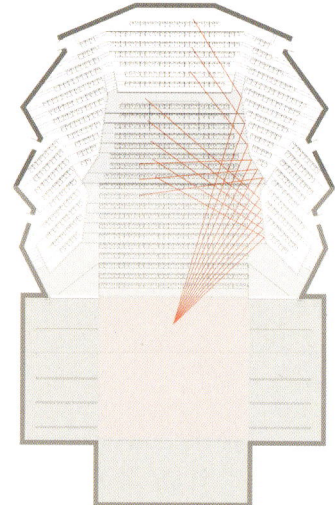

The wall structures of the "baskets" zone allows close proximity for ideal sound wave reflexions; the form that we propose assures excellent acoustic results.

VARIABILITY OF THE STAGE - CONCERT THEATRE

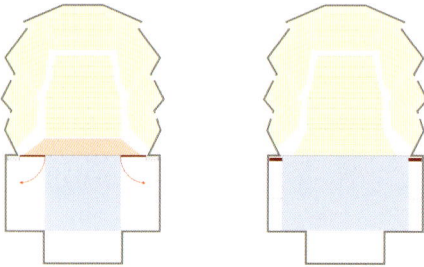

ACOUSTIC SLIDING PANNEL ALONG THE BALCONY TOWER

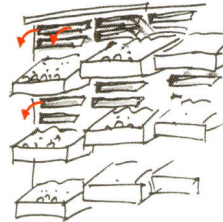

SLIDING PANNEL FOR WIDE FREQUENCY BAND ABSORPTION ADJUSTMENT

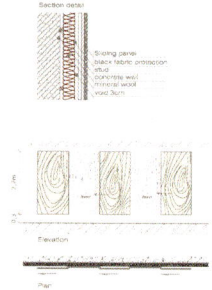

VARIABILITY OF THE STAGE - SMALL THEATRE

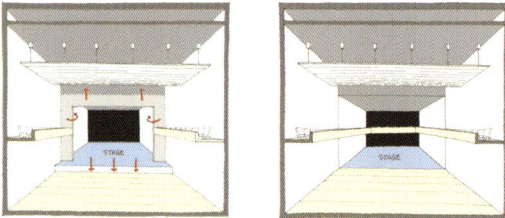

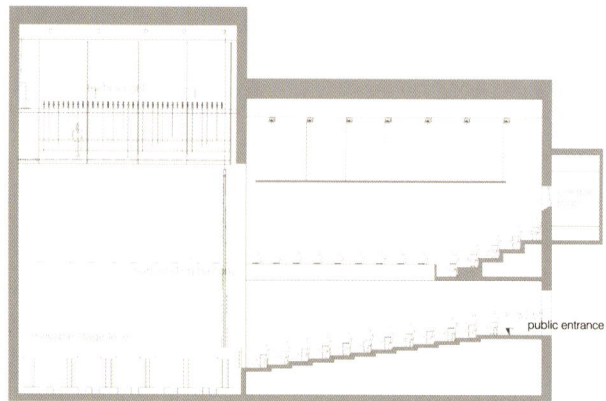

THE SMALL THEATRE

public entrance

STRUCTURAL CONCEPT

A simplicity of construction with vertical concrete walls
3 expansion joints at the separation between the large halls
Each blocks has its own structure
The concrete facade wall with large opening

One typical curved angle
in a gentle curve

Only one pannel type
of prefabricated concrete
large scale lattices
(2m x 2m)

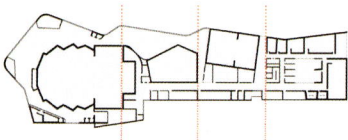

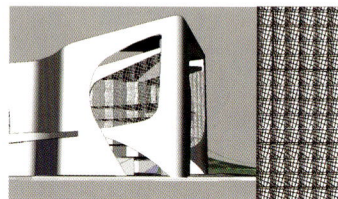

ECOLOGICAL CONCEPT

The garden on the roof
protects the building from
the heat and sun's rays

Photovoltaic pannels

Cooling air due to
the chimney effect

THE ENTRANCE - LOBBY

THE THEATER WITH TWO CONFIGURATIONS:

SUROUNDING AUDIENCE- CENTRAL STAGE

FRONT STAGE THEATER

THE LOBBY OF THE THEATER

THE CONCERT HALL

THE CONCERT HALL

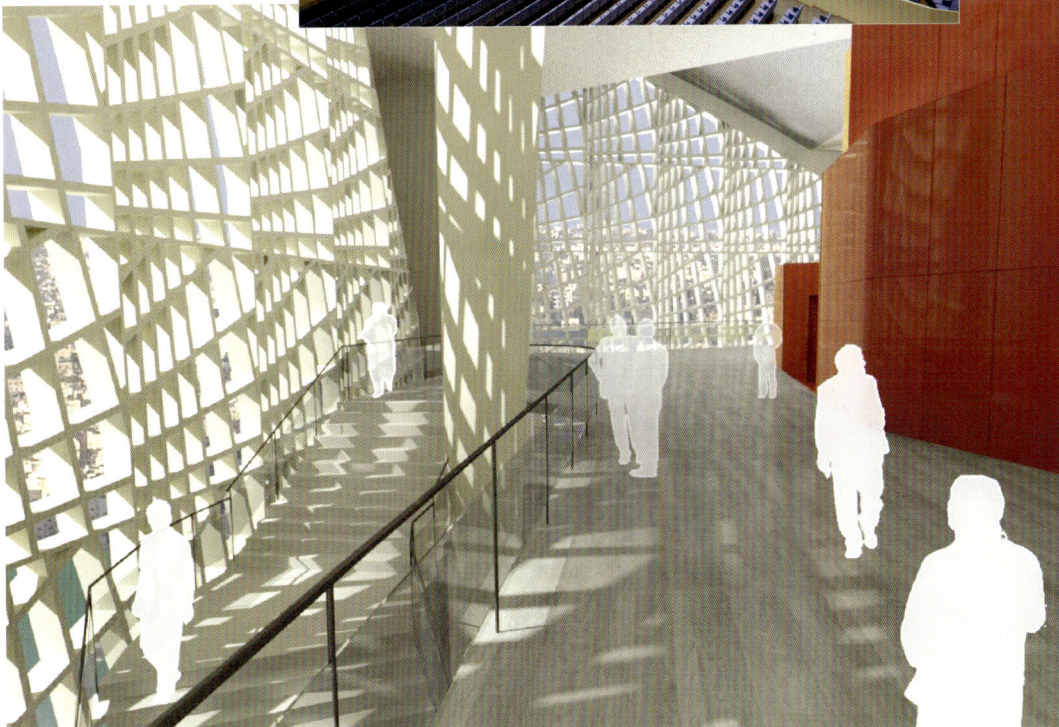

THE LOBBY OF THE CONCERT HALL

Atelier Christian de Portzamparc, Paris

Darat King Abdullah II, Amman

Kerry Hill Architects Singapore
Kerry Hill建筑事务所 新加坡

"The building is to be a stage for the performance, creation, and education of performing arts in Jordan – a building which welcomes people of all ages and backgrounds, that respects a rich architectural past, and aspires to innovation for the future."

"建筑物将成为约旦表演艺术的表演、创造和教育舞台——欢迎不同年龄和背景的人，尊重建筑丰富的历史，渴望未来的创新。"

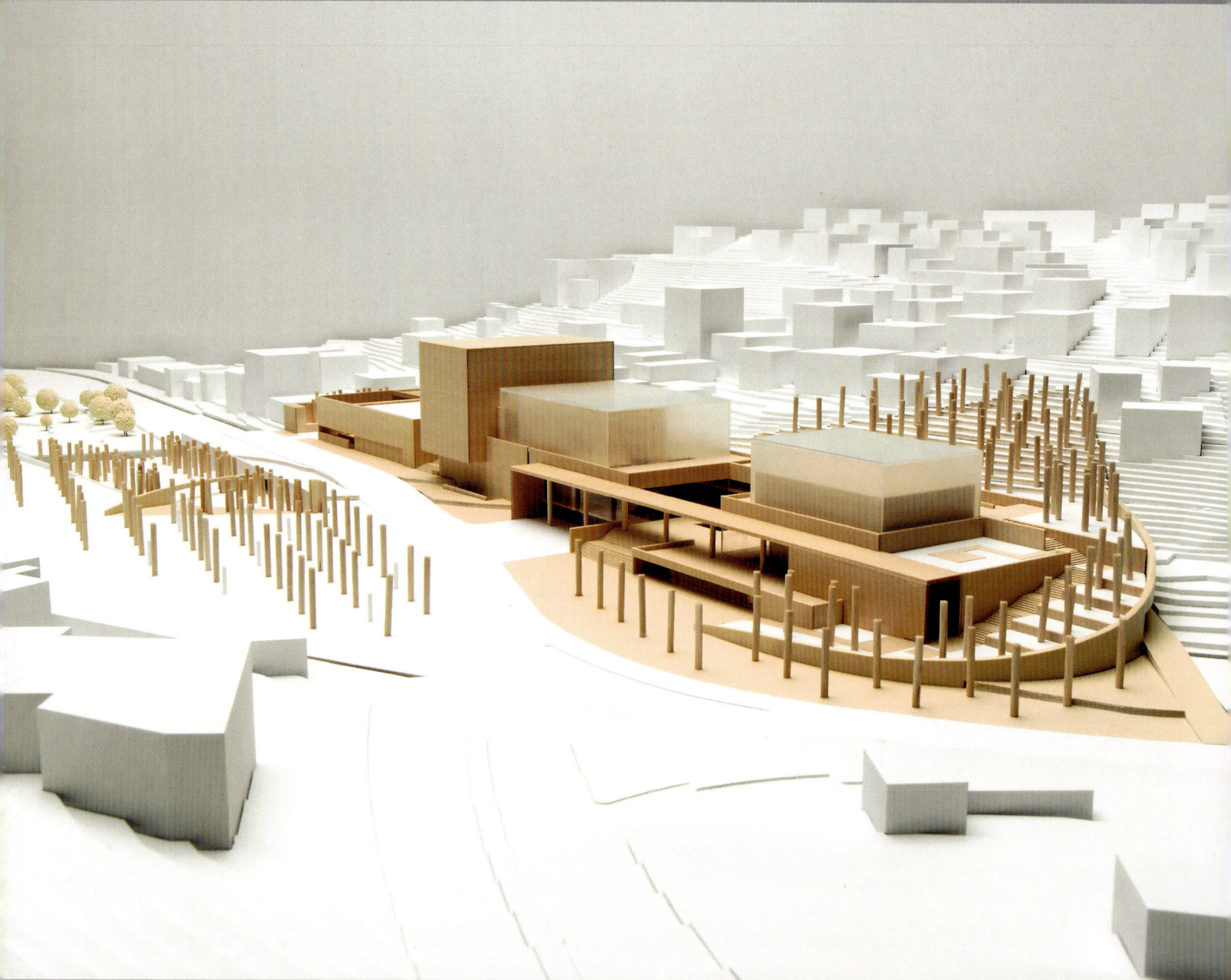

作者 Author A: Kerry Hill; Author B: Justin Hill   合作伙伴 Patrick Kosky, Cheng Ling Tan, Ken Lim, David Gowty, Bernard Lee, Chee Hong Lim, Duncan Payne, Alessandro Perinelli, Lidya Koes   专家 acoustics: Marshall Day Acoustics, Melbourne; landscape architecture: Tierra Design, Netherlands; lighting: The Flaming Beacon, Melbourne; ESD consultant: Arup, Singapore

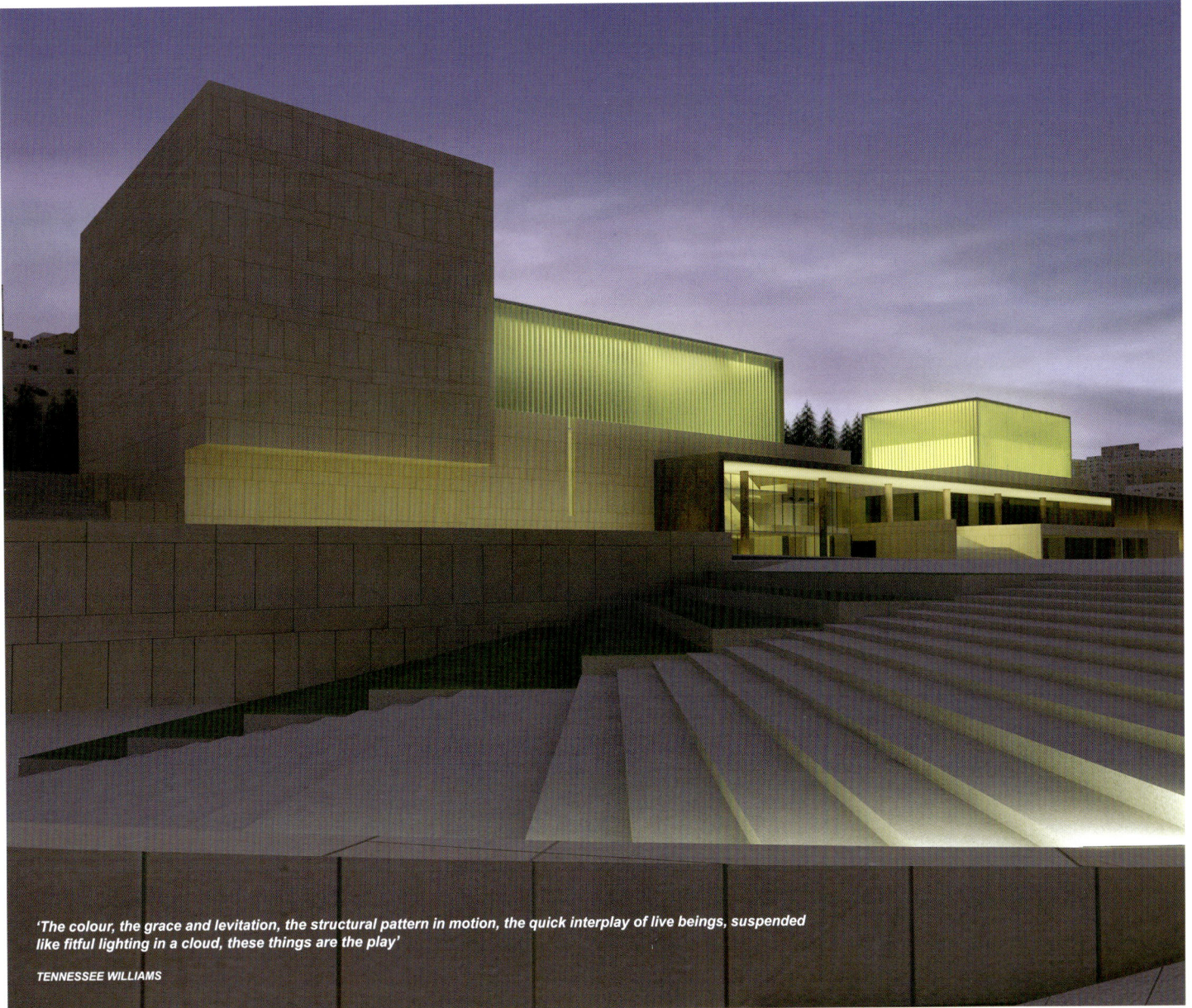

'The colour, the grace and levitation, the structural pattern in motion, the quick interplay of live beings, suspended like fitful lighting in a cloud, these things are the play'

TENNESSEE WILLIAMS

An important civic space generously accomodates the arrival and departure of large audiences

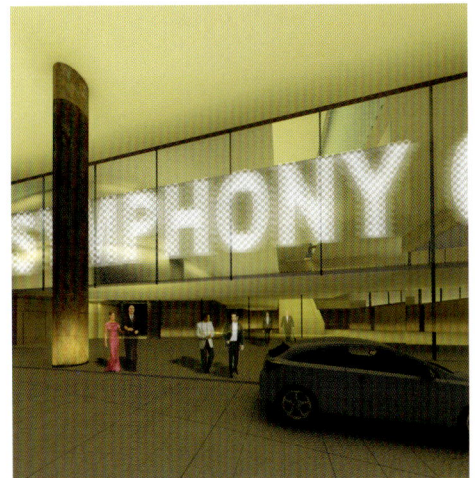

ILLUMINATED DIGITIZED SIGNAGE

Our vision is to proclaim "pride in the present without overwhelming the quiet history of the past". An architecture that clearly identifies itself through place, purpose and material. One that is contemporary, yet filtered through a sieve of traditional values.

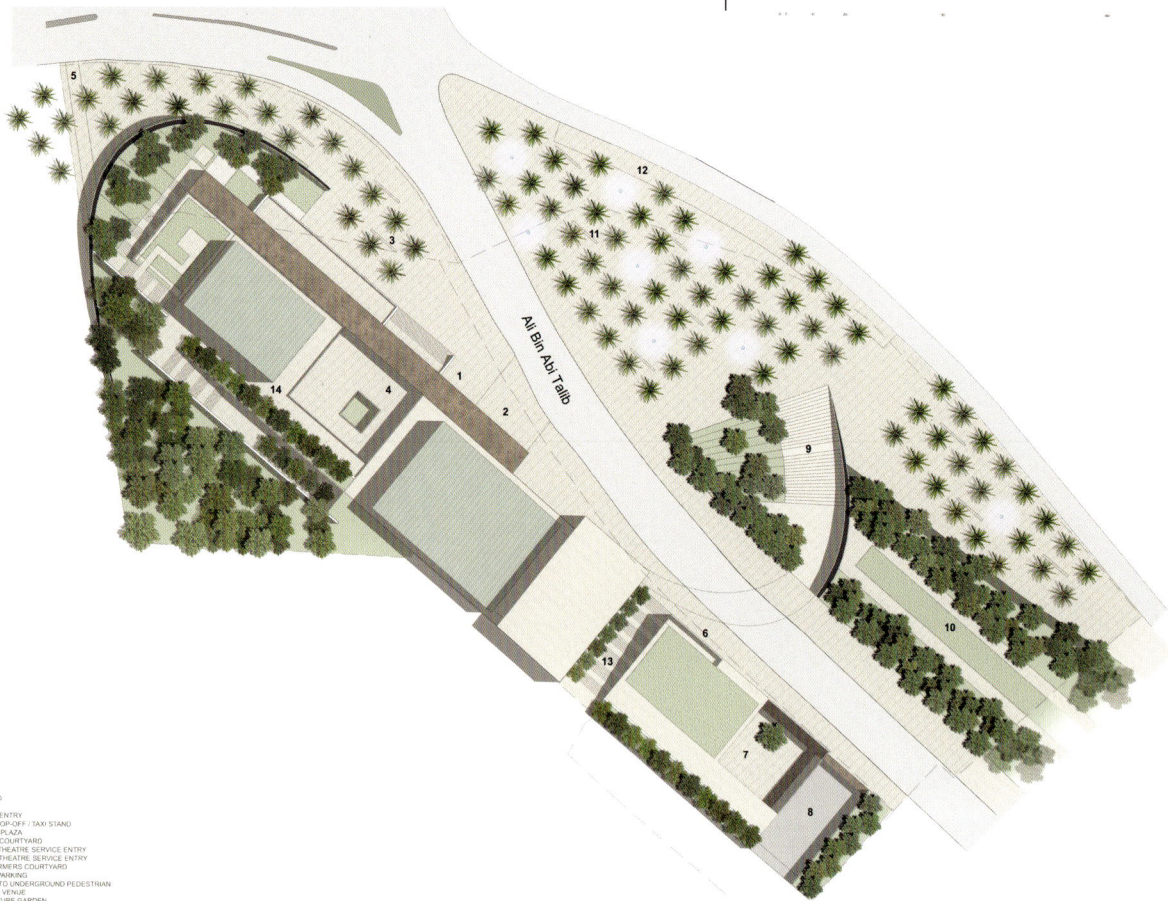

The stone podium merges with the landscape as a system of paths and stairways that evokes an urban experience distinct to Amman

APPROACH TO UNDERGROUND PEDESTRIAN LINK

STAIRS TO ROOF TERRACE

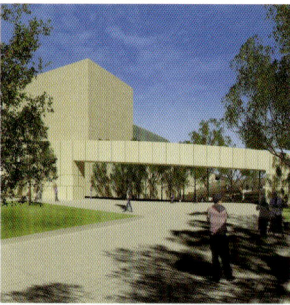

APPROACH FROM G.A.M PARK

LANDSCAPE ROOF

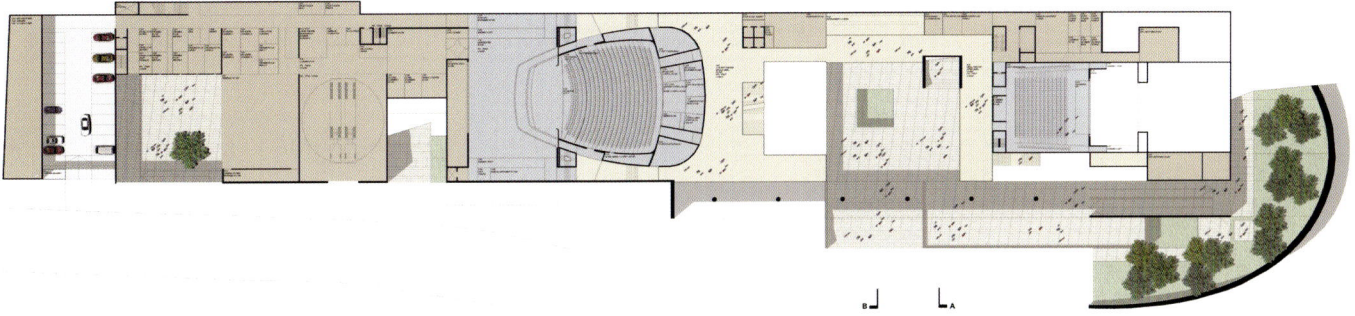

The education centre is given prime importance - located between the two theatres, in close proximity to the small theatre, it is entered past a cafe that spills onto the public plaza.

Ali Bin Abi Talib

1. Pre-cooling of makeup air provision via the concrete structure of the thermal labyrinth significantly reducing active mechanical cooling energy consumption.

2. Rainwater / grey water harvest and a/c condensate water collection for purification and subsequent soft usages (irrigation, cooling tower, etc...)

3. Exterior façade in stone clad pre-cast concrete panels provides thermal mass and insulation to inner theatre spaces; low velocity displacement air conditioning through plenum and 'swirl' diffusers.

4. Precast flooring system that allows cavities to be ventilated during the cooler night, purging the heat built up during the summer day.

5. Rooftop greenery to mediate micro-climate and provide insulation for building roofs.

6. A combination of photovoltaic panels and fritted glass to allow optimization of shading and natural illumination, while also generating electricity on site (building integrated photovoltaic panels) explore opportunities for a grid connect system to avoid large storage requirements. Circulation spaces designed to maximize daylight utilization and cross ventilation.

7. Hierarchy of air conditioned zones to maximize efficiency and minimize energy consumption.

NATURAL VENTILATION
MIXED MODE VENTILATION
FULLY AIR CONDITIONED

TYPICAL DETAIL THROUGH AUDITORIUM ROOF

PHOTOVOLTAIC PANEL

Intermission....

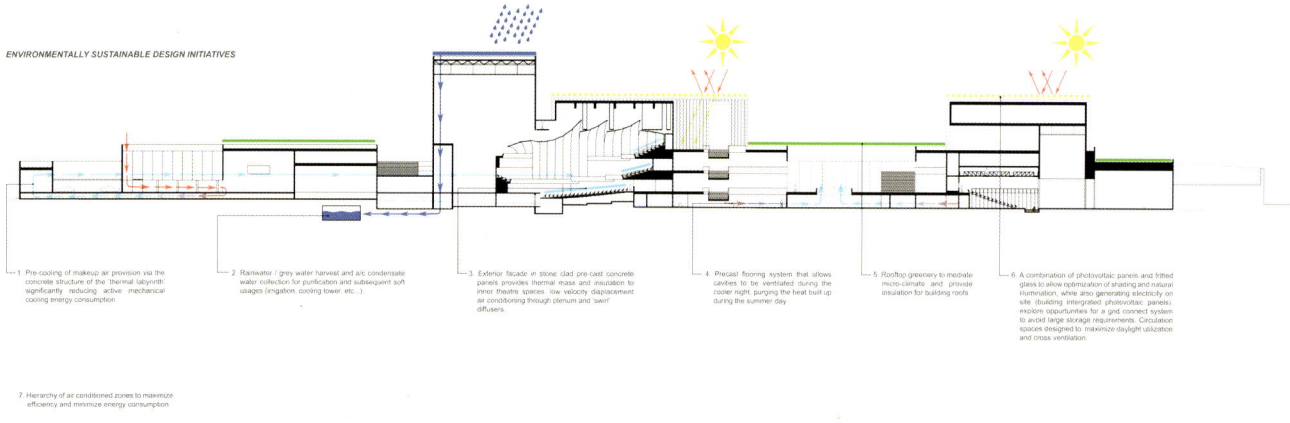

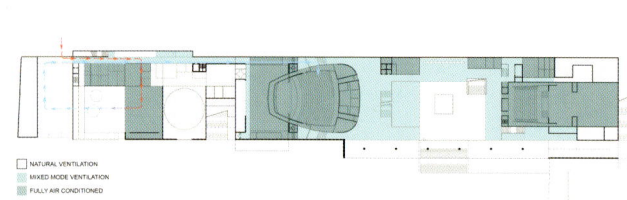

Theatre Acoustics

The Large Theatre

The functional brief gives highest priority to the symphonic use of the large theatre. The volumetric provision should reflect this. The design provides a gross symphonic volume of 19,250 m3 of which 2,624 m3 are on stage within the shell and 5,300 m3 are in the upper volume. That is a net volume per seat in the lower space of approximately 9 m³ per person.

Acoustical concept. In our realisation of the brief, the early reflected sound field beneath the visual ceiling provides the required clarity. Surfaces on the walls and soffits have been designed to ensure uniform high clarity of symphonic sound to all seats. Late sound is supplied by the upper volume outside these reflecting surfaces. Diffusion is guided throughout the space from a maximum intensity close the source to a minimum on remote surfaces.

The stage is to be equipped with an orchestral shell which provides an acoustical as well as a visual completion of the concert hall at the stage end. The shell is in two parts. An up-stage part is constructed as a single piece integral with its seating and top, and is able to be moved into storage as a unit at the rear of the stage. The sides and the remainder of the top are demountable elements flown in a conventional manner, or formed by wheeled towers. Stage shells of several configurations can therefore be matched to the size of the performing group.

The many reflecting surfaces produce lateral reflections required for the symphonic function. More precise localisation of the voice is required for Opera. This is achieved by a forestage reflector flown in front of the proscenium.

Variation of the reverberation time is achieved by deploying purpose designed acoustic barriers in the upper volume of the auditorium above the acoustically transparent ceiling.

The Small Theatre

Simplicity and adaptability are the twin themes of this intimate theatre.

The principal and perhaps most demanding function is drama, and care has been taken to ensure excellent speech clarity. The room has been designed to have a suitable reverberance to permit fullness in the sound. The volume of 7m³ per seat is rather higher than many conventional theatres. The high degree of clarity is assured by the provision of three early reflections to each seat. Even when in thrust stage mode, the same consideration of reflected sound distribution has been applied. The provision for reverberance will be particularly valued during the briefed chamber music/recital function. This will be an engaging room for performers.

The main reflecting surfaces are the side walls which bound the seating plane. These surfaces are supplemented by the overhead reflectors which are incorporated into the design under the catwalks, lighting bridges, and particularly the forestage reflector. The principal mode of varying the acoustics is by way of deployable absorption mounted on the upper side wall. This would be in use for all amplified functions and probably for Arabic and world music.

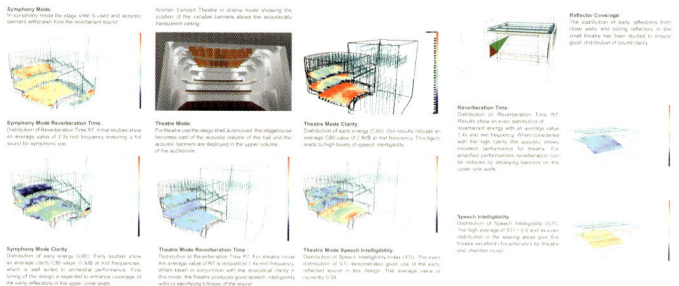

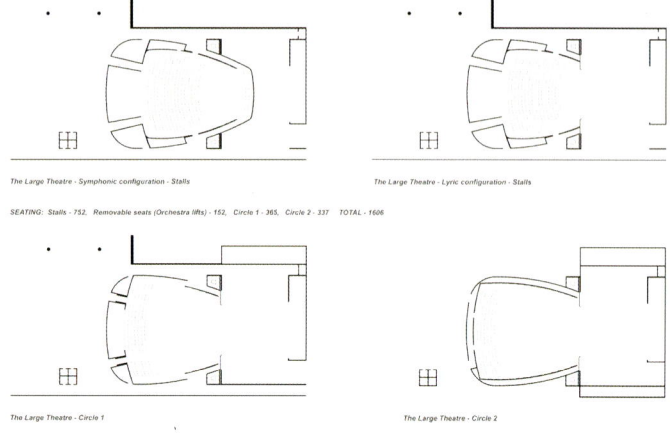

Symphony Mode.

Reflector Coverage.

Reverberation Time.

Speech Intelligibility.

The Large Theatre - Symphonic configuration - Stalls

The Large Theatre - Lyric configuration - Stalls

SEATING: Stalls - 752, Removable seats (Orchestra lifts) - 152, Circle 1 - 365, Circle 2 - 337 TOTAL - 1606

The Large Theatre - Circle 1

The Large Theatre - Circle 2

The pianist

The fabric of the theatre architecture and the sound it sustains melds seemlessly into one entity

A building that celebrates activity by day and by night. Allowing spectators to momentarily become spectacle

'From the start it has been theatres job to entertain people....it needs no other passport than fun,
BERTOLT BRECHT

Henning Larsens Tegnestue Copenhagen
Henning Larsens Tegnestue建筑事务所 哥本哈根

"The heart of the building is the magnificent amphitheatric foyer. Here all the qualities of the project are brought together in a lively, social meeting point."

"建筑物的灵魂是华丽高贵的圆剧场的休息厅。所有项目的品质都被一起带到一个充满活力的社交场所。"

作者 Troels Troelsen   合作伙伴 Viggo Haremst, Nina La Cour Sell, Charlotte Soderhamn Nielsen   专家 Buro Happold; Ove Arup & Partners, London

Approaching the new Darat King Abdullah II, after a stroll in the park, looking forward to a cultural evening.

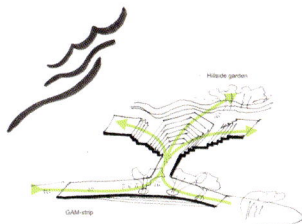

Hillside garden

GAM-strip

FLOW

The flow of the public starts in the GAM-strip and creates a crescendo culminating in the building complex thus creating a prominent landmark. Passers by are invited to take an informal stroll through the building to experience and partake in its life i.e. through the foyer amphitheatre to the hillside garden and further down to the park. Thus the flow does not terminate in the building, but divides further into several branches and creates circles and loops.

The continuous flow may include a wide underpass making the lower part of the park continue under the road to the site and further uphill to the garden at the top of the foyer.

The western end of the GAM-strip is proposed modelled to create a fluent integration between the valley landscape with the front plaza, overbuilding the road and extending into the 'stage' of the amphitheatric foyer. Further the landscape modelling may in the future extend to the areas to the west, creating extension of the park overlooking the bus station.

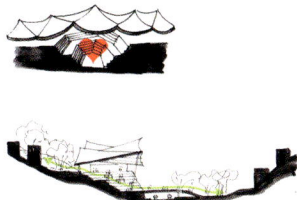

THE HEART OF THE BUILDING

The heart is the magnificent amphitheatric foyer. Here all the qualities of the place are brought together in a lively, social meeting point.

The public is invited to pass through the building from the GAM-strip to the hillside garden. Thus the vivid heart is the shadowy crossing point between public landscape promenade and the multi-level foyer.

The transparent foyer visually opens to the gardens at the hillside and to the valley park, thus creating a link between the communities of the two sides of the valley.

THE MEETING PLACE

The shape of the foyer unifies the characteristics of the place: On one side the theatrical spaces of the building, on the other side the amphitheatric character of the surrounding city.

The foyer is created as a terraced landscape giving access to the different levels of the auditoria via openings in the terraces.

From the terraces of the foyer you overview the GAM-strip below and the garden terraces at the hillside above.

The stage level of the foyer is extended across the street creating a front plaza unifying the amphitheatre with the GAM-strip.

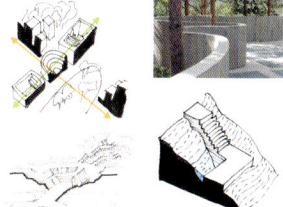

LANDSCAPE AND BUILDING

The landscape and the building are unified. Indoor and outdoor areas are merged to a continuous and coherent sequence of spaces. Contrast and interaction are accentuated between the terrain and the prismatic elements of the building with the amphitheatric foyer. Furthermore a contrast between the monolithic base and the floating curved roof like an artificial cloud is created.

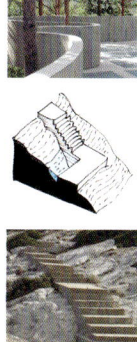

THE FIFTH FAÇADE

The roof structure is a modern transformation of native roof structures. It is something between inverted domes and solidified tents.

The curved roof seen from the hills, shining in the sun symbolizes the social meeting place linking the two sides of the city together. Thus the appearance from the southern hillside becomes appealing like from the GAM-strip.

THE ENTRANCE

The main access is from the GAM-strip with the underground parking structure. Here in the park, a drop-off can be reached from the left lane of the Ali Bin Abi Talib Street.

Additional / alternative drop off points can be created.

A street level entrance is created at Ali Bin Abi Talib Street. From here there is direct access to the education rooms and to the lower level of the small stage and the large rehearsal room as well as to the box office, lockers and toilets. Wide stairs lead up to the lower level of the amphitheatric foyer.

Accesses from the open spaces in the western and eastern ends of the site have built-in walkways at first floor along the street to the main foyer.

VIPs have direct access to the top level of the foyer via the hillside street (A-street).

Busses can be parked during performances at the eastern end of the site.

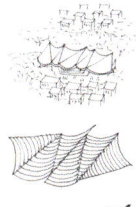

The main foyer amphitheatre
The new meeting point for Amman citizens

restaurant and view terrace

Road A

VIP and
hillside entrance

side entrance

pedestrian bridge

access to GAM Strip

Ali Bin Abi Talib Road

GAM Strip

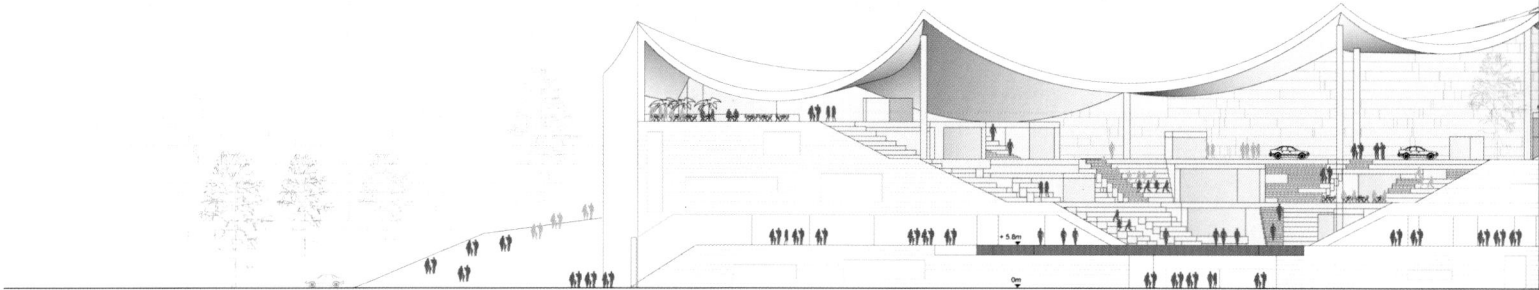

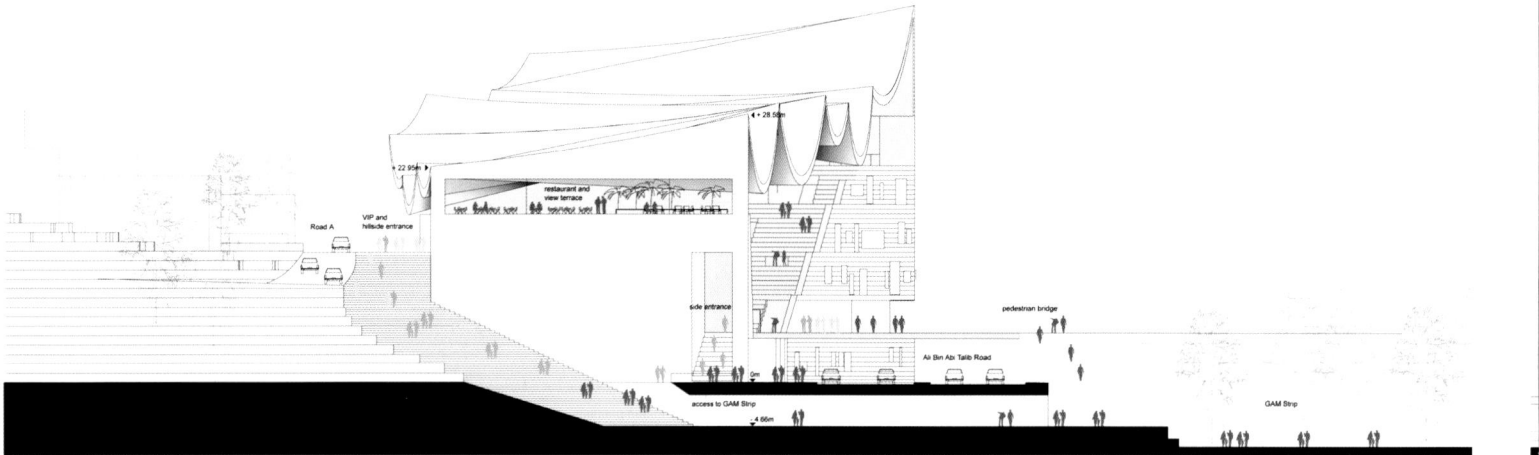

east elevation 1:200

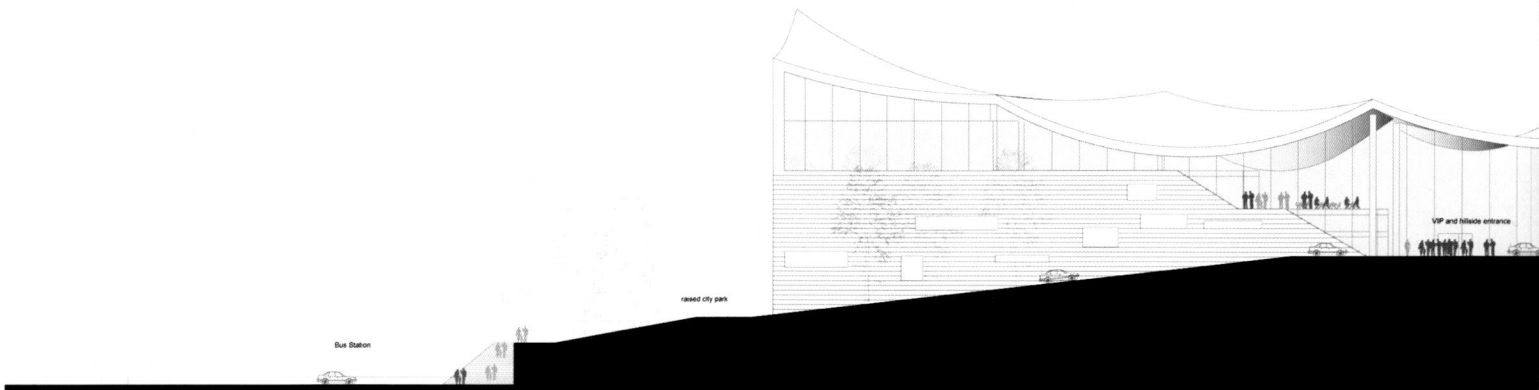

VIP and hillside entrance

raised city park

Bus Station

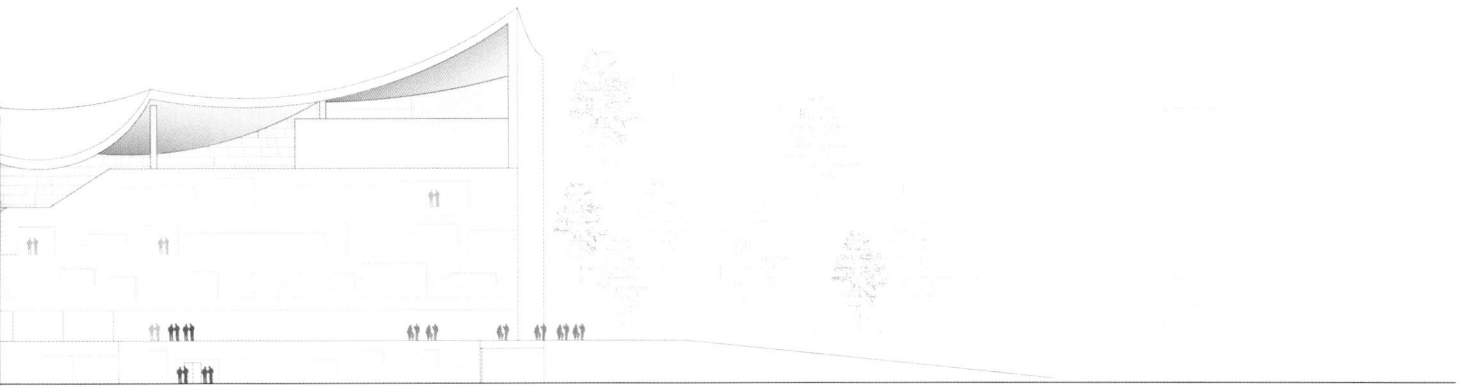

north elevation 1:200

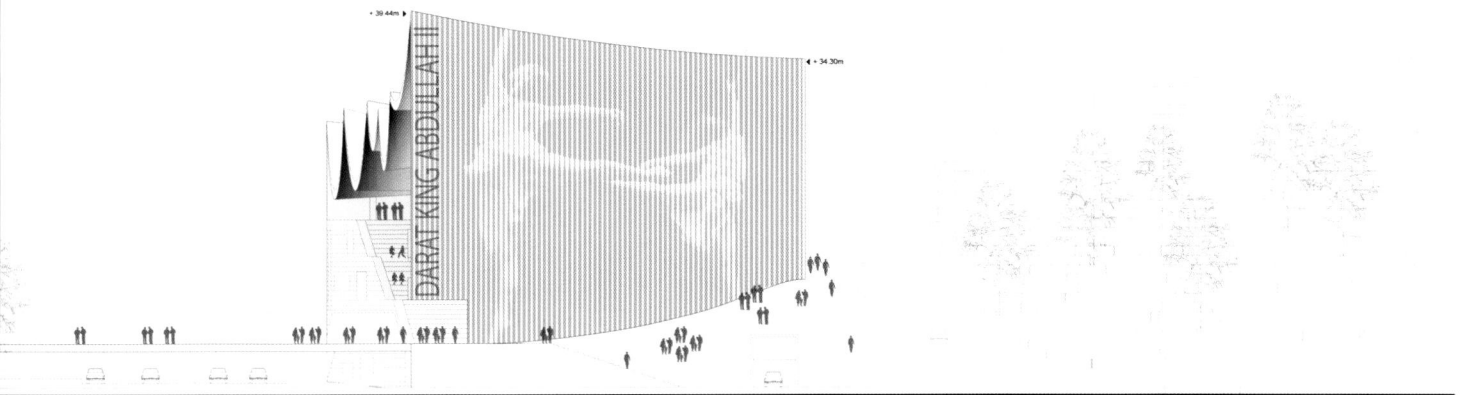

west elevation 1:200

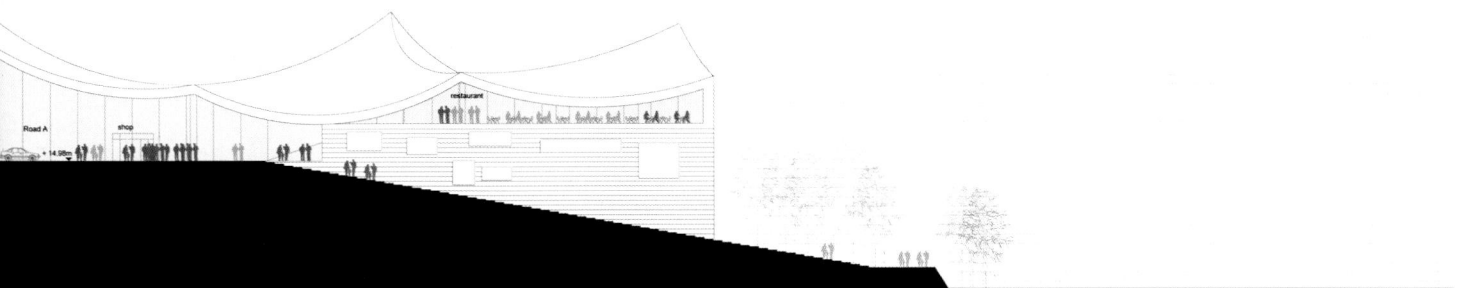

DARAT KING ABDULLAH II

+ 39.44m ▶

◀ + 34.30m

Road A

shop

restaurant

+ 14.98m

longitudinal section 1:200

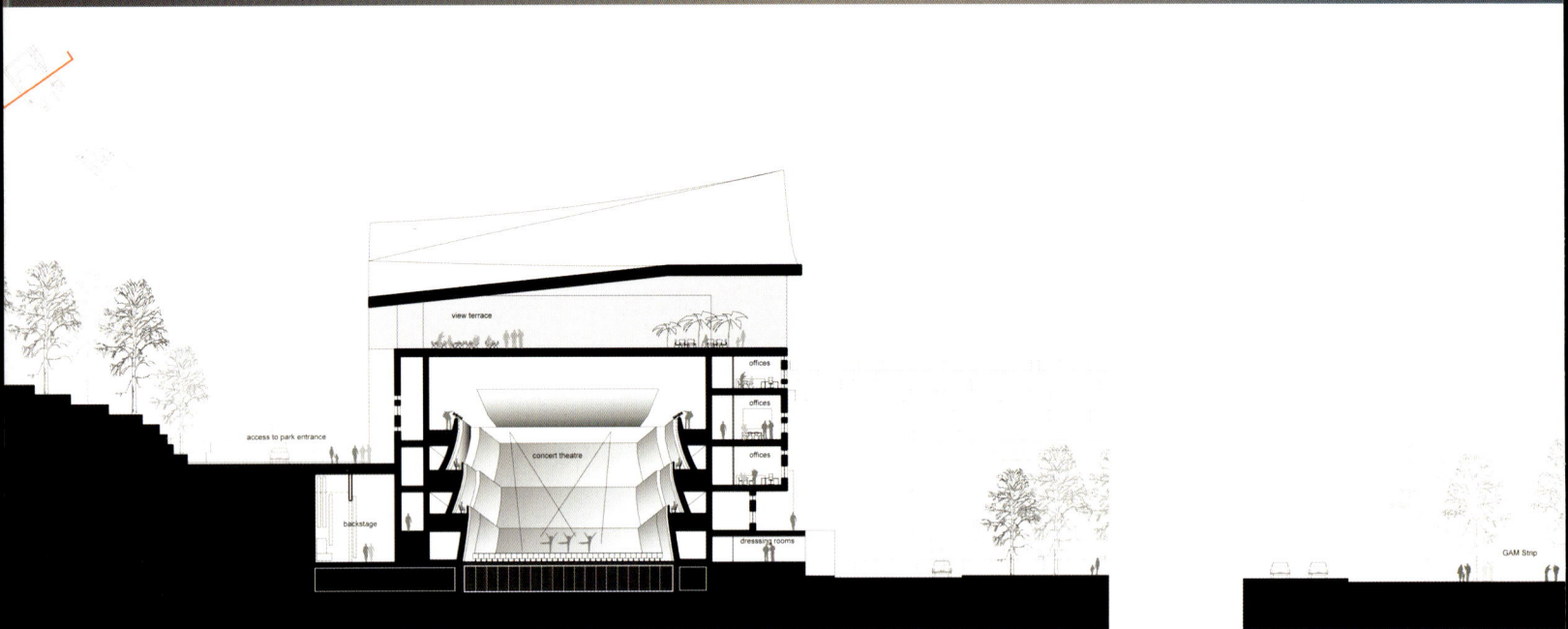

view terrace

access to park entrance

offices

offices

concert theatre

offices

backstage

dressing rooms

GAM Strip

section through concert theater 1:200

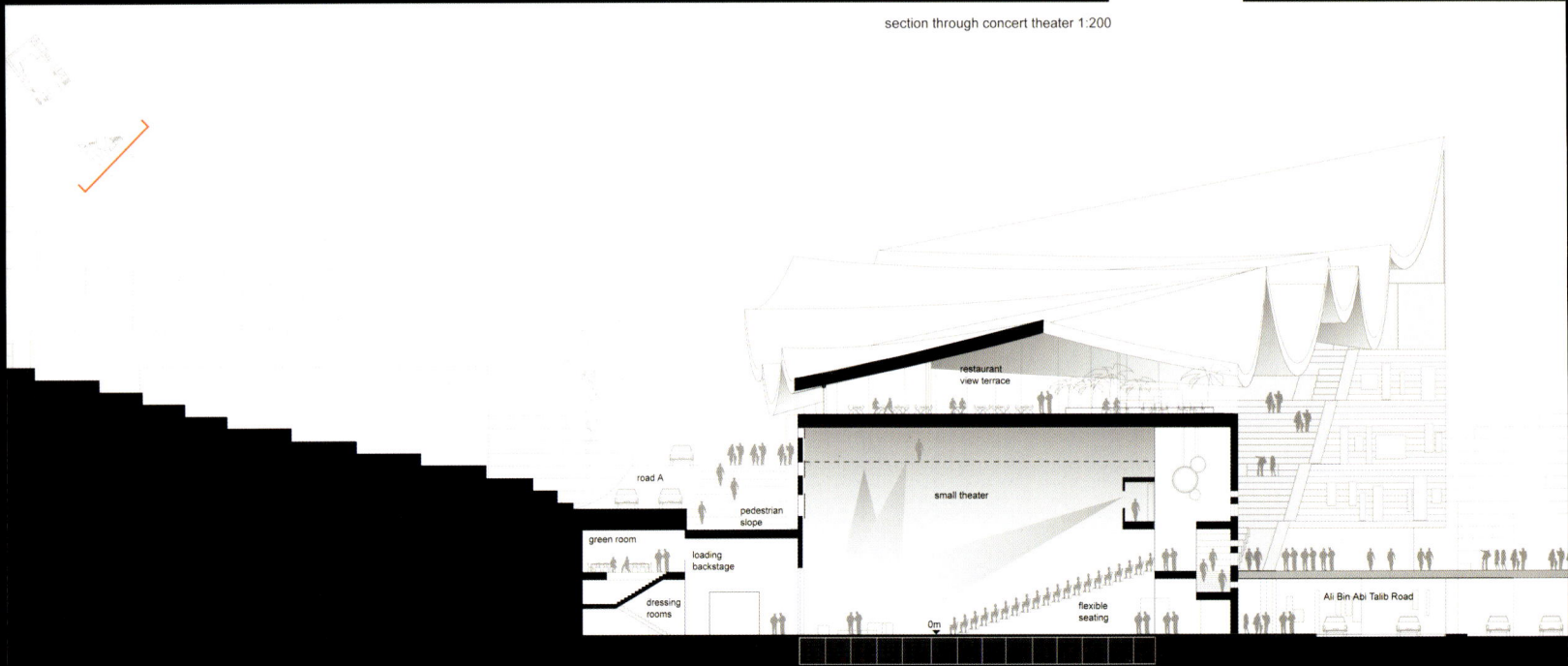

restaurant
view terrace

road A

small theater

pedestrian
slope

green room

loading
backstage

flexible
seating

Ali Bin Abi Talib Road

dressing
rooms

0m

The many informal stages in the foyer communicate the rich variety of cultural possibilities of Darat King Abdullah II

THE FOYER

The foyer becomes a focus for Amman's cultural life, a social, vivid meeting place with the audience as actors and also an informal amphitheatre for an occasional, large event or for several small performances at different levels of the foyer.

The terraced amphitheatric landscape divides between the public areas above and the private or semi-private spaces below.

A variety of recesses and projections create additional terraces with bars as well as enclosed spaces for music-shop, VIP room, education spaces etc. overlooking the grand foyer.

Like the roman theatre the foyer has sitting-steps combined with zones of normal steps, but in a more fluid and informal way.

Restaurant, VIP-rooms and café are in the upper part of the amphitheatre overlooking the foyer and the GAM-strip parkland and facing the hillside gardens to the south.

POSSIBLE FOYER SECTIONS

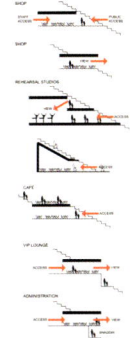

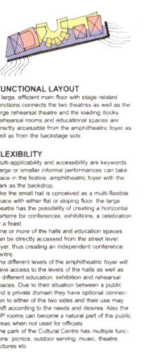

THE GROTTO

Under the stepped foyer a grotto-like space is created giving access to the theatre halls and to the education rooms, rehearsal spaces etc. exposed as separate volumes. Slots under the sitting-steps together with the larger openings and transparent volumes create dramatic light to the space.

THE MAGIC OF LIGHT

The interplay of light in the recesses and perforations of the stone material contrasts to the polished, shining surfaces of the roof structure and makes the building stand out like a sculpture or a piece of jewellery. At night, seen from the park, in the lit, amphitheatric foyer will stand out as a grand festive stage.

During the day, the play of the sun will emphasise the monolithic character, whereas from the inside, openwork parts will appear transparent and weightless. At night when viewed from the outside, the density of the monolith will dissolve from the artificial light from the lively inner house.

FUNCTIONAL LAYOUT

A large efficient main floor with stage related functions connects the two theatres as well as the large rehearsal theatre and the loading docks. Rehearsal rooms and educational spaces are directly accessible from the amphitheatric foyer as well as from the backstage side.

FLEXIBILITY

Multi-applicability and accessibility are keywords. Large or smaller informal performances can take place in the festive, amphitheatric foyer with the park as the backdrop.

Like the small hall is conceived as a multi-flexible space with either flat or sloping floor the large theatre has the possibility of creating a horizontal partiene for conferences, exhibitions, a celebration or a feast.

One or more of the halls and education spaces can be directly accessed from the street level foyer, thus creating an independent conference centre.

The different levels of the amphitheatric foyer will have access to the levels of the halls as well as to different education, exhibition and rehearsal spaces. Due to their situation between a public and a private domain they have optional connection to either of the two sides and their use may shift according to the needs and desires. Also the VIP rooms can become a natural part of the public areas when not used for officials.

The park of the Cultural Centre has multiple functions: picnics, outdoor serving, music, theatre, lectures etc.

THE LARGE HALL

A fluid visual coherence is created between hall and foyer via sliding gates, at stalls level opening the entire width of the hall. When closed these offer sound insulation to allow simultaneous use. The interior of the hall resembles a faceted grotto. There is a festive and intense atmosphere with the audience distributed in the stalls, the balconies and the side balconies that lean into the volume like terraces, creating intimacy and involving everyone.

They are designed in a dynamic staggered way like cone scales to emphasize the intimate atmosphere but are carefully sculpted to augment the acoustics. When the hall is used as a theatre, the "scales" closest to the proscenium can be moved/turned to enhance acoustics and mould the space to an even more intimate framing of the stage opening.

At concerts the "scales" are moved back to create a unity with the orchestra shell completing the hall with slanting wall and ceiling "scales".

SMALL THEATRE FLEXIBILITY

CONCERT HAL FLEXIBILITY

The concert theater embraces a wide variety of different events engaging all senses

Lightning, music and performance play together creating different worlds of experiences

[annex].
[附录].

Glossary
词汇表

Glossary
词汇表

A to Z of architectural competitions
建筑竞赛A——Z

Anonymity 匿名方式

A basic principle of architectural competitions, anonymity is to ensure that the jury decides solely on the basis of each entry's merit, i.e. without consideration of its author's reputation. Procedurally, during the otherwise anonymous process, anonymity may be lifted in favour of enhanced communication between sponsor and competitors in the cooperative phases.

Application procedure 申请程序

Restricted or partially restricted competitions are preceded by an application procedure where a selection committee picks a suitable number of competing candidates. Selection criteria must be unequivocal and non-biased.

Architectural Juror 建筑评奖委员会

The jury is composed of architectural jurors and technical jurors. The former are so designated because they must possess at least the same level of professional qualifications as the competitors. This is to ensure that the jury meets the standards required for its task. The distinction between architectural and technical jurors can be traced back to regulations of the German Public Contract Code, but in practice its importance has become less and less significant.

Author 作者

The project design is submitted by its author who is also the beneficiary of the promise of contract agreement. In this respect the author is the sponsor's contracting party. An author must meet the eligibility criteria set forth in the competition brief. In the case of a design team, all its members must meet these criteria and are considered as the project's authors.

Chamber of Architects 建筑师协会

National or regional professional association of architects and town planners within the relevant jurisdiction. The Chamber of Architects monitors architectural competitions for compliance with competition rules and, upon acceptance, issues a registration number. In Germany this is done by the State Competition Committees made up of members of the regional Chamber.

Competition Amount 竞赛费用

The overall monetary sum allotted by the sponsor to prize money, purchase awards, and payment for architectural services constitutes the competition amount. It is calculated on the basis of the fees normally charged, according to the official scale, for the services provided in the competition.

Competition Brief 竞赛任务说明

The competition brief unites, in the form of a brochure, all the textual information, plans, images, and tables required for undertaking the tasks of the competition. Its addenda include information drawings, working drawings, and other informational material as required.

Competition Cost Estimate 竞赛成本预算

Included in the offer that a competition manager makes to a (potential) sponsor for preparing and conducting a competition is an estimate of the overall cost of the procedure. Besides the remuneration due to the competition manager, this estimate includes the competition amount, the remuneration due to the jury, experts, and examiners and other costs incurred (third-party costs) such as printing cost, travel expenses, venue rental, etc. The cost estimate is continually updated as a basis for controlling costs during the course of the competition.

Competition Documents 竞赛文件

The documents made available to competitors as the basis for their design work comprise the brochure of the competition brief (with the programme of functions and spaces), the information drawings and the working drawings. Also included are a number of forms to be filled in and various tables, possibly the pilot plate for the environment model, the minutes of the online inquiry procedure, and the competitors' colloquium, if any.

Construction (GRW 1995), the rules for the
Award of Professional Services Contracts
(VOF), the Public Contract Code (VgV),
and the Law against Restraint of Trade (GWB).

Competition Homepage   竞赛主页

This webpage helps to disseminate information on the
competition. It forms the virtual space to distribute
and exchange data and other pieces of information
relevant to the project, to conduct an online forum,
and to document the outcome of the competition.
It facilitates the administration of major parts of the
competition (application procedure, response
to inquiries, document management) and thus
helps to keep down competition costs.

Competition Management   竞赛管理

Competition management (a.k.a. competition coordination) involves all the activities required for the preparation,
conduct, and documentation of an architectural competition.

Competitor   竞赛者

Participation in architectural competitions is open to
(a) all individuals or legal entities whose statutory purpose
includes pertinent planning services, and (b) design teams
consisting of such individuals or legal entities. Eligibility
may be restricted on the basis of regulations governing
the procedure in question (such as the rules for the
Award of Professional Services Contracts – VOF).

Competitors' Colloquium   参赛者专题座谈会

To provide a thorough understanding of the project task a
competitors' colloquium may be held before the midpoint
of the design period, normally at the project location so
that competitors can inspect the project site.
The jury, too, attends this colloquium. Its outcome is
recorded in the colloquium minutes and is considered part
of the competition task. The colloquium may be comple-
mented with or replaced by an online forum.

Contract Award Law   合同授予法

In Germany the legal basis for architectural competitions
and for awarding planning contracts are the EU Service
Directive, the Principles and Guidelines for Competitions in
Regional Planning, Town Planning and

Cooperative Competition   合作型竞赛

A competition where an immediate dialogue takes place
between the sponsor, the jury, the experts and the
examiners on one side and the competitors on the other.
This dialogue provides an opportunity for optimising one's
approach to problems arising from the designs proposed,
at stage of the competition most conducive to solutions.
A special feature of cooperative competitions is waiving
anonymity during at least parts of the procedure.

Data Bank   数据库

Once an applicant is approved to take part in a competition managed by [phase eins]., his data (name of
practice, address, authors' data, reference project data,
etc.) are stored in an online data bank. The planner
can access his data record at any time for updating
or reuse in case of a later application. This practice
ensures full and coherent presentation of the elements
of an application; in addition, it simplifies the (often very
time-consuming) application procedure and speeds up
subsequent processing to a high degree of precision.

Declaration of Authorship   原著声明

The anonymously submitted competition entry is
accompanied by a sealed envelope containing a declara-
tion that reveals the author's name and the names of
his collaborators and consulting experts. The sealed
envelopes are opened only at the end of the jury
meeting, i.e. after the short-listed designs have been
ranked and prizes have been awarded. The reading
out of each author's name concludes the meeting.

Design Elaboration, Suggestions as to...   设计详细说明

The jury's recommendations to the sponsor, prior to
revealing the identities of the designers, should include
suggestions for elaborating the short-listed entries. If
required, this document should also include suggestions as to
any necessary change to the project task and what conclu-
sions the sponsor should draw from the competition and
its outcome.

Documentation   文档资料

Upon conclusion of the competition, all the parties
involved are provided with an account of the
procedure and of the design proposals submitted.

This documentation adds to the transparency of the procedure and to the plausibility of the decision taken.

Downloads 下载

The Internet constitutes an ideal platform for the distribution of information, to as many recipients as required. Where competitions are concerned the applicants, competitors, and all interested parties can download various data files (forms, CAD files, illustrations, movie clips, text) from the competition homepage.

Eligibility for Participation 参赛资格

The eligibility for participation depends on the pertinent regulations of the public contract-award code and on the content of the task. Eligible to participate in an architectural competition are generally individuals who are entitled to designate themselves as an "architect", under the law of their country of origin. Analogous regulations govern which town planners and graduated engineers may participate in urban-planning competitions, engineering competitions, and integrated competitions.

Evaluation Round 评估会议

During the jury meeting, the informational round is followed by several evaluation rounds when jurors
(a) discuss which entries meet to a higher or lesser degree the requirements set forth in the competition brief and
(b) decide which entries are to remain for further consideration and which entries to eliminate from the selection procedure.
During the first evaluation round, one vote in favour suffices for the entry in question to be kept in the contest. In other words, rejection in round one requires a unanimous vote. In subsequent rounds decisions are taken simply by a majority. If required, between evaluation rounds discussions can be held that help to form opinions but where no actual decisions are taken.

Examination Drawings 测试设计图纸

In addition to presentation drawings, calculations, and other explanatory texts, the examination drawings form a part of the material to be submitted by each competitor. The examination drawings must contain all pieces of information (dimensions, distribution of uses, etc.) required to make the project intelligible and examinable in quantitative ways. To complement or replace the examination drawings, CAD data files as the basis for examination play an ever more important role.

Examiner 主考人

The examiners are professionals, i.e. generally architects or engineers from the fields covered by the competition. Upon receiving the entries, they appraise the designs for compliance with the competition requirements. The examiners are appointed by the sponsor, they attend to his interests and provide advice to the jury in their role as advocates of the project authors. The examiners should be involved with the competition throughout its course. Normally the competition manager coordinates their activity, and in particular during the quantitative preliminary examination the competition manager's team supports the examiners.

Exhibition 展览

In the interest of public acceptance of the project it is recommended that the submitted proposals be exhibited for at least two weeks, upon conclusion of the competition. The exhibition serves the double function of being a forum for discussion and an event celebrating the contributions made by each competitor.

Experts 专家

These are specialists from various fields that may be called up for setting up the competition and for contributing to the preliminary examination. Experts do not have a vote at the meeting of the jury.

Feasibility Study 可行性研究

A feasibility study may, for example, determine a site's potential for building development. It may be carried out in preparation of a competition or it may be included among the services to be rendered within the framework of project development.

Ideas Competition 观点竞赛

An ideas competition, as opposed to a project competition, seeks to obtain a variety of solutions to a given problem without the immediate intention of implementing the project under consideration. Consequently no promise of contract award is made, but in return a higher amount of money is allotted to prizes. An ideas competition may become the basis for a subsequent project competition.

Information Drawings 施工图

All information required (site plan, built environment, topography, existing vegetation, utility lines, programme of functions and spaces, etc.) is made available to all competitors on CD-ROM, by means of download, or through illustrations in the competition brochure, or, if required, as hardcopy.

Information drawings (as opposed to working drawings) are not a graphic model to be followed for the representation of each competitor's design.

Informational Round 资料会议
The jury begins design appraisal with an informational round where the examiners present a detailed, unbiased explanation of each entry. In this round all the designs are inspected, but judgement is withheld for the time being.

Inquiry 咨询
At the start of the period allotted to the task, competitors may ask for clarifications by sending written inquiries by mail, fax, or e-mail and, if a competitors' colloquium is held, by asking questions at that event. The sponsor will answer these questions in the shortest time possible, after consulting with jurors or experts. Prepared and moderated by the competition manager, this inquiry process is often conducted as an online forum on the competition webpage and concluded before the midpoint of the design period by distributing the forum minutes to all interested parties. The questions asked and the answers given define requirements that are considered as parts of the competition task.

Integrated Competition 综合竞赛
In an integrated competition (a.k.a. multidisciplinary competition) the concern is to integrate issues arising from various different disciplines. Consequently, the entries present design approaches and planning services that transcend the borders of any single area of expertise. An integrated competition is the best option in cases where the complexity of the task calls for the collaboration of experts from various fields.

International Competition 国际竞赛
Originally adopted as recommendations by UNESCO in 1956, the "Standard Regulations for International Competitions in Architecture and Town Planning" were last revised in November 1978. This document defines the basic rules for international competitions, i.e. any competition in which architects or town planners of more than one country are invited to participate. Among the issues settled are the various types of competitions, the composition of the jury, the commissioning process subsequent to the competition, the announcement and publication of competition results, the language or languages to be used in the brief and in deliverables, the anonymity of the competition procedure, and the obligation to appoint a professional adviser and supervisor. The body UNESCO entrusted with the preparation of a detailed set of rules and supervision of international competitions is the "union internationale des architectes" (uia). Thus, the "uia Guide for International Competitions in

Architecture and Town Planning" constitutes the recognised basis for competitions organised on an international level.

Issuing of Competition Documents 竞赛文件的发布
Describes the formal procedure for issuing competition documents to competitors or making the data files required for the task available for download from the competition homepage. This takes place on a given date.

Juror 评奖委员会
Jurors exercise their duties independently, in person, and are guided solely by their professional expertise. Due to their professional qualifications, architectural jurors must match the requirements placed on competitors at a particular high level. Technical jurors should be familiar with the content of the competition task and with local conditions in particular. In addition, architectural and technical alternates with voting capacity should be appointed so that they can replace regular jurors who may be absent.

Jurors' Colloquium 评奖委员会的专题座谈会
Jurors, alternate jurors, experts, and examiners should participate in the formulation of the competition brief. For this purpose a jury colloquium (a.k.a. jury "kick-off") is held where the project content and the competition task are discussed, possibly reformulated and ultimately adopted.

Jury 评奖委员会
The jury's remit is to decide which submitted proposals are to be admitted to the competition, to appraise the admitted entries, to pick those proposals that best comply with the competition requirements, and to suggest the direction of the process. The jury is appointed by the sponsor. It makes its decisions solely on the basis of the criteria listed in the competition brief. Its voting members are architectural jurors on the one hand and technical jurors on the other. They are supported in their decision-making by non-voting alternate jurors, experts, and examiners.

Jury Meeting 评奖委员会会议
Chaired by an experienced architectural juror appointed by consensus amongst its members, the jury meets in a closed session to select the competition entry that best responds to the task. Jury meetings are jointly moderated by the jury chair and the competition manager. Their proceedings and outcome are recorded in the jury meeting minutes. Meetings are attended by jurors, experts, examiners, representatives of the sponsor, by the competition

manager, and his agents. Guests too may be invited. All attendees must declare not to have discussed the competition task with any of the competitors and must undertake to keep deliberations confidential.

Life-cycle Cost Analysis 生命周期成本分析

This cost analysis relates planning and construction costs to those incurred over the lifetime of a (proposed) building: management costs of the real estate, the costs of consumables, of conversion and adaptation, of renovation and restoration plus costs associated with demolition and disposal. This provides the basis for appraising (initial) investment costs against consequential costs and for assessing the impact of planning decisions on costs arising during the overall (expected) lifetime of a building.

Minutes 会议记录

All meetings held in the course of the competition and during its preparation are documented in the form of minutes. The minutes of the colloquia and the jury meeting are automatically distributed to all the parties involved.

Official Scale of Fees for Services by Architects and Engineers 支付建筑师和工程师服务酬金的官方标准

In short referred to as HOAI, this by-law (a) defines the official profile of services rendered by architects, interior architects, landscape architects, town planners, and graduated engineers from various fields and (b) regulates their remuneration. The HOAI as a price-regulatory regime has binding force upon the contracting parties. However, remuneration for the consulting services provided by a competition manager is not covered by the HOAI.

Online-Forum 在线论坛

Clarification of issues raised may take the form of an Internet-based online forum. Questions about the project task can be posted anonymously in a password-protected area of the competition homepage and coordinated answers are made available, so as to ensure a uniform level of information on the part of all the competitors. Online forums improve the level of dialogue with competitors as they allow exchanges to take place over a longer period of time. They also contribute to cost saving as they may replace colloquia, or make colloquia more efficient.

Open Competition 公开竞赛

A participant in an open competition may be any individual with the required professional qualification. The number of competitors is not restricted by a candidature procedure or any similar scheme. In order to reduce the burden on

the parties involved it is customary to conduct an open competition in two stages.

Ordnung SIA 142 SIA142条令

This standard established by the Swiss Institute of Architects and Engineers (SIA) lists the rules governing Swiss planning and comprehensive-services competitions in architecture, engineering, and similar fields. The more detailed annexes ("guidelines") complementing it deal with determining the amounts of prize money, the selection of competitors, and procedural options. In 2007, SIA 143 was issued as a separate body of rules governing the commissioning of studies.

Period Allotted to the Task 任务分配阶段

The period allowed to the competitors for preparing their design proposals, i.e. the space of time between issuing of competition documents and projects due-date. Customarily competitors are allowed about eleven weeks or, in the instance of a two-stage competition, five weeks in stage one and seven weeks in stage two.

Preliminary Examination 初试

This term covers two different notions, as it designates
(a) the examiners as a group entity, together with the competition manager's support team, and
(b) the activity exercised by these individuals.
The preliminary examination — as a group — is responsible for appraisal of the competition entries and processing the relevant data and pieces of information, prior to the meeting of the jury.
The outcome of the preliminary examination — as an activity — is recorded in the Report on the Preliminary Examination. At the meeting of the jury it is up to the examiners to point out to the jury the essential functional and economic characteristics of the competition entries and to make the jury aware of any aspect that in their opinion the jury should not disregard.

Preparation of a Competition 竞赛的准备工作

The tasks in the run-up to a competition include formulating the task, drafting the competition materials, preparing, conducting, moderating, and documenting all preparatory meetings and colloquia. Next are scheduling and preparing all meetings and colloquia to be held during the competition proper, advice as to the appointment of the jury, and coordinating the procedure with the competent permit authorities and the Chamber of Architects. This includes monitoring costs and controlling schedule during this stage and, in the case of a restricted open competition, carrying out an application procedure resulting in the selection of a suitable number of competitors.

Presentation Drawings 示意图

The materials to be submitted by the competitors normally include presentation drawings, examination drawings, calculations, and documents such as explanatory texts and building specifications. The presentation drawings are the original drawings that are to be presented to the jury and are ultimately exhibited.

Press Conference 记者招待会

Upon conclusion of a competition, the outcome is normally announced at a press conference that may coincide with an exhibition opening for all the designs submitted.

Principles and Guidelines for Competitions in Regional Planning, Town Planning, and Architecture 地区规划、城市规划和建筑竞赛的原则和指导方针

This refers to the German Public Code governing the preparation and conduct of competitions in the fields specified (short reference: GRW 95). Its preamble states that it "should form the basis [...] for a fair cooperation, in a spirit of partnership, of all the parties involved in a competition and help to develop architectural culture [...] in pursuit of societal, economic, ecologic, and technologic progress." Other regulations that apply to public authorities in Germany are those of the Public Code for Contracting Services from the Liberal Professions (VOF).

Prizes 奖金

The competition amount is split into prize money, money for purchase awards, and (possibly) remuneration for services rendered. The amounts allotted to prizes are mentioned in the public announcement of the competition. The first prize is to be awarded to the entry that best satisfies the problem described by the sponsor. If a prize winner is commissioned with the elaboration of his proposal, remuneration for services rendered during the competition will be reduced by the amount of prize money received, provided that essential parts of the submitted design form the basis for the detailed plan.

Programme of Functions and Spaces 功能与空间的规划

This is one of the task-defining elements of a competition. When not provided by the sponsor, it is up to the competition manager to prepare, in cooperation with the sponsor, the programme of functions and spaces.

Project Competition 项目竞赛

A project competition is intended to demonstrate, on the basis of a clearly defined programme and unambiguous service requirements, the feasibility of a project in terms of planning and to help select the party to be awarded the planning contract.

Project Cost Estimate 项目成本预算

A prerequisite for calculating competition costs is the project cost estimate based on the available characteristic data. This cost estimate may also be a part of the problem posed to the competitors or it may be included in the preliminary examination.

Project Development 项目开发

This term designates all the activities intended to integrate the location elements, project idea, and capital in a way that ensures the creation and sustained profitable utilisation of a competitive, socially acceptable, and environmentally compatible built object. In a more restricted sense project development comprises the stage from project kick-off (or conception of the project idea) to the award of planning commissions, while in a wider sense it covers the whole life cycle of the project.

Project Due-date 项目到期日

The moment when the time period allowed for the task ends and all competitors must hand in their completed proposals in full compliance with competition requirements. As a rule, it is a given point in Central European Time by which a competitor must deliver his entry to a courier service or to the postal service. This practice ensures that each competitor is accorded the same amount of time regardless of geographical location. On demand, the competitors must provide proof of timely delivery.

Project Management 项目管理

This term describes the unbiased, independent exercise of (delegable) technical, economic, and legal activties otherwise exercised by the sponsor.

Promise of Contract Award 合同判授承诺

A competition constitutes a contractual relationship between the sponsor and each competitor. An essential element of this relationship is the pledge to award one of the prize-winners, in return for the mostly unremunerated services rendered by the competitors, the planning commission for the project. The promise of contract award must state accurately and in a legally sound manner the scope of services offered.

Purchase Award 购买奖金

In open or (partially) restricted competitions, the jury customarily reserves part of the prize money (usually 20 percent) for purchase awards. These are given to entries that, while not realisable in full, are distinguished by outstanding partial solutions. The purchase award implies that the sponsor may make use of that partial solution.

RAW 2004 RAW2004

Derived from the GRW 1995, the Rules for the Conduct of Competitions in Regional Planning, Town Planning and the Construction Industry (RAW 2004) are applicable law in only three German states (Bremen, Lower Saxony, and North Rhine-Westphalia). The discrepancies reflected by this state of affairs, outcome of many years of dispute over a reform of GRW, are to be resolved with the upcoming revision of GRW. In contrast to GRW, RAW rules seek to introduce less stringent requirements, an attempt that partly succeeds and partly results in ambiguities or, to give it a positive spin, more leeway.

Remuneration for Services Rendered 支付提供服务的酬金

In the case of competitions by invitation and in stage two of some two-stage competitions, the competitors are paid a portion of the competition amount as remuneration for the services rendered. The remainder of the competition amount is distributed, as in any ordinary competition, in the form of prizes and purchase awards.

Report on the Preliminary Examination 初试报告

Jury sessions begin with the delivery of the report on the preliminary examination. In the first round, the procedure to be followed during preliminary examination is explained and the outcome of the formal inspection of entries is announced. During the subsequent informational round, the examiners present the designs in detail and in an unbiased manner. They explain the results of the preliminary examination for each entry so as to acquaint the jury, within a short span of time, with the range of problems posed and the set of solutions provided.

Service Phase 使用阶段

The official scale of fees for services by architects and engineers (HOAI) divides the performance profiles of German architects and engineers into service phases, to each of which is assigned a portion of the remuneration due. The service phases describe a regular planning and construction process with reference to the different activities performed by an architect or engineer. Service phase one, establishing the basis of the project, normally pertains to the competition manager, while service phase two, preliminary design, pertains to the competitors.

Specialist Consultant 专家咨询人员

The competitor submits his design to the competition, and in return receives sponsor's promise of contract award for one of the entries submitted. The design may call for input from specialist consultants, experts from various disciplines such as landscaping and engineering. The sponsor may even suggest the participation of specialists with specific expertise. These experts are not required to furnish the same evidence or level of eligibility as that demanded of applicants to the competition, in the procedural prerequisites.

Sponsor 主办方

As a rule, a competition's sponsor is the party that subsequently will award the planning contract for the project emerging from the competition. Concerning the promise of contract award, the sponsor is the contracting party for each competitor.

Suggestions for Work 设计作品建议

Essential elements of the competition brief are its precision and inclusiveness. To ensure uniformity, the brief should comprise clear statements as to what items are to be included in the design proposal and how they are to be presented.

Technical Juror 专家评奖委员会

The jury comprises architectural jurors and technical jurors. What is required of the latter is particular familiarity with the content of the competition task and with local conditions. Generally, representatives of the sponsor and of the local community act as technical jurors.

Two-Stage Competition 两个阶段的竞赛

To limit overall cost and improve the competition results, a competition may be conducted in two stages, thus allowing for a more intensive dialogue with the competitors and for phased elaboration of designs. At the end of phase one, the jury picks, from a comparatively large number of not highly developed entries, the ones that respond best to the requirements and that are to be elaborated upon during stage two, at the end of which the jury takes its ultimate decision.

Urban-planning Competition 城市规划竞赛

An urban-planning competition, as opposed to a building project competition, aims at the solution of town-planning tasks. Its outcome may form the basis for an urban land-use planning procedure or a building project competition.

Working Drawings 施工图

As a basis for their design work, the competitors are provided with all the required planning documents, digitally formatted. Working drawings (as opposed to information drawings) are a graphic model to be followed for the representation of each competitor's design. They contain such elements (e.g. neighbouring buildings and borders of the competition site) as are required to ensure that the plans submitted are at a uniform scale and depict identical elements.

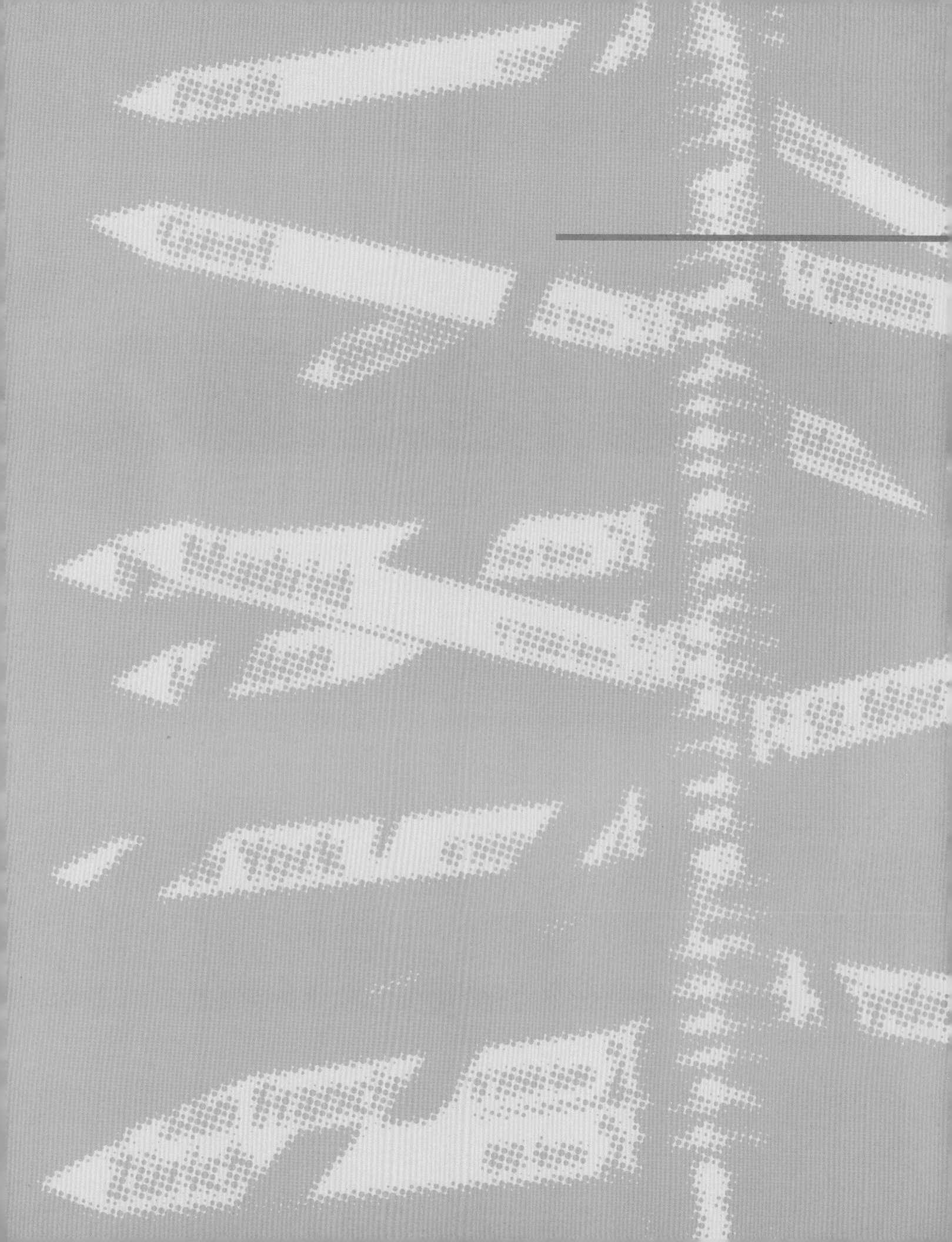

Links to Platforms, Organisations, and Legal Regulations

与论坛、机构和法律条文的相关链接

Links to Platforms, Organisations, and Legal Regulations
与论坛、机构和法律条文的相关链接

Topical links to websites covering competitions, the building industry, and pertinent regulations
与竞赛、建筑业和有关法律条文的代表性链接

Architectural competitions 建筑竞赛
- www.BauNetz.de/arch/_wettbewerbe
  The competition section of Baunetz
- www.archi.fr/EUROPAN
  Europan
- www.archinform.de
  Literature, works of reference, biographies
- www.newitalianblood.com
  Interactive platform for the announcement of competitions and of their outcome
- www.europaconcorsi.com
  Competitions in Europe (I)
- www.nextroom.at
  Architecture and competitions in Austria (AT)

Platforms for the building industry 建筑业平台
- www.baunetz.de
  Internet portal of Bertelsmann AG
- www.archi.fr
  French portal to topics architectural
- www.archguide.com
  Architecture network in Belgium
- www.irish-architecture.com
  Architecture network in Ireland
- www.archicool.com
  Architecture network in France
- www.archined.nl
  Architecture network in The Netherlands
- www.architettura.it
  Architecture network in Italy
- www.architectenwerk.nl
  Architecture network in The Netherlands
- www.architektura.info
  Building portal in Poland
- www.vitruvio.ch
  Architecture network in Switzerland
- www.europaconcorsi.com
  Competition portal from Italy
- http://archnet.org/lobby.tcl
  Architects' portal focusing on Islamic architecture
- www.archijob.co.il
  Israeli architecture portal
- www.abbs.com.cn
  Chinese architecture portal

Trade journals 贸易杂志
- www.wettbewerbe-aktuell.de
  Current competitions (Germany)
- www.BauNetz.de/arch/bauwelt
  Bauwelt (Germany)
- www.baumeister.de
  Baumeister (Germany)
- www.dbz.de
  DBZ (Germany)
- www.tema.de/mass/news/cl4.htm
  AIT (Germany)
- www.detail.de
  Detail (Germany)
- www.sia.ch
  Tec21 (Switzerland)
- www.hochparterre.ch
  Hochparterre (Switzerland)
- www.penrose-press.com/IDD/pub/cards/J11726.html
  International directory of design (Switzerland)
- www.archis.org
  Archis (Netherlands)
- www.arplus.com
  Architectural Review (UK)
- www.elcroquis.es
  El Croquis (Spain)
- www.competitions.org
  Competition (USA)

Chambers of architects and federations 建筑师协会和建筑师联盟
- www.bak.de
  Bundesarchitektenkammer Germany
- www.architekt.de/bdia
  Bund Deutscher Innenarchitekten (BDIA)
- www.ak-berlin.de
  Architektenkammer in Berlin
- www.aknw.de
  Architektenkammer Norh Rhine-Westphalia
- www.architekten-thueringen.org
  Architektenkammer Thuringia
- www.akh.de
  Hessische Architektenkammer (Hesse)
- www.byak.de
  Bayerische Architektenkammer (Bavaria)
- www.akrp.de
  Architektenkammer Rhineland-Palatinate
- www.akbw.de
  Architektenkammer Baden-Württemberg

Organisations 组织机构

- www.bda-architekten.de
  Bund Deutscher Architekten
- www.architekten-ueber-grenzen.de
  Architekten über Grenzen (D)
- www.planned-in-germany.de
  Planned in Germany
- www.vbi.de
  Verband Beratender Ingenieure

- www.ak-hh.de
  Hamburgische Architektenkammer
- www.architektenkammer-bremen.de
  Architekten- und Ingenieurkammer Bremen
- www.ak-brandenburg.de
  Brandenburgische Architektenkammer Potsdam
- www.aksachsen.org
  Architektenkammer Saxony
- www.architektenkammer-mv.de
  Architektenkammer Mecklenburg-Western-Pomerania
- www.aknds.de/htm/start.htm
  Architektenkammer Lower Saxony
- www.ak-lsa.de
  Architektenkammer Saxony-Anhalt
- www.archi.fr/UIA
  Union internationale des architectes
- www.sia.ch
  Schweizer Ingenieurs- und Architektenverein (Switzerland)
- www.bna.nl
  Bond van Nederlandse Architecten (Netherlands)
- www.architecture.com
  Royal Insitute of British Architects (UK)
- www.aik-sh.de/
  Architekten- und Ingenieurkammer Schleswig-Holstein
- www.oai.lu
  Luxemburgische Architektenkammer (Luxembourg)
- www.rias.org.uk
  Royal Institute of Architects in Scotland
- www.riai.ie
  Royal Institute of Architects in Ireland
- www.aia.org
  American Institute of Architects
- www.architecture.com.au
  The Royal Australian Institute of Architects
- www.comarchitect.org
  The Commonwealth Association of Architects
- www.aua-architects.com
  Africa Union of Architects

Legal regulations 法律条文

- "Grundsätze und Richtlinien für Wettbewerbe auf den Ge-
  bieten der Raumplanung, des Städtebaus und des Bauwesens",
  novel version of the 22nd December 2003 – GRW 1995
  (Principles and Guidelines for Competitions in
  Regional Planning, Town Planning, and Construction)
- "Verdingungsordnung für freiberufliche Leistungen" of the 26th
  August 2002 – VOF (Rules for the Award of Professional Services
  Contracts)
- "Gesetz gegen Wettbewerbsbeschränkungen" of the 26th August
  1998 – GWB (Law against Restraint of Trade)
- Council Directive 92/50/EEC of the 18th June 1992 relating to
  the Coordination of Procedures for the Award of Public
  Service Contracts
- "Verordnung über die Vergabe öffentlicher Aufträge" of the
  9th January 2001 (novel version of the 11th February 2003) –
  VgV (Ordinance on the Award of Public Contracts)

http://www.phase1.de/forum_links.htm

This page of the [phase eins]. website provides active links
to the above legal regulations and other topical web pages.

Partners and Collaborators 2006–2008

合作人与共事人 2006 – 2008

Partners and Collaborators 2006–2008
合作人与共事人2006 – 2008

The people behind [phase eins].
[phase eins].公司背后的工作人员

The partners 合作伙伴
Benjamin Hossbach (co-founder, since 1998),
Christian Lehmhaus (since 2001)

Christine Eichelmann (since 2008), Martin Linz (since 2008)

The permanent staff 正式员工
Alexander Bulgrin, Uwe "Matt" Dahms, Barbara Frei,
Julia Grahl, Stefan Haase, Raschid Hafiz, Marc Havekost,
Michaela "Svea" Heinemann, Sebastian Illig, Maja Kastaun,
Ronny Kutter, Susanne Mocka, Brigitte Panek,
Birgit Pfisterer, Angela Salzburg, Björn Steinhagen,
Harald Theiss, Silke Wischhusen

The freelancers, students, and trainees 自由建筑师、建筑系在校生和培训生
Mogdeh Ali, Katrin Bade, Sameh Balo, Uwe Barsch,
Mario Bär, Anna-Luisa Bories, Martin Bütow, Annika Bleckat,
Jewgeniy Borshchevskiy, Max Dölling, Marc Dufour-Feronce,
Philippe Dufour-Feronce, Lana Eichelmann, Nicole Erbe,
Oliver Gassner, Philipp Haas, Michael Kandel, Jens Kärcher,
Patrick Kutterolf, Kornelia Klimmeck, Paul Maiwald,
Jurgen Middelberg, Paul-Merlin Muller, Viet Dung Nguyen,
Florian Pacher, Michael Pawelzick, Anne Peters,
Andreas Reeg, Ceva Sahinarslan, Ina Schoof, Ben Tullin,
Yakup Vardar, Lisamarie Villegas Ambia, Anina Wagner,
Alexander G. Williams

The independent examiners 独立的主考官
Annette Bresinsky, Heinrich Burchard, Friedhelm Gülink,
Helmut Hanle, Roland Kuhn, Birgit Petersen

The third-party specialists 第三方专家
- Michael Rädler (exhibition building, display panels), Berlin, www.mraedler.de
- Klaus Rupprecht + Bernard D. Wilmot (translations), Berlin
- Olaf Schreiber, Fa. Raecke & Schreiber (Internet and data bank solutions), Berlin, www.raecke-schreiber.de
- Sirko Sparing, Fa. dBusiness (repro and printing), Berlin, www.dbusiness.de
- Olaf Thiede, Fa. Jack-in-the-Box (hardware and server administration), Berlin, www.jack-in-the-box.de
- Hans-Joachim Wuthenow, (photography), Berlin, www.wuthenow-foto.de

Acknowledgement 致谢

We extend our thanks to all our fellow-architects who with their designs contributed to this book, and our particular gratitude is due to our clients, including those whose projects did not fit into the limited space of this book. Their commitment to architectural excellence and their aspiration to make contributions of their own to a high-quality built environment are at the very basis of the work performed by [phase eins].

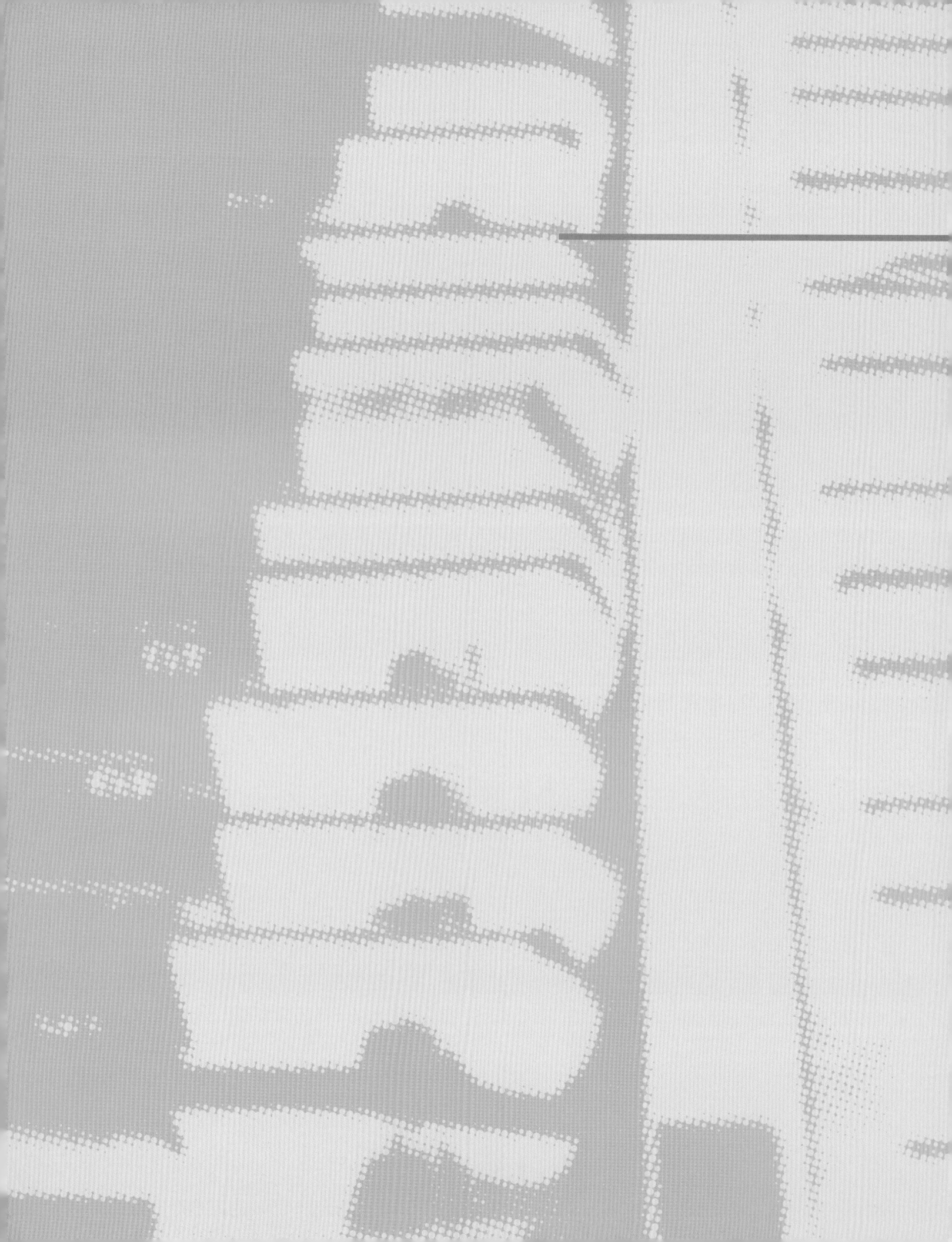

Index
索引

Index
索引

Complete index of authors and offices
作者及建筑事务所的全部索引

A

Abbas, Mohammad	434
ACXT - IDOM Group	111, 113, 124–126
Adjaye, David	270
Adjay Associates	263, 270–275
Aedas	451, 480, 481–483
Agirbas/Winstroer	183, 187
Agoston, Boran	480
agps	6, 353, 355, 376–379
Albert Speer und Partner	111, 353
Ammeter, Laurent	192
Anderhalten Architekten	307
Angélil, Dr. Marc	376
Architekturbüro Kühn	307
ARGE	183, 187, 222, 330, 398
AS&P	219, 353
ASP Schweger Assoziierte Gesamtplanung	339
ASTOC Architects & Planners	353, 355
Atelier Christian de Portzamparc	489, 512–517
Ateliers Lion architectes urbanistes	31, 111, 113, 114–119, 405, 426–429
Auer+Weber+Assoziierte	339, 344–345, 353, 355

B

Bandi, Christian	183, 187
Bär, Friedrich	356
Bär, Stadelmann, Stöcker Architekten	353, 355, 356–359
Bates, Donald	102
Behet, Martin	330
behet bondzio lin architekten	321, 330–333
BEHNISCH ARCHITEKTEN	339
Benthem Crouwel	73, 98–101, 353
Besson, Adrien	192
Blom, Marcel	98
Bofinger & Partner	217, 221
Bofinger, Prof. Helge	217, 221
BOLLES + WILSON	263
Bondzio, Roland	330
Bothe, Jens	74
Brands, Bart	98
Broenimmann, Tarramo	192
BRT Bothe Richter Teherani	73, 74–79, 219
Bründler, Andreas	188
Brüning, Arndt	230
Brüning Klapp Rein	217, 221, 230–235, 263
Buchner Bründler	183, 187, 188, 189, 191

Bunyamin, Derman	132
Burckhardt + Partner	405, 414, 415, 417, 419
Burger, Stefan	364
Burger Rudacs Architekten	353, 355, 364–367
Buro Happold	142, 162–169, 524

C

Centola & Associati	208
Centola, Luigi	208
Chaix & Morel et Associés	58, 221, 222–229
Chaix, Philippe	222
Chipperfield, David	263, 264–269
Consolidated CE - Jafar Tukan Architects	120–123, 405, 434–437
constructconcept	389
COOP HIMMELB(L)AU	451, 476–479
Crone, Daniel	183

D

Dagher, Fadhallah	114
Daher, Rami	120
Dar Al Handasah Nazih Tahleb & Partners	111
Darwish, Yasser	120, 434
David Chipperfield Architects	263, 264–269
Davidson, Peter	102
DB Architecture & Consulting	111, 113, 132–135
Delugan, Roman	498
Delugan-Meissl, Elke	498
DELUGAN MEISSL	489, 498–505

E

Engel, Jürgen	94, 246, 340
Engstrom, Peter	480

F

Faust, Joachim H.	254
Feiner, Henry	472
ff-Architekten	353, 355
Fidanza, Alain	198
Frank, Charlotte	290, 380, 442
Freiesleben, Antje	312
Frisk, Oscar	192
Frank, Charlotte	290, 380, 442

Freiesleben, Antje   312
Frisk, Oscar   192
Fuksas, Massimiliano   466
Fürst Architects   217, 221

G

Gasser, Prof. Markus   350
Gastier, Victor   206
Gerber, Prof. Eckhard   150, 260, 322, 372
Gerber Architekten   141, 150–157, 217, 321, 322–325, 353,
                                       355, 372–375, 389
Gessert, Martin   250
Gigon, Annette   276
Gigon Guyer Architekten   263, 276–279
gmp   353, 405, 425
Grashug, Walter   222
Greenwood, Robert   90, 506
Grüntuch, Armand   314
Grüntuch Ernst Architekten   307, 314–315
Guyer, Mike   276

H

h4a Gessert + Randecker   217, 221, 250–253
Hadid, Zaha   236, 284, 438, 490
Hagopian, Manuel Der   192
Harry Seidler and Associates   451, 472–475
HASCHER JEHLE Architektur   339, 346–347, 353
Hascher, Prof. R.   346
Hauswirth Keller Branzanti   183
Hayawan, Ibrahim El   90
Henning Larsen Architects   353, 489, 524–529
Herzog + Partner   353
Hijjas Kasturi Associates Sdn.   111, 113, 128–131
Hill, Justin   518
Hill, Kerry   518
Hillmer, Jürgen   170, 420
HILMER & SATTLER und ALBRECHT   263
Hirst, Peter   472
Höhne, Thilo   360
HPP Hentrich – Petschnigg & Partner   217

I

Ilhan, Cem   132
Ingenhoven & Ingenhoven   219
Ingenhoven, Christoph   204
Ingenhoven Architekten   204–205, 339
Isa, Azmilhram Md.   128

J

J.S.K.-SIAT International   217, 221
Jabr, Abdul Halim   31, 114
Jafar Tukan Architects   111, 113, 120–123, 405,
                                           434 –437
Jaspert, Konstantin   217, 222
Jehle, Prof. S.   346
JSWD Architekten + Planer   58, 217

K

Kaehne + Lange   307
Kasturi, Hijjas   128
Kellenberger, Christoph   183, 187
Keller, Christian   310
keller mayer wittig   307, 310–311
Kerry Hill Architects   489, 518–523
Kirsten Schemel Architekten   353, 355
Kisho Kurokawa architect & associates   405, 430–433
Klapp, Eberhard   230
Kleihues + Kleihues   217, 451–458
Kleihues, Jan   452
Königs, Ilse   298
Königs, Ulrich   298
Königs Architekten   263, 298–301
Kretzschmar & Weber Architekten   307
KSP Engel und Zimmermann Architekten 73, 94–97, 217, 221,
                                 246–249, 339, 340–343, 353, 355
Kurokawa, Kisho   430

L

L'OEUF   210–211
Lab Architecture Studio   73, 102–105
Leclercq, Francois   426
lee + mundwiler architects   183, 187, 196–197
Legner, Prof. Klaus   217, 250
Lehmann, Philipp   198
Lehmann Fidanza   183, 187, 198, 199
Lehrecke, Jakob   308
Lehrecke Architekten   307, 308–309
Léon, Hilde   406, 448, 486
Léon Wohlhage Wernik Architekten   405, 406–413
Lin, Yu-Han Michael   330
Lion, Yves   114, 138, 426
Llewelyn Davies Yeang London   111
LOVE architecture and urbanism   219, 389, 396–397
Lussi+Halter   183

M

Machado, Rodolfo	12–13, 138, 402
Maltby, Elliott	392
Mancuso, Mark	392
Mandrelli, Doriana	466
Marignac, Francçois de	192
Massimiliano Fuksas architetto	451, 466–471
Mechs, Martin	389, 398–399
Memhood, Khalid	124
mischa badertscher architekten	183
Modersohn & Freiesleben	307, 312–313
Modersohn, Johannes	312
Moriyama & Teshima Architects	141, 142–149
Müller, Harald	264
Müller, Philippe	183
Müller, Thomas	321, 326, 327
Mundwiler, Stephan	196
MVRDV	263

N

Nagel, Manfred	217
Nandan, Gita	392
netzwerkarchitekten	217
netzwerkarchitekten	353, 355, 360–363
Neumann, Matthias	392
Nieuwenhove, Remy van	222
Nishizawa, Ryue	294
normaldesign	389, 392–395

O

Oester, Hanspeter	376
Olivier, Bernard	210
oos ag	183, 187
Ortner & Ortner Baukunst	73, 84–89, 217, 353
Ortner, Prof. Laurids	84
Ortner, Prof. Manfred	84

P

Pasquier, Grégoire Du	192
Paul, Joop	98
Pearl, Daniel S.	210
Peia, Giampero	460
Peia Associati	451, 460–465
Peter Kulka Architektur	353, 355
Petzinka Pink Architekten	339
Pfenninger, Reto	376
Pidoux, Christophe	192
Piguet, Claire	114, 426
Plasma Studio	353, 355
Pocaterra, Isabel Cecilia	206
Pocaterra, Maria Ines	206
Poddubiuk, Mark	210
Portzamparc, Christian de	489, 512, 512–517
Prix, Dreibholz&Partner	451, 476
Prix, Prof. Dr. Wolf D.	476
Proyectos Arqui 5	206, 207
Puy, Juan Pablo	124

R

Racz, David	102
Randecker, Albrecht	250
Randel, Mark	264
re-urbanism	183
Reimann, Ivan	326, 368
Rein, Volker	230
Rhode Kellermann Wawrowsky	339
Richter, Kai	74
Riitano, Mariagiovanna	201, 208, 209
Robert Greenwood	90, 506
Rogers, Paul	162
Rudacs, Birgit	364

S

SANAA	263, 294–297
Saucier + Perrotte Architects	111
Scharabi, Karim	360
Schiffer, Karim	360
schmiedeknecht architekten planschmiede	217
schneider + schumacher Architekturgesellschaft	339
Schuh, Jochen	360
Schuhmacher, Dr. Patrick	236, 284, 438, 490
Schultes, Axel	290, 380, 442
SCHULTES FRANK ARCHITEKTEN	23, 263, 290–293, 353, 355, 380–383, 405, 442–445
Schwieger, Marcus	360
Seidler, Penelope	472
Sejima, Kazuyo	294
Shiraishi, Hiromi	472
Slapeta, Dr. Sc. Vladimír	14–17
Snøhetta	73, 90–93, 489, 506–511
Soonets, Silvia	206
Staab, Volker	280
Staab Architekten	263, 280–283
Stauffenegger + Stutz	196
struhk architekten Planungsgesellschaft	339

T

T. R. Hamazah & Yeang Sdn.Bhd.	73, 80–83
Taylor, Mark	162
Tebroke, Stefan	217
Teherani, Hadi	74
Thilo Höhne	360
Thomas Lussi	183
Thomas Müller Ivan Reimann	321, 326–329, 353, 355, 368–371
thread collective	389, 392–395
Tinner, Mathis Simon	414
tp bennett LLP	219
Troelsen, Troels	524
Tukan, Jafar	18–19, 120, 138, 183, 405, 434, 486

V

Valerio, Juan Carlos   124
Van Bergen, Paul   98
Velvet Creative Office   183
von Gerkan, Marg und Partner   170–177, 353, 405, 420–425
von Gerkan, Meinhard   420

W

Werner, Roloff   242
Wernik, Siegfried   406
Weuthen, Volker   254
Witan, Oliver   360
Wohlhage, Konrad   406

Y

Yeang, Dr. Kenneth   80

Z

Hadid, Zaha   6
Zaha Hadid Architects   217, 221, 236–241, 284–289, 405, 438–441, 489, 490–497
Zamarbide, Daniel   183, 192
Zimmermann, Michael   94
Zlonicky, Prof. Peter   20–23, 214, 402